**NORTH-HOLLAND
PERSONAL LIBRARY**

LEPTONS AND QUARKS

LEPTONS AND QUARKS

L. B. OKUN

Translated from the Russian by
V. I. Kisin

NORTH-HOLLAND
AMSTERDAM · OXFORD · NEW YORK · TOKYO

© Elsevier Science Publishers B.V. 1982

All rights reserved. No part of this publication may be reproduced, stored in a retrieval system, or transmitted, in any form or by any means, electronic, mechanical photocopying, recording or otherwise, without the prior permission of the publisher, Elsevier Science Publishers B.V. (North-Holland Physics Publishing Division), P.O. Box 103, 1000 AC Amsterdam, The Netherlands.

Special Regulations for readers in the USA: This publication has been registered with the Copyright Clearance Center Inc. (CCC), Salem, Massachusetts. Information can be obtained from the CCC about conditions under which photocopies of parts of this publication may be made in the USA.

All other copyright questions, including photocopying outside of the USA, should be referred to the publisher.

ISBN: 0 444 86924 7

Hardbound edition 1982
Second printing (paperback) 1984

Published by:
North-Holland Physics Publishing
a division of
Elsevier Science Publishers B.V.
P.O. Box 103
1000 AC Amsterdam
The Netherlands

Sole distributors for the U.S.A. and Canada:
Elsevier Science Publishing Company, Inc.
52 Vanderbilt Avenue
New York, N.Y. 10017
U.S.A.

QC
793.5
.L422
O3813
1984

Library of Congress Cataloging in Publication Data

Okun', Lev Borisovich.
  Leptons and quarks.

  Bibliography: p.
  1. Leptons. 2. Quarks. 3. Weak interactions
  (Nuclear physics) I. Title.
QC793.5.L42203813   539.7'211   80-24254
ISBN 0-444-86002-9
ISBN 0-444-86924-7 (North-Holland
  Personal Library, Pbk)

Printed in The Netherlands

Preface

This book is an introduction to the theory of weak interactions of elementary particles. It is based on the course of lectures that I give regularly at the Institute of Theoretical and Experimental Physics to graduate students of the Moscow Physico-Technical Institute, who specialize in experimental high-energy physics.

The book was meant to outline the current situation in weak interaction theory and to discuss the prospects for the coming decade. In addition, I wanted to familiarize the reader with simple theoretical techniques and with the calculation of decay rates, interaction cross sections, angular and spin correlations, and so on.

The first part of the book (Chapters 1 through 17) can be regarded, in a certain sense, as a theoretical commentary to the well-known tables of elementary particles, the so-called Rosenfeld tables, excerpts from which are given at the end of the book (Chapter 30). During the lectures, students kept these tables on their desks, and the results of calculations were compared with the corresponding experimental data. The reader is advised to follow the same procedure.

The second part of the book is mostly an introduction to unified gauge theory of electromagnetic and weak interactions (Chapters 18 through 24). This part describes the Higgs mechanism of spontaneous symmetry breaking and discusses the expected properties of the intermediate and Higgs bosons. The main emphasis is on the physical, not on the formal mathematical, side of gauge theory. In particular, a detailed discussion is given of the role of conserved currents and scalar bosons in the elimination of the unacceptable growth of amplitudes (typical for the longitudinal components of vector bosons).

Two chapters are devoted to attempts to construct a unified theory of all known interactions: the electromagnetic, weak, strong (Chapter 25), and even the gravitational interaction (Chapter 26). Chapter 27 contains a short digression to astrophysics and cosmology in relation to the properties of elementary particles. This is followed by an annotated bibliography (Chapter 28) and a mathematical appendix (Chapter 29) which the reader is advised to look through before undertaking to read the book.

The book was written for experimental physicists who are familiar with relativistic quantum mechanics and with some elements of quantum field theory, but who possibly have forgotten some of the finer details (this last condition is by no means necessary).

I tried mostly not to resort to the operator techniques of quantum field theory, and to work only in terms of Feynman graphs. Most calculations in this book can be handled after reading Feynman's monographs "Fundamental processes" and "Quantum electrodynamics". My purpose was to demonstrate how much can be calculated, explained, and even predicted, by using the simplest tools. Originally the intention had been to write a book so simple that it could be readable even in a town bus. In this respect the attempt seems to have failed, with some of the chapters being difficult for student readers.

As for the physics approach, the title of the book, "Leptons and quarks", clearly shows that the contents are based on the quark model. In fact, the book was meant to be the second volume of a two-volume textbook on the theory of elementary particles, of which the first volume, "Hadrons and quarks", was to give an elementary introduction to the quark-gluon theory of strong interactions. Unfortunately, the first volume is not yet completed. On the other hand, quarks and gluons have already made the grade of newspaper coverage, not to mention numerous popular articles, so that hardly any physics major can be found who is unaware of color symmetry.

I am deeply grateful to Yu. Yu. Kloss and S. V. Semenov who kept detailed records of the lectures, to E. G. Gulyayeva, G. M. Karaseva, and I. A. Terekhova for their help in preparing the manuscript for publication, to I. S. Tsukerman who compiled the tables of experimental data, and to the translator of the book V. I. Kisin for valuable suggestions. I am grateful to those of my friends and colleagues who have read the manuscript, for their encouragement and comments.

I owe special thanks to my co-authors in scientific publications: A. D. Dolgov, S. S. Gershtein, V. N. Gribov, B. L. Ioffe, I. Yu. Kobzarev, V. A. Novikov, B. M. Pontecorvo, M. A. Shifman, A. I. Vainshtein, M. B. Voloshin, V. I. Zakharov, and Ya. B. Zeldovich. Our discussions helped me in clarifying many of the subjects discussed in the book.

L. B. Okun

January, 1980

Contents

Preface v

1. *Introduction* *1*
   1.1. Quark currents 3
   1.2. On the color of weak currents 4
   1.3. Currents and processes 5
   1.4. About the outline of this book 7

2. *Structure of weak currents* *9*
   2.1. Left-handed charged currents 10
   2.2. Breaking of P- and C-invariance 12
   2.3. Universality of the charged current 12
   2.4. The neutral current 13

3. *Muon decay* *15*
   3.1. Decay amplitude 15
   3.2. Decay probability 17
   3.3. Decay of polarized muon 19
   3.4. Qualitative discussion 20

4. *Strangeness-conserving leptonic decays of hadrons. Properties of the ud-current* *22*
   4.1. Isotopic properties of the *ud*-current 23
   4.2. Relationship between the vector current *ud* and the isovector electromagnetic current 23
   4.3. Weak charge 24
   4.4. Chiral invariance 25

5. *Leptonic decays of pions and nucleons* ...27
   - 5.1. $\pi \to \ell\nu$ decays ...27
   - 5.2. $\pi^+ \to \pi^0 e^+ \nu$ decay ...29
   - 5.3. β-decay of the neutron ...30
   - 5.4. Vector form factors ...31
   - 5.5. Axial form factors ...32
   - 5.6. The probability of β-decay. Angular correlations ...35
   - 5.7. $\Sigma^\pm \to \Lambda e^\pm \nu$ decays ...36

6. *Leptonic decays of K-mesons and hyperons* ...38
   - 6.1. $|\Delta S| = 1$ and $\Delta Q = \Delta S$ rules ...38
   - 6.2. SU(3) and SU(2) properties of the us current ...39
   - 6.3. $K_{\ell 2}$ decays ...40
   - 6.4. $K_{\ell 3}$ decays ...41
   - 6.5. $K_{\mu 3}$ decays ...42
   - 6.6. K_{e4} decays ...43
   - 6.7. Leptonic decays of hyperons ...44

7. *Strangeness-changing non-leptonic interactions* ...48
   - 7.1. Properties of the bare non-leptonic lagrangian ...48
   - 7.2. The inclusion of hard gluons ...50
   - 7.3. The gluonic monopole ...52
   - 7.4. Effective non-leptonic lagrangian ...56

8. *Phenomenology of non-leptonic decays of hyperons* ...58
   - 8.1. Relativistically invariant amplitude ...58
   - 8.2. Non-relativistic form of the amplitude ...59
   - 8.3. Spin correlations in hyperon decays ...59
   - 8.4. Isotopic amplitudes and the $\Delta T = \frac{1}{2}$ rule ...61
   - 8.5. Phases of the S- and P-wave amplitudes ...63
   - 8.6. SU(3) relation between amplitudes of hyperon decays ...65

9. *Dynamics of non-leptonic decays of hyperons* ...67
   - 9.1. Quark graphs ...67
   - 9.2. Factorization of external diagrams for the decay $\Lambda^0 \to p\pi^-$ ...69
   - 9.3. Enhancement of the contribution of right-handed quarks ...71
   - 9.4. $\Lambda^0 \to n\pi^0$ decay ...72
   - 9.5. Ω-hyperon decays ...74

10. *Non-leptonic decays of K-mesons* 76
    10.1. K_1^0 and K_2^0 mesons 76
    10.2. Isotopic relations for $K \to 2\pi$ decays 77
    10.3. Quark graphs for $K^\pm \to \pi^\pm \pi^0$ decays 79
    10.4. $K \to 3\pi$ decays 82

11. *Neutral K-mesons in vacuum and in matter* 85
    11.1. $K^0 \leftrightarrow \overline{K}^0$ transitions and the K_1^0–K_2^0 mass difference 85
    11.2. Glashow-Iliopoulos-Maiani mechanism 88
    11.3. Oscillations of strangeness 89
    11.4. Regeneration 91

12. *Violation of CP invariance* 94
    12.1. $K_L^0 \to \pi^+ \pi^-$ decay 94
    12.2. Other observed *CP* violating effects 95
    12.3. Superweak mixing 97
    12.4. On decays of the K_S^0 meson 98
    12.5. Violation of time-reversal invariance and the neutron dipole moment 99
    12.6. Gedanken experiments 101

13. *Decays of the τ-lepton* 103
    13.1. ν_τ neutrino 103
    13.2. Decays $\tau^- \to \mu^- \bar{\nu}_\mu \nu_\tau$ and $\tau^- \to e^- \bar{\nu}_e \nu_\tau$ 104
    13.3. Semi-hadronic decays. General remarks 104
    13.4. Decay $\tau \to \pi \nu_\tau$ 105
    13.5. Decay $\tau \to \rho \nu_\tau$ 107
    13.6. Decays $\tau \to \nu_\tau + 2n\pi$ 110
    13.7. Decays $\tau \to \nu_\tau + (2n+1)\pi$ 114
    13.8. Decays $\tau \to \nu_\tau + K + n\pi$ 116
    13.9. Summary of results 116

14. *Decays of charmed particles* 117
    14.1. Leptonic and non-leptonic decays 117
    14.2. The role of virtual gluons 118
    14.3. Selection rules for *T*-, *U*-, and *V*-spin 121

Contents

15. *Weak decays of b- and t-quarks* *123*
    - 15.1. Unitary $n \times n$ matrix 123
    - 15.2. Angles θ_1, θ_2, θ_3 and phase δ 124
    - 15.3. Metastability of the b-quark 126
    - 15.4. Contribution of t-quarks to $K^0 \leftrightarrow \overline{K}^0$ transitions 127
    - 15.5. Cascade decays of b- and t-hadrons 129

16. *Neutrino–electron interactions* *130*
    - 16.1. Kinematics of $\nu + e \to \nu + \ell$ reactions 130
    - 16.2. Cross section of the $\nu_\mu e^- \to \nu_e \mu^-$ reaction 133
    - 16.3. Cross section of the $\bar{\nu}_e e^- \to \bar{\nu}_\mu \mu^-$ reaction 134
    - 16.4. Elastic νe and $\bar{\nu} e$ scattering induced by charged current 135
    - 16.5. Cross sections of the νe and $\bar{\nu} e$ scattering: general formulas 138
    - 16.6. Other effects of the νe interaction 139
    - 16.7. Creation of muon pairs by neutrinos in the Coulomb field of the nucleus 140

17. *Neutrino–nucleon interactions* *143*
    - 17.1. Kinematics 143
    - 17.2. Quasi-elastic scattering 144
    - 17.3. Partons 145
    - 17.4. Kinematics of lepton–parton collisions 146
    - 17.5. Lepton–parton collision cross sections 147
    - 17.6. Parton distributions 149
    - 17.7. Cross sections of deep-inelastic processes 151
    - 17.8. Production of strange and charmed particles 152
    - 17.9. Phenomenology of deep-inelastic processes 154
    - 17.10. The parton model and quantum chromodynamics 155

18. *Renormalizability* *156*
    - 18.1. Why do we need renormalizability? 156
    - 18.2. Unitarity limit 157
    - 18.3. Intermediate bosons 158
    - 18.4. Wave function of the vector boson 160
    - 18.5. Digression on the photon mass 161
    - 18.6. Vector boson propagator 162

19. *Gauge invariance* *164*
    - 19.1. Global abelian symmetry $U(1)$ 164
    - 19.2. Global non-abelian symmetry $SU(2)$ 165

	19.3.	Local abelian symmetry 165
	19.4.	Digression on baryonic and leptonic photons 166
	19.5.	Local SU(2) symmetry 167
	19.6.	Panegyric to Yang-Mills theory 170
	19.7.	How to take masses into account? 171
20.	*Spontaneous symmetry breaking* *174*
	20.1.	Spontaneous breaking of discrete symmetry 174
	20.2.	Spontaneous breaking of global U(1) symmetry 177
	20.3.	Spontaneous breaking of global SU(2) symmetry 179
	20.4.	Spontaneous breaking of abelian gauge symmetry 180
	20.5.	On the conservation of electric charge 181
	20.6.	Spontaneous breaking of local SU(2) symmetry 183
21.	*Standard model of the electroweak interaction* *185*
	21.1.	Main features of the model 185
	21.2.	Nine terms of the lagrangian 187
	21.3.	Masses of W- and Z-bosons 189
	21.4.	Relation between electric charge and the constants g and g' 191
	21.5.	Relation between the vacuum-expectation value η and the Fermi constant G 192
	21.6.	More about the masses of the W- and Z-bosons 193
	21.7.	Electron mass 194
	21.8.	Introduction of other leptons and quarks 195
22.	*Neutral currents* *199*
	22.1.	Scattering of ν_e and $\bar{\nu}_e$ on the electron 199
	22.2.	Scattering of ν_μ and $\bar{\nu}_\mu$ on the electron 201
	22.3.	Annihilation $e^+ e^- \to \mu^+ \mu^-$ 202
	22.4.	Neutral currents and neutrino–nucleon interaction 206
	22.5.	Isotopic properties of the neutral current 208
	22.6.	Parity non-conservation in scattering of electrons by nucleons 209
	22.7.	Parity non-conservation in atoms 211
	22.8.	P-odd nuclear forces 213
23.	*Properties of intermediate bosons* *214*
	23.1.	Decays of W-bosons 214
	23.2.	Decays of Z-bosons 216
	23.3.	Production of Z-bosons 217

	23.4.	Production of W-bosons 218
	23.5.	Colliding beam projects 220
	23.6.	DUMAND 221
24.	*Properties of Higgs bosons* *223*
	24.1.	On the H-boson mass 223
	24.2.	The role of H-bosons at high energies 225
	24.3.	Coupling of H-bosons to heavy quarks 227
	24.4.	Coupling of H-bosons to gluons 228
	24.5.	"Higgs charge" of nucleons 230
	24.6.	Digression on the trace of the energy-momentum operator .231
	24.7.	Coupling of H-bosons to W- and Z-bosons 232
	24.8.	Coupling of H-bosons to photons 233
	24.9.	General remarks on Higgs bosons 235
25.	*Grand unification* *236*
	25.1.	Three generations of fermions 236
	25.2.	Quintet and decuplet in SU(5) 237
	25.3.	24 vector bosons 239
	25.4.	Running coupling constants 240
	25.5.	Unstable proton 243
	25.6.	Grand Higgs bosons 246
	25.7.	The SO(10) group and other orthogonal groups 248
	25.8.	The exceptional groups 249
26.	*Superunification* *251*
	26.1	Supersymmetry 251
	26.2	Sub-quarks? 252
27.	*Particles and the universe* *254*
	27.1.	Hot universe 254
	27.2.	Upper bound on the neutrino mass 257
	27.3.	On the number of possible types of neutrinos 258
	27.4.	Concentration of relic quarks 260
	27.5.	On baryonic asymmetry of the universe 261
28.	*Bibliography* *263*
	28.1.	Monographs. Proceedings of conferences 265
	28.2.	Decays of leptons and hadrons 267
	28.3.	Neutrino reactions 278
	28.4.	Weak interactions at high energies 281

	28.5.	Models of grand and super unification 288
	28.6.	Particles and the universe 290
	28.7	Supplementary bibliography (autumn 1980) 296
29.	*Appendix (some useful formulas)* *308*
	29.1.	Pseudo-euclidean metric 308
	29.2.	Groups 309
	29.3.	Properties of Dirac matrices 317
	29.4.	Rules for the calculation of probabilities 325
30.	*Tables of experimental data* *329*
	30.1.	Physical and numerical constants and parameters 330
	30.2.	Tables of particle properties 332
	30.3.	Weak decays and the $\Delta I = \frac{1}{2}$ rule 343
	30.4.	CP- and CPT-invariances 350
	30.5.	Conservation of leptonic numbers 352
	30.6.	Selection rules for the weak current 353

CHAPTER 1

Introduction

The weak interaction is responsible for a large number of physical processes: nuclear β-decay, numerous decays of elementary particles, reactions induced by neutrinos from accelerators and nuclear reactors, and also some subtle effects involving parity violation in γ-decays of nuclei and in atomic optical spectra. All known leptons and hadrons are subject to the weak interaction. It plays an important role in such astrophysical phenomena as the sun's burning and supernova explosions. Some of the weak processes were already put to use (for example, the angular asymmetry in the muon decay is a promising new tool in chemistry). Mainly, however, our interest in the weak interaction is rooted not in its possible applications but in the hope that its study will ultimately yield a unified theory of elementary particles and of the interactions between them. And although it would be very difficult today to predict any practical consequences of such a unified theory, there can be no doubt of their utmost importance.

In contrast to "stronger" interactions, namely the strong and electromagnetic, the weak interaction violates a number of conservation laws. Among the quantum numbers that are not conserved are space parity P, charge conjugation parity C, combined inversion parity CP, strangeness, charm, and some others.

The standard theory of weak interactions is based on the analogy with the electromagnetic interaction which is produced by the electromagnetic current coupled to the photon (see fig. 1.1). Likewise, the weak interaction is postulated to result from weak currents being coupled to the so-called intermediate bosons W^+, W^-, Z. Intermediate bosons have not yet been found experimentally; however, this does not point to a defect in the theory since the expected masses are of the order of 100 GeV, the energies of the existing accelerators being well below their production thresholds. The W^+ and W^- bosons are coupled to charged currents which change the charges of particles involved. Such are the currents $\bar{e}\nu_e$ or $\bar{\mu}\nu_\mu$ and their hermitian conjugate currents $\bar{\nu}_e e$ and $\bar{\nu}_\mu \mu$. The last two currents, for instance, interact

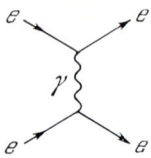

Fig. 1.1.

by exchanging a virtual W-boson and yield the muon decay $\mu \to e\bar{\nu}_e \nu_\mu$ (fig. 1.2). The Z^0 bosons are created by neutral currents of the types $\bar{e}e$, $\bar{\nu}_\mu \nu_\mu$, $\bar{\mu}\mu$, and so on, involving identical ingoing and outgoing particles. Neutral currents are responsible, for example, for the scattering $\nu_\mu e \to \nu_\mu e$ (fig. 1.3). Both the charged and the neutral currents include a leptonic and a hadronic part. At present we know six leptons which are naturally grouped into three pairs:

$$\begin{array}{ccc} \nu_e & \nu_\mu & \nu_\tau \\ e^- & \mu^- & \tau^- \end{array},$$

so that each lepton has its neutrino counterpart. Each lepton enters the charged current j with the appropriate neutrino:

$$j_\ell = \bar{e}\nu_e + \bar{\mu}\nu_\mu + \bar{\tau}\nu_\tau.$$

This current emits W^+ bosons and absorbs W^- bosons. The hermitian conjugate current j_ℓ^+

$$j_\ell^+ = \bar{\nu}_e e + \bar{\nu}_\mu \mu + \bar{\nu}_\tau \tau$$

emits W^- bosons and absorbs W^+ bosons. The neutral leptonic current j_ℓ^0 contains six terms: $\bar{\nu}_e \nu_e$, $\bar{\nu}_\mu \nu_\mu$, $\bar{\nu}_\tau \nu_\tau$, $\bar{e}e$, $\bar{\mu}\mu$, $\bar{\tau}\tau$.

The two leptonic currents given above are responsible for the processes involving both leptons ($e^-, \mu^-, \tau^-, \nu_e, \nu_\mu, \nu_\tau$) and antileptons ($e^+, \mu^+, \tau^+, \bar{\nu}_e, \bar{\nu}_\mu, \bar{\nu}_\tau$). This follows from the properties of the relevant operators. For example, the operator $\bar{e}$ creates an electron and annihilates a positron, while the operator e creates a positron and annihilates an electron. Operators of other particles act in a similar manner.

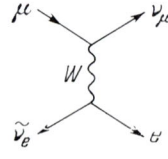

Fig. 1.2.

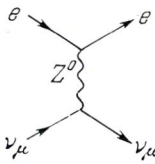

Fig. 1.3.

1.1. Quark currents

Hadrons are represented in weak currents by quarks. According to quark theory, all known hadrons consist of quarks of five types (five flavors): u, d, s, c and b. Theoretical arguments, however, point to the existence of a sixth quark t, so that in analogy to the six leptons, the six quarks form three pairs:

u   c   t
d   s   b.

We recall that the charges of quarks u, c, and t are $+\frac{2}{3}$, and those of quarks d, s, and b are $-\frac{1}{3}$. The quark structure is uud for the proton, udd for the neutron, $u\bar{d}$ for the π^+ meson, and so on. Strange particles include s-quarks (for instance, Λ = uds, $K^+ = u\bar{s}$), and charmed particles include c-quarks (for example, $D^+ = c\bar{d}$, $F^+ = c\bar{s}$). Particles with hidden charm, such as the J/ψ meson, are represented by $c\bar{c}$. The structure of the Υ-meson is $b\bar{b}$.

No hadrons with single b-quarks have so far been found★, and only very scant indirect information is available on the weak interaction of the b- and t-quarks (see Chapter 15). Our knowledge of the first two pairs of quarks is, on the other hand, quite substantial. We know, first of all, that a quark may enter charged currents both with its paired partner and with a partner from another pair. For instance, in addition to the currents $\bar{u}d$ and $\bar{c}s$, the current $\bar{u}s$ exists as well. If it did not, strange particles would be absolutely stable whereas in fact they undergo decays; for example, the current $\bar{u}d$ is responsible for neutron decay (fig 1.4), while the $\bar{u}s$ current is responsible for the decay of the Λ-hyperon (fig. 1.5).

If it is assumed that each of the upper quarks can go over to any of the lower quarks, then in the general case the charged hadronic current j_h must contain nine terms: $\bar{u}d$, $\bar{u}s$, $\bar{u}b$, $\bar{c}d$, $\bar{c}s$, $\bar{c}b$, $\bar{t}d$, $\bar{t}s$, $\bar{t}b$; the picture is similar for

---

★B mesons containing single b-quarks were discovered in 1980, see footnote on p. 5.

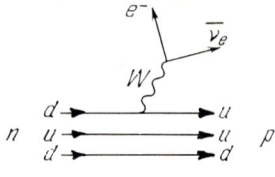

Fig. 1.4.

the hermitian conjugate current j_h^+. As for the neutral hadronic current j_h^0, it must have six terms: $\bar{u}u$, $\bar{d}d$, $\bar{s}s$, $\bar{c}c$, $\bar{b}b$, $\bar{t}t$. There are no neutral currents transforming quarks of one type into quarks of a different type, such as $\bar{d}s$, $\bar{u}c$, and so on (see Chapter 2).

1.2. On the color of weak currents

In addition to flavor, quarks are characterized by color. The same flavor characterizes three non-identical quarks differing in a quantum number quoted as color, so that we shall refer to yellow (y), blue (b), and red (r) u-quarks, d-quarks, and so on. The total number of quarks is therefore 18. Physical hadrons are singlets in color space, and are termed colorless or white. Color symmetry is absolute, and weak quark currents are, just as hadrons, white. This means, for example, that $\bar{u}d$ is in fact a sum of three terms:

$$\bar{u}d = \bar{u}^i d_i = \bar{u}^1 d_1 + \bar{u}^2 d_2 + \bar{u}^3 d_3,$$

where suffices 1, 2, 3 stand for yellow, blue, and red colors, respectively. The same is true for other quark currents. Hereafter, unless stated otherwise, summation over color suffices in quark currents is omitted.

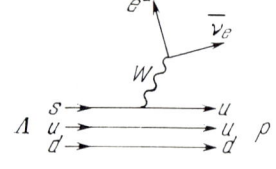

Fig. 1.5.

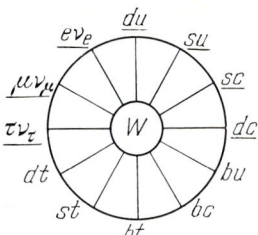

Fig. 1.6.

1.3. Currents and processes

The theory thus contains twelve charged currents coupled to W-bosons (see fig. 1.6) and twelve neutral currents coupled to Z-bosons (fig. 1.7). Underlined in figs. 1.6 and 1.7 are the currents verified by experiment*.

A symbol du in fig. 1.6 denotes either the current $\bar{d}u$ or the hermitian conjugate current $\bar{u}d$; the same is true for other currents in this figure. In some cases this condensed notation proves convenient.

As each of the twelve currents can interact with each current of the diagram, the total number of possible interactions must equal 78 both for fig. 1.6 and for fig. 1.7. So far only fourteen such current × current interactions are experimentally detected for charged currents, and seven for

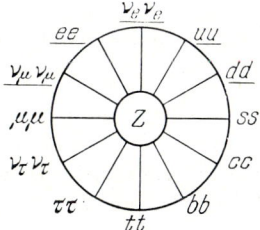

Fig. 1.7.

---

*Decays of Υ''' mesons into B and $\bar{\text{B}}$ mesons containing single b-quarks were observed in the reaction $e^+e^- \to \Upsilon''' \to B\bar{B}$ in 1980, with charmed mesons dominating the decay products of B-mesons. The observed decays of B mesons seem to be caused by the interactions $(bc)(e\nu)$, $(bc)(\mu\nu)$, and $(bc)(ud)$. This means that one more current, namely, bc, should be underlined in fig. 1.6.

neutral ones. Let us make a list of these interactions, indicating in square brackets the processes in which they are experimentally observed. These processes are usually classified into three groups: pure leptonic, semi-leptonic (involving both leptons and hadrons), and non-leptonic (involving hadrons only). Four leptonic, seven semi-leptonic, and three non-leptonic interactions were found for charged currents:

$(e\nu_e)(e\nu_e)$ $[\bar{\nu}_e e \to \bar{\nu}_e e]$,

$(e\nu_e)(\mu\nu_\mu)$ $[\mu \to e\nu\nu]$, $(e\nu_e)(\tau\nu_\tau)$ $[\tau \to e\nu\nu]$, $(\mu\nu_\mu)(\tau\nu_\tau)$ $[\tau \to \mu\nu\nu]$,

$(e\nu)(ud)$ $[n \to pe\nu]$, $(e\nu)(us)$ $[K \to e\nu\pi]$, $(e\nu)(cs)$ $[D \to e\nu K]$,
$(\mu\nu)(ud)$ $[\pi \to \mu\nu]$, $(\mu\nu)(us)$ $[K \to \mu\nu]$, $(\mu\nu)(cs)$ $[D \to \mu\nu K]$,
$(\tau\nu)(ud)$ $[\tau \to \rho\nu]$,

$(ud)(du)$ $[P\text{-odd nuclear forces}]$,
$(ud)(su)$ $[\Lambda \to p\pi]$, $(ud)(cs)$ $[D \to K\pi\pi]$.

(Where the subscripts of neutrinos and the antiparticle bar ¯ were not essential, they were dropped in order to avoid overloading the formulas).

In the case of neutral currents, two leptonic, four semi-leptonic, and three non-leptonic interactions were found:

$(\nu_\mu\nu_\mu)(ee)$ $[\nu_\mu e \to \nu_\mu e]$, $(\nu_e\nu_e)(ee)$ $[\bar{\nu}_e e \to \bar{\nu}_e e]$,

$(\nu_\mu\nu_\mu)(dd), (\nu_\mu\nu_\mu)(uu)$ $[\nu_\mu + \text{nucleus} \to \nu_\mu + \text{hadrons}, \nu_\mu p \to \nu_\mu p]$,
$(ee)(uu), (ee)(dd)$ $[e + p \to e + \text{hadrons, photons in bismuth}]$,

$(dd)(dd), (uu)(uu), (dd)(uu)$ $[P\text{-odd nuclear forces}]$.

In two of the above cases, the same physical phenomenon results both from charged and from neutral currents. These are the scattering process $\bar{\nu}_e e \to \bar{\nu}_e e$, and the nuclear parity-violating interaction between nucleons. A special analysis is required to separate the contributions of charged and neutral currents to these processes.

Each of the 156 interactions shown in figs. 1.6 and 1.7 generates a number of related processes. We have already mentioned that according to quantum field theory, field operators describe the production and annihilation of both particles and antiparticles. Hence, the same interaction results in several processes. Thus, the interaction $(\mu\nu)(ud)$ is responsible for muon capture $\mu^- p \to n\nu$, decay of the π-meson $\pi^+ \to \mu^+ \nu$, neutrino reactions of the type ν_μ + nucleon $\to \mu^-$ + hadrons, and so on.

Figs. 1.6 and 1.7 obviously invite the following questions:
(i) What are the couplings between different currents?
(ii) What is the space-time structure of the different currents and what are its experimental corollaries?
(iii) What properties are anticipated for W- and Z-bosons?

1.4. About the outline of this book

The first two questions are discussed in Chapters 1 through 17, devoted mostly to a phenomenological analysis of a number of weak processes at low energies (below the production threshold of W- and Z-bosons). The second part of the book treats the high-energy physics of weak interactions (at energies above the production threshold of W- and Z-bosons, i.e. in the range that the experimenters plan to reach in the eighties). The idea of a unified electroweak interaction, as stated in the standard model of electroweak interaction, makes the backbone of this part of the book.

In the low-energy limit, the fact that current interactions are mediated by intermediate bosons, is immaterial. We are thus justified in referring to an effective local interaction between currents. For instance, for muon decay, the non-local graph reduces to a local four-fermion interaction graph (see fig. 1.8). This local (i.e. occurring in a single world-point) four-fermion interaction is characterized by the Fermi constant $G \simeq 10^{-5} m_p^{-2}$. (The system of units employed throughout the book assumes $\hbar = c = 1$.) The same coupling constant characterizes (although with some qualifications) the remaining 155 four-fermion interactions. The next chapter deals with the form of the effective four-fermion lagrangian of the weak interaction in more detail.

The first part of the book is thus devoted to the applications of the theory of four-fermion weak interactions. The second part starts with exposing the

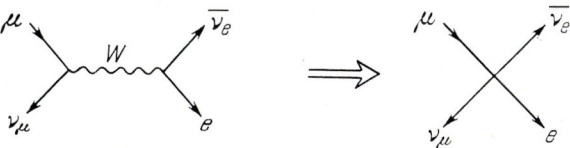

Fig. 1.8.

weak spots of the four-fermion theory and presents the gauge theory of the unified electroweak interaction. Then we discuss the corollaries of this last theory.

From the purely theoretical point of view it would be more logical, of course, to choose another, deductive approach. Namely, the book should have started with presenting the $SU(2) \times U(1)$ symmetry and its spontaneous breaking, writing down the lagrangian of the electroweak interaction and discussing the properties of intermediate bosons, and only then should the results of the theory for low-energy processes have been described. I am of the opinion, however, that the presentation must be based on the phenomena which have been investigated not only theoretically but experimentally as well.

With the plan of the book as chosen here, the reader gradually assimilates important elements of the theory, such as particle states with a definite helicity. He will learn about the experiments from which the main parameters of the theory have been extracted. He will recognize the inherent contradictory nature of the four-fermion theory and the need to introduce the vector and scalar intermediate bosons.

Naturally, the book will unavoidably take on an archaic appearance after the W-, Z-, and especially H-bosons will have been found experimentally. But if it helps to bring us closer to these discoveries, I shall consider its mission fulfilled.

CHAPTER 2

Structure of weak currents

In the introduction we gave a qualitative "bird's eye view" of the weak interaction. Let us discuss now the quantitative description of weak currents. It was mentioned above that the low-energy behavior of weak processes is represented very well by an effective four-fermion lagrangian comprising two terms:

$$\mathcal{L}^w(x) = \mathcal{L}^{ch}(x) + \mathcal{L}^n(x).$$

Here $\mathcal{L}^{ch}$ is the Lagrangian of the interaction of charged currents, and $\mathcal{L}^n$ is that for neutral currents:

$$\mathcal{L}^{ch}(x) = \sqrt{\tfrac{1}{2}}\, Gj^{\alpha+}(x) j_\alpha(x),$$

$$\mathcal{L}^n(x) = \rho\sqrt{\tfrac{1}{2}}\, Gj^{0\alpha}(x) j_\alpha^0(x),$$

where $j^\alpha(x)$ is the charged current (it decreases the particle charge by unity), $j^{+\alpha}(x)$ the hermitian conjugate current (increasing the particle charge by unity), and $j^{0\alpha}$ a neutral current (it is diagonal, i.e. transforms the particle into itself). As mentioned in the introduction, $G \simeq 10^{-5} m_p^{-2}$, and the factor $\sqrt{\tfrac{1}{2}}$ is a tribute to tradition. According to the standard theory of the electroweak interaction (see Chapter 22) the dimensionless parameter ρ equals unity. This prediction agrees with the experimental data within $\simeq \pm 3\%$ accuracy.

The interaction of currents in the lagrangians $\mathcal{L}^{ch}$ and $\mathcal{L}^n$ is local, that is the interaction takes place only when the operators of the two currents act at the same world-point. The four-fermion weak interaction is sometimes referred to as the universal interaction; this implies that different currents interact with the same coupling constant G. This statement of universality calls for serious qualifications which will be given later in this chapter.

2.1. Left-handed charged currents

Whenever the space-time structure of a charged current yielded to experimental determination, it was found to be of the type

$$\bar{f} O_\alpha^L i,$$

where i and f, the operators of the initial and final particles, are Dirac bispinors, $\bar{f} = f^+ \gamma_0$, and the matrix O_α^L is of the form

$$O_\alpha^L = \gamma_\alpha(1 + \gamma_5), \qquad \alpha = 0, 1, 2, 3.$$

The reader is reminded that $\gamma^\alpha = (\gamma^0, \boldsymbol{\gamma})$, $\gamma_\alpha = (\gamma_0, -\boldsymbol{\gamma})$, where

$$\gamma^0 = \gamma_0 = \begin{pmatrix} 1 & 0 \\ 0 & -1 \end{pmatrix}, \qquad \boldsymbol{\gamma} = \begin{pmatrix} 0 & \boldsymbol{\sigma} \\ -\boldsymbol{\sigma} & 0 \end{pmatrix},$$

$$\gamma_5 = \gamma^5 = -i\gamma^0\gamma^1\gamma^2\gamma^3 = -\begin{pmatrix} 0 & 1 \\ 1 & 0 \end{pmatrix}$$

(the properties of Dirac matrices γ are given in the appendix). The quantity $\bar{f} O_\alpha^L i$ is the difference between a vector $V(\gamma_\alpha)$ and an axial vector $A(\gamma_5\gamma_\alpha)$. Consequently, the interaction of charged currents is referred to as the universal $V - A$ interaction.

By virtue of the anticommutation of the matrices γ_5 and γ_α,

$$O_\alpha^L = \gamma_\alpha(1 + \gamma_5) = \tfrac{1}{2}(1 - \gamma_5)\gamma_\alpha(1 + \gamma_5).$$

We can write, therefore,

$$\bar{f} O_\alpha^L i = 2 \bar{f}_L \gamma_\alpha i_L,$$

where

$$i_L = \tfrac{1}{2}(1 + \gamma_5)i, \qquad \bar{f}_L = \bar{f}\tfrac{1}{2}(1 - \gamma_5).$$

The suffix L indicates that i_L and f_L are the so-called left-handed components of Dirac bispinors. In order to clarify this term, consider a wave function

$$\psi_L = \tfrac{1}{2}(1 + \gamma_5)\psi,$$

where ψ satisfies the Dirac equation

$$(\hat{p} - m)\psi = (E\gamma_0 - \boldsymbol{p}\cdot\boldsymbol{\gamma} - m)\psi = 0$$

(as usual, $\hat{a} \equiv a^\alpha \gamma_\alpha = a^0 \gamma_0 - \boldsymbol{a}\cdot\boldsymbol{\gamma}$ for any 4-vector a^α). Expressing the bispinor ψ via two-component spinors φ and χ,

$$\psi = \begin{pmatrix} \varphi \\ \chi \end{pmatrix},$$

we obtain that the Dirac equation yields a relationship between χ and φ:

$$\chi = \frac{\boldsymbol{\sigma} \cdot \boldsymbol{p}}{E + m} \varphi.$$

Now the explicit form of the matrix γ_5 yields that

$$\psi_L = \tfrac{1}{2}(1 + \gamma_5)\psi = \tfrac{1}{2}\begin{pmatrix} 1 & -1 \\ -1 & 1 \end{pmatrix}\begin{pmatrix} \varphi \\ \chi \end{pmatrix} = \tfrac{1}{2}\begin{pmatrix} \varphi - \chi \\ \chi - \varphi \end{pmatrix},$$

that is ψ_L contains the combination $\varphi - \chi$ and not the combination $\varphi + \chi$. Taking the above expression for χ, we obtain that

$$\varphi - \chi = \left(1 - \frac{\boldsymbol{\sigma} \cdot \boldsymbol{p}}{E + m}\right)\varphi.$$

In the ultra-relativistic limit, that is for $E \gg m$ and $v = |\boldsymbol{p}|/E \to 1$, we obtain

$$\varphi - \chi \cong (1 - \boldsymbol{\sigma} \cdot \boldsymbol{n})\varphi,$$

where $\boldsymbol{n} = \boldsymbol{p}/|\boldsymbol{p}|$ is a unit vector in the direction of the particle momentum. If a coordinate axis z is directed along $\boldsymbol{n}$, then

$$(\varphi - \chi) \cong (1 - \sigma_z)\varphi = \begin{pmatrix} 0 & 0 \\ 0 & 1 \end{pmatrix}\varphi.$$

Now let us take into account that the two-component spinor $\varphi = \begin{pmatrix} 1 \\ 0 \end{pmatrix}$ represents a particle with a spin along z, and $\varphi = \begin{pmatrix} 0 \\ 1 \end{pmatrix}$ a particle with the opposite direction of spin. Consequently, $\psi_L = 0$ if $\varphi = \begin{pmatrix} 1 \\ 0 \end{pmatrix}$, and $\psi_L \neq 0$ if $\varphi = \begin{pmatrix} 0 \\ 1 \end{pmatrix}$, so that ψ_L represents a particle with spin in the direction opposite to that of its momentum. Such particles are said to possess left-handed helicity, or left-handed polarization. A particle is said to possess right-handed helicity, or polarization, if its spin is directed along its momentum. The concept of helicity is not Lorentz invariant if the particle mass is non-zero. The helicity of such a particle depends upon the motion of the observer's frame of reference. For example, it will change sign if we try to catch up with the particle at a speed above its velocity. Overtaking a particle is the more difficult, the higher its velocity, so that helicity becomes a better quantum number as velocity increases. It is an exact quantum number for massless particles (neutrinos and photons).

The above space-time structure of charged currents means that only left-handed components of bispinor operators are involved. This means further that at $v \to 1$, particles have only left-handed helicity, and antiparticles only right-handed helicity.

2.2. Breaking of P- and C-invariance

The $V - A$ nature of the charged current implies that the weak interaction is not invariant with respect to space inversion. Indeed, this operation affects the signs of V- and A-terms in an opposite manner, so that the terms VV^+ and AA^+ in a lagrangian $(V - A)(V - A)^+ = VV^+ + AA^+ - VA^+ - AV^+$ do not change sign while VA^+ and AV^+ do. P-violation in weak interactions is manifested by a number of physical effects, and among them by P-odd $\boldsymbol{J \cdot p}$-type correlations between the particle momentum $\boldsymbol{p}$ and its spin or the spin of another particle, $\boldsymbol{J}$. We recall that under space inversion, $\boldsymbol{p} \to -\boldsymbol{p}$, $\boldsymbol{J} \to +\boldsymbol{J}$, P-odd correlations of this type were first found experimentally in 1956 in β-decay and in decays of elementary particles.

Furthermore, the product $(V - A)(V - A)^+$ is not invariant with respect to charge conjugation since this operation does not affect the signs of the terms in $VV^+ + AA^+$, but reverses those of the terms in $VA^+ + AV^+$. Violation of charge conjugation invariance is best seen in the neutrino: charge conjugation invariance demands that a weak process emitting a left-handed neutrino be transformed to a weak process emitting a left-handed antineutrino; this latter, however, is stated by the $V - A$ theory to be non-existent.

Breaking of CP invariance is a more delicate problem (see Chapter 12). CP may be broken by the complex coefficients with which the twelve currents of fig. 1.6 enter the total charged current j.

2.3. Universality of the charged current

It has been mentioned above that according to the theory, three leptonic pairs enter the weak current in a symmetrical manner, and the total leptonic current has the form

$$j_\alpha^\ell = \bar{e} O_\alpha^L \nu_e + \bar{\mu} O_\alpha^L \nu_\mu + \bar{\tau} O_\alpha^L \nu_\tau.$$

A quark current can be written in a similar manner:

$$j_\alpha^h = \bar{d}' O_\alpha^L u + \bar{s}' O_\alpha^L c + \bar{b}' O_\alpha^L t,$$

where d', s', b' are orthonormalized linear combinations of d-, s-, and b-quarks. Nine coefficients characterizing these combinations can be expressed, in the most general form, in terms of three Euler angles θ_1, θ_2, θ_3 and one phase factor δ. These coefficients will be written in their explicit form in Chapter 15; now it is sufficient to mention that if $b' = b$, that is if transitions $\bar{b}u$ and $\bar{b}c$ can be neglected, we are left with a single angle θ

characterizing mixing of d- and s-quarks:
$$d' = d\cos\theta + s\sin\theta,$$
$$s' = -d\sin\theta + s\cos\theta.$$

The angle θ is called the Cabibbo angle. (Sometimes we shall replace θ with θ_C.) Experimental data on decays of strange particles yield $\theta \cong 13°$.

The following correspondence is therefore found between leptons and quarks in weak interactions:

$$\begin{pmatrix} \nu_e & \nu_\mu & \nu_\tau \\ e & \mu & \tau \end{pmatrix} \leftrightarrow \begin{pmatrix} u & c & t \\ d' & s' & b' \end{pmatrix};$$

(obviously, the correspondence

$$\begin{pmatrix} \nu_e & \nu_\mu & \nu_\tau \\ e & \mu & \tau \end{pmatrix} \leftrightarrow \begin{pmatrix} u' & c' & t' \\ d & s & b \end{pmatrix},$$

where u', c', t' are linear combinations of u-, c-, t-quarks, would be absolutely equivalent). The variety of weak decays of hadrons stems from the fact that quark doublets do not include particles with definite (and non-identical) masses, but rather, their linear combinations. There is no such mixing in the case of leptons if the neutrinos are massless.

Experimenters can go a long way to test the details of the theoretical scheme as presented above. For instance, the coefficient for the term $\bar{\tau} O_\alpha^L \nu_\tau$ was not measured, and very little is known about experimental values of the parameters θ_2, θ_3 characterizing the admixture of b-quarks to d- and s-quarks, and about the parameter δ characterizing the breaking of CP in the lagrangian $\mathcal{L}^{ch}$.

2.4. The neutral current

Two peculiarities of neutral currents stand out when one compares them with charged currents: first, they are diagonal, that is transform a particle into itself, and secondly, they contain, in addition to left-handed components of spinors, right-handed ones as well.

The absence of non-diagonal terms in a neutral current is born out by the fact that experiment does not detect processes which would occur with high decay rates if such terms existed. Here we mean the process $\mu \to e e \bar{e}$ and a related process $\mu \to e\gamma$, which would reveal the current μe, and also the processes $K \to \pi e\bar{e}$, $K \to \pi\mu\bar{\mu}$, $K \to \pi\nu\bar{\nu}$ which would reveal the current ds. (The decay $K^0 \to \mu\bar{\mu}$ recorded with the relative probability of the order of

10^{-8}, results from the joint action of the non-leptonic interaction of charged currents $(du)(us)$ and the electromagnetic interaction.)

That neutral currents comprise both the left-handed and right-handed components, is also an experimental fact (see Chapter 22). In the general case, the neutral current may be written as a sum of twelve terms:

$$j_\alpha^0 = \sum_i (g_L^i \bar{\psi}_i O_\alpha^L \psi_i + g_R^i \bar{\psi}_i O_\alpha^R \psi_i),$$

where

$$i = \nu_e, \nu_\mu, \nu_\tau, e, \mu, \tau, u, c, t, d, s, b,$$

$$O_\alpha^L = \gamma_\alpha(1 + \gamma_5), \qquad O_\alpha^R = \gamma_\alpha(1 - \gamma_5),$$

and g_L^i and g_R^i are numerical coefficients. As shown in Chapter 22, the unified theory of weak and electromagnetic interactions relates coefficients g_L^i and g_R^i to the charges of particles:

$$g_L^i = \tfrac{1}{2}, g_R^i = 0 \qquad \text{for } i = \nu_e, \nu_\mu, \nu_\tau,$$

$$g_L^i = -\tfrac{1}{2} + \xi, g_R^i = +\xi \qquad \text{for } i = e, \mu, \tau,$$

$$g_L^i = \tfrac{1}{2} - \tfrac{2}{3}\xi, g_R^i = -\tfrac{2}{3}\xi \qquad \text{for } i = u, c, t,$$

$$g_L^i = -\tfrac{1}{2} + \tfrac{1}{3}\xi, g_R^i = +\tfrac{1}{3}\xi \qquad \text{for } i = d, s, b,$$

where $\xi = \sin^2\theta_W$, and θ_W is the Weinberg angle. Experiments give:

$$0.20 \lesssim \xi \lesssim 0.25.$$

Chapter 22 will show how this value of ξ is derived from the experimental data on elastic and inelastic scattering of ν_μ and $\bar{\nu}_\mu$ by nucleons, and from the cross section of deep-inelastic scattering of electrons on nucleons measured as a function of the longitudinal polarization of electrons. This value of ξ is in agreement with the data on the scattering processes $\nu_\mu e \to \nu_\mu e$ and $\bar{\nu}_e e \to \bar{\nu}_e e$.

This concludes a brief review of the properties of weak currents, and we are now ready to start systematic calculations of weak processes.

CHAPTER 3

Muon decay

The muon decay $\mu \to e\nu\bar{\nu}$ is a process which traditionally opens calculations of weak decays in textbooks. The reason is two-fold. First, this is a pure leptonic process not involving hadrons, and therefore is easily calculated to the end. Second, experimentally this is one of the most thoroughly investigated decays of elementary particles. In this chapter we calculate the electron spectrum, the total decay probability, and finally the angular and spin correlations in polarized muon decays.

3.1. Decay amplitude

Let us consider the decay $\mu^-(p) \to e^-(k)\bar{\nu}_e(q_1)\nu_\mu(q_2)$ (quantities in parentheses denote 4-momenta of the particles). The Feynman graph of this process given in fig. 3.1a is equivalent to that of fig. 3.1b. In this second case the matrix element is of the form

$$M = \sqrt{\tfrac{1}{2}}\, G \bar{\nu}_\mu O^\alpha \mu\, \bar{e} O_\alpha \nu_e,$$

where $O_\alpha \equiv O_\alpha^L = \gamma_\alpha(1 + \gamma_5)$, and particle symbols stand for their wave functions. By using the Fierz transformation (see appendix) this matrix element can be recast in the form

$$M = -\sqrt{\tfrac{1}{2}}\, G \bar{e} O^\beta \mu\, \bar{\nu}_\mu O_\beta \nu_e.$$

The complex conjugate of this last expression is

$$M^* = M^+ = -\sqrt{\tfrac{1}{2}}\, G \nu_e^+ O^{\beta+} \gamma_0 \nu_\mu \mu^+ O_\beta^+ \gamma_0 e$$

$$= -\sqrt{\tfrac{1}{2}}\, G \nu_e^+ \gamma_0 \gamma_0 O^{\beta+} \gamma_0 \nu_\mu \mu^+ \gamma_0 \gamma_0 O_\beta^+ \gamma_0 e = -\sqrt{\tfrac{1}{2}}\, \bar{\nu}_e O^\beta \nu_\mu \bar{\mu} O_\beta e.$$

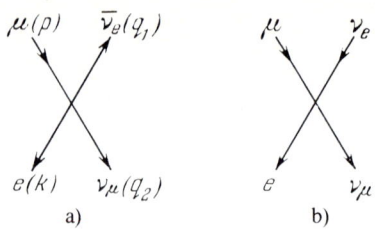

Fig. 3.1.

[We made use of the fact that $\gamma_0 O^{\beta+} \gamma_0 = O^\beta$, because $\gamma_0 [\gamma_0 (1 + \gamma_5)]^+ \gamma_0 = \gamma_0 (1 + \gamma_5) \gamma_0 \gamma_0 = \gamma_0 (1 + \gamma_5)$, and also

$$\gamma_0 [\boldsymbol{\gamma}(1 + \gamma_5)]^+ \gamma_0 = \gamma_0 (1 + \gamma_5)(-\boldsymbol{\gamma}) \gamma_0 = (1 - \gamma_5) \boldsymbol{\gamma} = \boldsymbol{\gamma}(1 + \gamma_5)].$$

Similarly, $\gamma_0 O_\beta^+ \gamma_0 = O_\beta$. Therefore,

$$|M|^2 = MM^* = -\tfrac{1}{2} G^2 \bar{\nu}_\mu O^\alpha \mu \bar{e} O_\alpha \nu_e \bar{\nu}_e O^\beta \nu_\mu \bar{\mu} O_\beta e$$

$$= -\tfrac{1}{2} G^2 \bar{\nu}_\mu O^\alpha \mu \bar{\mu} O_\beta e \bar{e} O_\alpha \nu_e \bar{\nu}_e O^\beta \nu_\mu.$$

As demonstrated in the appendix, the density matrix for an unpolarized Dirac particle has the form

$$\sum_s u^k(s) \bar{u}_i(s) = (\hat{p} + m)_i^k,$$

where the summation is over the polarization states of the particle; p is the particle 4-momentum and m its mass; and i, k are bispinor suffices, $i, k = 1, 2, 3, 4$. Using this we obtain

$$\overline{|M|^2} = -\tfrac{1}{2} G^2 \operatorname{Tr} \hat{q}_2 O^\alpha (\hat{p} + m_\mu) O_\beta (\hat{k} + m_e) O_\alpha \hat{q}_1 O^\beta.$$

The bar over $|M|^2$ denotes summation over spin states; Tr denotes the trace of the product of matrices to the right of it. Since $\gamma_5 \gamma_\alpha = -\gamma_\alpha \gamma_5$, and since $(1 + \gamma_5)^2 = 2(1 + \gamma_5)$,

$$\overline{|M|^2} = -8 \cdot \tfrac{1}{2} G^2 \operatorname{Tr} \hat{q}_2 \gamma^\alpha \hat{p} \gamma_\beta \hat{k} \gamma_\alpha \hat{q}_1 \gamma^\beta (1 + \gamma_5)$$

$$= 8 G^2 \operatorname{Tr} \hat{q}_2 \hat{k} \gamma_\beta \hat{p} \hat{q}_1 \gamma^\beta (1 + \gamma_5) = 32 G^2 (pq_1) \operatorname{Tr} \hat{q}_2 \hat{k} (1 + \gamma_5)$$

$$= 128 G^2 (pq_1)(kq_2).$$

(In deriving this equality, we took into account that $\gamma^\alpha \hat{A} \hat{B} \hat{C} \gamma_\alpha = -2 \hat{C} \hat{B} \hat{A}$, $\gamma^\alpha \hat{A} \hat{B} \gamma_\alpha = 4AB$, see Chapter 29.)

3.2. Decay probability

It is now possible to calculate the probability of the muon decay by means of the general expression (see Chapter 29, sect. 4)

$$d\Gamma = \frac{|M|^2}{2 \cdot 2m} d\Phi.$$

Here $d\Gamma$ is the differential probability of decay per unit volume of phase space $d\Phi$, and the factor $\frac{1}{2}$ appears because of the averaging (and not summation) over muon polarizations; the factor $2m$ (where m is the muon mass) appears because of the chosen normalization of wave functions of particles. As shown in the appendix,

$$d\Phi = (2\pi)^4 \delta^4(p - k_1 - q_1 - q_2) \frac{d\mathbf{k}}{2E(2\pi)^3} \frac{d\mathbf{q}_1}{2\omega_1(2\pi)^3} \frac{d\mathbf{q}_2}{2\omega_2(2\pi)^3},$$

where E, ω_1, ω_2 are the energies of the electron, $\bar{\nu}_e$, and ν_μ; $\mathbf{k}$, $\mathbf{q}_1$, and $\mathbf{q}_2$ stand for their momenta. Neutrinos being not observed, let us integrate over their momenta. This means that we must calculate the integral

$$I_{\alpha\beta} = \int q_{1\alpha} q_{2\beta} \frac{d\mathbf{q}_1}{\omega_1} \frac{d\mathbf{q}_2}{\omega_2} \delta^4(q_1 + q_2 - q),$$

where $q = p - k$. The anticipated result can be written as a sum of two mutually orthogonal terms:

$$I_{\alpha\beta} = A(q^2 g_{\alpha\beta} + 2q_\alpha q_\beta) + B(q^2 g_{\alpha\beta} - 2q_\alpha q_\beta).$$

Here $g_{\alpha\beta}$ is a metric tensor (see Chapter 29), q is the total 4-momentum of two neutrinos, and A and B are dimensionless coefficients that we shall now determine. By multiplying both sides of the equality by $q^2 g^{\alpha\beta} - 2q^\alpha q^\beta$, we obtain

$$B \cdot 4q^4 = \int q_{1\alpha} q_{2\beta} (q^2 g^{\alpha\beta} - 2q^\alpha q^\beta) \cdots$$

$$= \int \left[q^2(q^1 q^2) - 2(qq_1)(qq_2) \right] \cdots = 0,$$

since $q^2 = (q_1 + q_2)^2 = 2q_1 q_2$, $qq_1 = (q_1 + q_2, q_1) = q_1 q_2$, $qq_2 = (q_1 + q_2, q_2) = q_1 q_2$, $q_1^2 = q_2^2 = 0$ (the neutrino is massless). By multiplying both sides of the tensor equality by $q^2 g^{\alpha\beta} + 2q^\alpha q^\beta$, we obtain

$$A \cdot 12q^4 = q^4 \int \frac{d\mathbf{q}_1}{\omega_1} \frac{d\mathbf{q}_2}{\omega_2} \delta^4(q_1 + q_2 - q) = q^4 \int \frac{d\mathbf{q}_1}{\omega_1 \omega_2} \delta(2\omega_1 - \omega)$$

$$= q^4 \cdot 4\pi \cdot \tfrac{1}{2},$$

so that $A = \frac{1}{6}\pi$. (The integral above is taken in the center-of-mass frame of two neutrinos). Finally,

$$I_{\alpha\beta} = \tfrac{1}{6}\pi\bigl(q^2 g_{\alpha\beta} + 2q_\alpha q_\beta\bigr),$$

where $q = p - k$. Substitution of this result into the expression for the decay width yields

$$d\Gamma = \frac{G^2}{2\cdot 2m}\frac{128}{(2\pi)^5 2\cdot 2}\tfrac{1}{6}\pi p^\alpha k^\beta\bigl[q^2 g_{\alpha\beta} + 2q_\alpha q_\beta\bigr]\frac{d\mathbf{k}}{2E}$$

$$= \frac{G^2}{48\pi^4 m}\bigl[q^2(pk) + 2(qp)(qk)\bigr]\frac{d\mathbf{k}}{E}.$$

As the electron mass is negligibly small compared to its energy, we have $qk = (p - k, k) = pk = mE$, $q^2 = (p - k)^2 = p^2 - 2pk = m^2 - 2mE$. Integration over the electron direction yields 4π, and we obtain

$$d\Gamma = \frac{G^2}{12\pi^3 m}(pk)\bigl(p^2 - 2pk + 2p^2 - 2pk\bigr)E\,dE$$

$$= \frac{G^2}{12\pi^3}(3m^2 - 4mE)E^2\,dE = \frac{G^2 m^5}{96\pi^3}(3 - 2\varepsilon)\varepsilon^2\,d\varepsilon,$$

where $\varepsilon = E/E_{\max} = 2E/m$.

The electron spectrum in muon decay, for the most general form of the four-fermion interaction, can be shown to take the form

$$\Gamma(\varepsilon)\,d\varepsilon = 12\Gamma\bigl[(1-\varepsilon) - \tfrac{2}{9}\rho(3 - 4\varepsilon)\bigr]\varepsilon^2\,d\varepsilon,$$

where the coefficient ρ is called the Michel parameter (the term proportional to ρ gives zero contribution when integrated over the whole range $0 \leq \varepsilon \leq 1$). Clearly, the spectrum obtained above corresponds to $\rho = 0.75$. Integration over the electron energy yields the total decay width

$$\Gamma = \frac{G^2 m^5}{96\pi^3}\int_0^1 (3 - 2\varepsilon)\varepsilon^2\,d\varepsilon = \frac{G^2 m^5}{192\pi^3}.$$

After correction for virtual photons, one obtains (see references in Chapter 28)

$$\Gamma = \frac{G^2 m^5}{192\pi^3}\left[1 - \frac{\alpha}{2\pi}\left(\pi^2 - \tfrac{25}{4}\right)\right].$$

Comparison of this expression with the experimentally measured muon lifetime gives the familiar value of the coupling constant G.

3.3. Decay of polarized muon

Let the spin of a muon in its rest frame be directed along the unit vector $\boldsymbol{\eta}$. The muon density matrix then becomes (see appendix)

$$u^k(s)\bar{u}_i(s) = \tfrac{1}{2}\left[(\hat{p}+m)(1-\gamma_5)\hat{s}\right]_i^k,$$

where a 4-vector s^α has the following properties: $s^2 = -1$, $sp = 0$. In the muon rest frame, $s^0 = 0$, $\mathbf{s} = \boldsymbol{\eta}$. In the reference frame in which the muon moves with momentum $\mathbf{p}$,

$$s^0 = \frac{\boldsymbol{\eta}\cdot\mathbf{p}}{m}, \qquad \mathbf{s} = \boldsymbol{\eta} + \frac{\mathbf{p}(\boldsymbol{\eta}\cdot\mathbf{p})}{m(E+m)}.$$

If we substitute in the above calculations

$$(\hat{p}+m) \to \tfrac{1}{2}(\hat{p}+m)(1-\gamma_5\hat{s}),$$

we will have to substitute $p - ms$ for p in the earlier result

$$d\Gamma = \frac{G^2}{48\pi^4 m}\left[q^2(pk) + 2(qp)(qk)\right]\frac{d\mathbf{k}}{E}.$$

[This is readily confirmed by considering that part of the trace which comprises the muon density matrix and its neighbor terms

$$\mathrm{Tr}\cdots(1+\gamma_5)(\hat{p}+m)(1-\gamma_5\hat{s})\gamma_\beta(1+\gamma_5)\cdots$$
$$= \mathrm{Tr}\cdots(1+\gamma_5)(\hat{p}-m\gamma_5\hat{s}+m+\gamma_5\hat{p}\hat{s})(1-\gamma_5)\gamma_\beta\cdots$$
$$= \mathrm{Tr}\cdots(1+\gamma_5)(\hat{p}-m\gamma_5\hat{s})\gamma_\beta(1+\gamma_5)\cdots$$
$$= \mathrm{Tr}\cdots(1+\gamma_5)(\hat{p}-m\hat{s})\gamma_\beta(1+\gamma_5)\cdots.$$

The terms $m + \gamma_5\hat{p}\hat{s}$ cancelled out since they were multiplied by $(1+\gamma_5) \times (1-\gamma_5)$. As for the factor $\tfrac{1}{2}$, it was already taken into account in the expression for the probability indicating averaging over muon polarizations, so that we need not do it a second time.] With muon polarization taken into account,

$$d\Gamma = \frac{G^2}{48\pi^4 m}\left\{\left[q^2(pk)+2(qp)(qk)\right] - m\left[q^2(sk)+2(qs)(qk)\right]\frac{d\mathbf{k}}{E}\right\}$$
$$= \frac{G^2}{48\pi^4 m}\left\{(pk)\left[(p-k)^2 + 2(p^2-pk)\right]\right.$$
$$\left. - m(sk)\left[(p-k)^2 - 2pk\right]\right\}\frac{d\mathbf{k}}{E}$$
$$= \frac{G^2 m^5}{384\pi^4}\left[(3-2\varepsilon) + \boldsymbol{\eta}\cdot\mathbf{n}(1-2\varepsilon)\right]\varepsilon^2\,d\varepsilon\,d\Omega,$$

where $\mathbf{n} = \mathbf{k}/E$ is a unit vector in the direction of the emission of the electron, $d\Omega = d\cos\theta\, d\varphi$ is an element of the solid angle, and $\cos\theta = \mathbf{n}\cdot\mathbf{\eta}$. The above derivation uses that $qs = (p - k, s) = -ks$ since $ps = 0$; moreover, we take into consideration that in the muon rest frame, $ks = k^0 s_0 - \mathbf{k}\cdot\mathbf{s} = -E\mathbf{n}\cdot\mathbf{\eta}$.

The angular distribution integrated over the electron spectrum takes the form

$$\frac{d\Gamma(\cos\theta)}{\Gamma} = \tfrac{1}{2}\left(1 - \tfrac{1}{3}\cos\theta\right) d\cos\theta.$$

The above formulas cover the case when the electron polarization is not measured. If we were interested in the decay probability as a function of electron polarization, k in the expression for $|M|^2$ would have to be replaced by $\tfrac{1}{2}(k - m_e s_e)$ where s_e is a vector characterizing the electron polarization: $s_e^2 = -1$, $s_e k = 0$. Explicit expressions for the components of s_e, $s_e^0 = \mathbf{k}\cdot\mathbf{\zeta}/m_e$, $\mathbf{s}_e = \mathbf{\zeta} + (\mathbf{k}\cdot\mathbf{\zeta})\mathbf{k}/m_e(E + m_e)$, demonstrate that the components of s_e normal to $\mathbf{n}$ can be neglected, in the ultrarelativistic limit, in comparison to s_e^0 and $\mathbf{n}\cdot\mathbf{s}_e = (E/m)(\mathbf{\zeta}\cdot\mathbf{n})$. We obtain therefore that for $v \to 1$,

$$\tfrac{1}{2}(k_\alpha - m_e s_{e\alpha}) \Rightarrow \tfrac{1}{2} k_\alpha(1 - \mathbf{\zeta}\cdot\mathbf{n}).$$

We have thus reproduced a well-known property of the weak interaction: the left-handed polarization of the emitted relativistic leptons. The probability of the decay $\mu^- \to e^- \nu_\mu \bar{\nu}_e$, with the electron polarization taken into account, is then written as

$$d\Gamma = \Gamma \tfrac{1}{2}(1 - \mathbf{\zeta}\cdot\mathbf{n})\left[(3 - 2\varepsilon) + \mathbf{\eta}\cdot\mathbf{n}(1 - 2\varepsilon)\right]\varepsilon^2 d\varepsilon\, d\cos\theta.$$

Similar calculations for the decay $\mu^+ \to e^+ \nu_e \bar{\nu}_\mu$ would give

$$d\Gamma = \Gamma \tfrac{1}{2}(1 + \mathbf{\zeta}\cdot\mathbf{n})\left[(3 - 2\varepsilon) - \mathbf{\eta}\cdot\mathbf{n}(1 - 2\varepsilon)\right]\varepsilon^2 d\varepsilon\, d\cos\theta,$$

where ε is the positron energy divided by its maximum energy, and $\mathbf{n}$, $\mathbf{\zeta}$, $\mathbf{\eta}$ are unit vectors in the directions of momentum of the positron, its spin, and the muon spin, respectively.

3.4. Qualitative discussion

A number of results derived in this chapter are very easy to interpret qualitatively, without any calculations. First of all, the fact that $\Gamma \sim G^2 m^5$ follows from dimension-based arguments, since $[\Gamma] = \mu$, and $[G] = \mu^{-2}$, where μ has the dimension of mass. We only use that the probability is

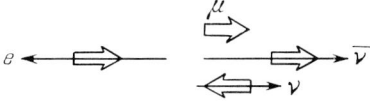

Fig. 3.2.

proportional to G^2 and that the only parameter determining the dynamics of muon decay is the muon mass, since the electron mass can be safely neglected.

Turning to the angular asymmetry, we readily notice that it is neither P- nor C-invariant: its signs are different in the left-handed and right-handed reference frames, as well as for the electron and positron in decays of μ^- and μ^+, respectively. Characteristic features of this asymmetry can also be explained. Let us consider an electron with nearly maximum energy ($\varepsilon \sim 1$). In this situation neutrinos (ν and $\bar{\nu}$) must be emitted in the direction opposite to that of the electron (see fig. 3.2) (the phase-space volume of the configuration in which one of them has low momentum, is small). Recalling that the helicities of ν and $\bar{\nu}$ are of opposite signs, we have to conclude that the angular momentum carried by neutrinos is zero. Consequently, the electron must be emitted with its spin parallel to that of the muon. But, the electron helicity being negative, its momentum must be predominantly directed opposite to the muon spin. This is in agreement with the formula derived above for the angular distribution of electrons, which for $\varepsilon \sim 1$ is proportional to $(1 - \boldsymbol{\eta} \cdot \boldsymbol{n})$.

For $\varepsilon \ll 1$, the neutrino and antineutrino move in the opposite directions, with the total spin equal to unity. This time, conservation of angular momentum demands that electrons be emitted along the muon spin direction (see fig. 3.3). Similar arguments will be valid for decays of positive muons.

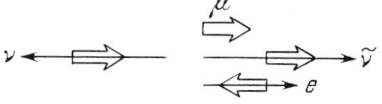

Fig. 3.3.

CHAPTER 4

Strangeness-conserving leptonic decays of hadrons. Properties of the *ud* current

The interaction between quark and leptonic currents leads to the leptonic decays of hadrons, sometimes referred to as semileptonic decays. In particular, the interaction between currents of light leptons,* $e\nu$ and $\mu\nu$, and the current of light quarks, ud, is responsible for decays of charged π-mesons, for β-decay of the neutron and atomic nuclei, for the capture of μ^- by nuclei, and for the decay $\Sigma \to \Lambda e\nu$. The same interaction operates in reactions involving neutrinos (see chapter 17). In the present chapter we shall discuss the general properties of the ud current. The coefficient with which the ud current enters the total charged current is $\cos\theta$, where θ is the Cabibbo angle: $\cos\theta \simeq 0.97$. When writing expressions for amplitudes of specific processes, we shall single out the factor $\cos\theta$ explicitly, and give them in the form

$$M = \sqrt{\tfrac{1}{2}}\, GL_\alpha H^\alpha \cos\theta,$$

where L_α is a leptonic matrix element, and H_α a hadronic matrix element. For instance, we obtain in the case of the process $\nu_\ell + i \to f + \ell$**

$$L_\alpha = \langle \ell | \bar{\ell} O_\alpha \nu_\ell | \nu_\ell \rangle e^{-ikx} = \bar{u}_\ell O_\alpha u_{\nu_\ell},$$

$$H_\alpha = \langle f | \bar{u} O_\alpha d | i \rangle e^{-iqx},$$

where i is the initial and f the final hadronic state,

$$k = p_\ell - p_\nu, \qquad q = p_f - p_i.$$

---

*We recall that the symbol $e\nu$ stands for one of two currents, $e\nu = \bar{e}O_\alpha \nu_e$ or $\bar{\nu}_e O_\alpha e$, and similarly, $\mu\nu = \bar{\mu}O_\alpha \nu_\mu$ or $\bar{\nu}_\mu O_\alpha \mu$, $ud = \bar{u}O_\alpha d$ or $\bar{d}O_\alpha u$, where $O_\alpha = \gamma_\alpha(1 + \gamma_5)$. This contracted notation is used when the discussion is of general character.
**We shall denote, in this chapter and hereafter, charged leptons by the symbol ℓ (for e and μ), using also the symbol ν_ℓ for ν_e and ν_μ.

In the general case, H_α is a sum of a polar and an axial vector:
$$H_\alpha = V_\alpha + A_\alpha,$$
where
$$V_\alpha = \langle f|\bar{u}\gamma_\alpha d|i\rangle e^{-iqx},$$
$$A_\alpha = \langle f|\bar{u}\gamma_\alpha\gamma_5 d|i\rangle e^{-iqx}.$$
The matrix elements V_α and A_α are expressed in terms of wave functions of hadrons in i and f states, and of their 4-momenta.

4.1. Isotopic properties of the *ud* current

The *ud* current is an isovector, since the u- and d-quarks are isospinors. Its vector component is G even, like the ρ-meson, and the axial component is G odd, like the A_1 or π meson. Matrix elements of these currents (V_α and A_α) must have the same properties. (We recall that the G-transformation is a product of charge conjugation and rotation by 180° around the y-axis in isotopic space. The G-parity of an isomultiplet is readily found by means of the formula $G = C_0(-1)^T$, where C_0 is the charge conjugation parity of the truly neutral component of the multiplet, and T is its isospin.)

4.2. Relationship between the vector current *ud* and the isovector electromagnetic current

There is a profound relationship between the vector currents $\bar{u}\gamma_\alpha d$, $\bar{d}\gamma_\alpha u$, and the isovector electromagnetic current of hadrons, $\bar{u}\gamma_\alpha u - \bar{d}\gamma_\alpha d$. The relationship stems from the fact that the three currents form a single isotopic triplet. This becomes clearer when we denote the isodoublet ($\begin{smallmatrix}u\\d\end{smallmatrix}$) by q. Then
$$\bar{u}\gamma_\alpha d = \bar{q}\gamma_\alpha \tau^+ q, \qquad \bar{d}\gamma_\alpha u = \bar{q}\gamma_\alpha \tau^- q,$$
$$\bar{u}\gamma_\alpha u - \bar{d}\gamma_\alpha d = \bar{q}\gamma_\alpha \tau_3 q,$$
where
$$\tau^+ = \begin{pmatrix} 0 & 1 \\ 0 & 0 \end{pmatrix}, \quad \tau^- = \begin{pmatrix} 0 & 0 \\ 1 & 0 \end{pmatrix}, \quad \tau_3 = \begin{pmatrix} 1 & 0 \\ 0 & -1 \end{pmatrix}.$$
(For these currents to have identical normalization, the last of them must be divided by $\sqrt{2}$. Sometimes $\tau^\pm$ are defined with an additional factor $\sqrt{2}$.)

We recall that the electromagnetic interaction between u- and d-quarks and the electromagnetic field A_α is of the form

$$eA_\alpha\left(\tfrac{2}{3}\bar{u}\gamma_\alpha u - \tfrac{1}{3}\bar{d}\gamma_\alpha d\right) = eA_\alpha\left(\tfrac{1}{2}\bar{q}\gamma_\alpha\tau_3 q + \tfrac{1}{6}\bar{q}\gamma_\alpha q\right).$$

The isoscalar current $\bar{q}\gamma_\alpha q$ is not the only isoscalar hadronic current. Indeed, there are isoscalar electromagnetic currents of other quarks: $-\tfrac{1}{3}\bar{s}\gamma_\alpha s$, $+\tfrac{2}{3}\bar{c}\gamma_\alpha c$, $-\tfrac{1}{3}\bar{b}\gamma_\alpha b$, and so on. The current $\bar{q}\gamma_\alpha\tau_3 q$ is, on the other hand, the only isovector current which defines isovector electromagnetic form factors of both the hadrons containing only u- and d-quarks and of all other hadrons, strange, charmed, and so on. In fact, the very existence of isotopic multiplets with non-zero isospin ($T \neq 0$) is caused by the presence of the light quarks, u and/or d, in the corresponding hadrons. Matrix elements of the currents $\bar{q}\tau^+\gamma_\alpha q$ and $\bar{q}\tau_3\gamma_\alpha q$ must be identical because they are in the same isotopic multiplet and also because the strong interaction is isotopically invariant. Specifically, the current $\bar{u}\gamma_\alpha d$ must be conserved like the electromagnetic current:

$$\frac{\partial}{\partial x_\alpha}\bar{u}(x)\gamma_\alpha d(x) = 0.$$

Consequently, all matrix elements of the *ud* current must be transverse:

$$q^\alpha V_\alpha = 0.$$

Before discussing numerous specific manifestations of the conserved vector current, CVC, let us analyze one important result of this conservation: non-renormalizability of the weak charge.

4.3. Weak charge

When the initial and final hadrons of a semi-leptonic decay belong to the same isotopic multiplet (such as the neutron and proton in the β-decay n $\to$ pe$^-\nu$), we can introduce the concept of the weak charge. The weak charge is defined as the time-like component of the 4-vector V_α for $q = 0$. The weak charge determines the strength of the weak vector interaction in the static limit. Evidently, the weak charge is a matrix element of the integral

$$\int u^+(x)d(x)\,\mathrm{d}x.$$

The integral is the generator $\hat{T}^+$ of the isotopic group SU(2) and its magnitude for a transition $T_3 \leftrightarrow T_3 + 1$ is determined completely by the hadron isospin T and its projection T_3. In analogy to a well-known relationship for the angular momentum operator, this integral is equal to

$$\sqrt{T(T+1) - T_3(T_3+1)},$$

where T_3 refers to the lower of the two components of the isotopic multiplet. This means, in particular, that the weak charge is unity in the β-decay of a neutron, and equals $\sqrt{2}$ in the β-decay of the π-meson, $\pi^+ \to \pi^0 e^+ \nu_e$. The weak charge has the same value in the so-called superallowed β-transitions in ^{14}O, ^{26m}Al, ^{34}Cl, ^{38m}K, ^{42}Sc, ^{46}V, ^{50}Mn, ^{54}Co. The levels involved in these transitions have $J^P = 0^+$ and constitute components of isotriplets. It is essential that in the limit of unperturbed isotopic invariance, the magnitude of the weak charge is unaffected by the structure of the decaying hadron. As a result, the weak charges of the nucleon and quark are identical, just as those of the π-meson and cobalt nucleus (provided we neglect small isotopic non-invariant corrections).

This property of the weak charge is of the same nature as the familiar (and absolutely non-trivial!) additivity of the electric charge. The charge additivity implies that the total electric charge of a system is equal to the sum of charges of the constituent particles, regardless of the form of interaction between the particles.

And although free quarks are never encountered, their weak interaction constant (that of the vector quark current ud) can be found experimentally, precisely because the weak charges of the nucleon and quark are identical. Otherwise we would hardly be justified in factoring out $\cos\theta$.

4.4. Chiral invariance

If the terms representing quark masses in the total quark-gluon lagrangian are omitted, the left-handed and right-handed quark fields become independent, and the degree of symmetry of the lagrangian increases. In particular, the lagrangian is invariant under isotopic transformations separately for the left-handed doublet and for the right-handed doublet,

$$q_L = \begin{pmatrix} u_L \\ d_L \end{pmatrix}, \quad q_R = \begin{pmatrix} u_R \\ d_R \end{pmatrix},$$

respectively, if we neglect the masses m_u and m_d (this approximation is especially good since both masses are small: $m_u \sim 4$ MeV, $m_d \sim 7$ MeV). Namely,

$$q_L \to e^{i\tau \cdot \omega_L/2} q_L, \quad q_R \to e^{i\tau \cdot \omega_R/2} q_R$$

and the relevant group is $SU(2)_L \times SU(2)_R$. This is the so-called chiral symmetry. The strong interaction lagrangian is said to be chiral invariant. The absolute chiral invariance implies conservation both of the vector currents forming the triplet $\bar{q}\gamma_\alpha \tau q$, and of the axial currents forming the triplet $\bar{q}\gamma_\alpha \gamma_5 \tau q$. The axial current is conserved, because by definition it is the difference between left-handed and right-handed currents, conserved due to chiral invariance.

It will be shown below that conservation of the axial current could be realized by introducing a triplet of massless mesons with the quantum numbers of π-mesons. The chiral generators $\hat{T}_5 = \int \bar{q}\gamma_0 \gamma_5 \tau q \, dx$ thereby transform the states with different numbers of such massless zero-energy π-mesons into one another: nucleon $\rightarrow$ nucleon $+ \pi \rightarrow$ nucleon $+ 2\pi \cdots$. Such a nonlinear realization of symmetry is usually referred to as the spontaneous breaking of symmetry.

In reality, chiral symmetry is not absolute since pion masses are distinct from zero. Nevertheless, pions are much lighter than other hadrons: $m_\pi^2 / m_p^2 \cong 2 \cdot 10^{-2}$, so that the observed deviations from chiral symmetry are proportional to a small parameter (m_π^2 is proportional to $m_u + m_d$). Instead of the rigorously conserved axial current (CAC), we have to deal with the partially conserved axial current (PCAC). Some of the corollaries of PCAC are discussed in chapter 5.

CHAPTER 5

Leptonic decays of pions and nucleons

The following leptonic decays of hadrons will be considered in this chapter: decays of π-mesons ($\pi^+ \to \ell^+ \nu$, where $\ell =$ e or μ, $\pi^+ \to e^+ \nu \pi^0$), β-decay of the neutron n $\to$ pe$^- \bar{\nu}$, and muon capture $\mu^- $p $\to$ nν. In addition, we shall briefly discuss the decays $\Sigma \to \Lambda e \nu$ which are also produced by the ud current.

5.1. $\pi \to \ell \nu$ decays

We start with $\pi^+ \to \ell^+ \nu$ decay. The quark graph of this decay is shown in fig. 5.1. The vertex marked in the graph by a black dot symbolizes that π^+ consists of u and $\bar{\text{d}}$. The vector component V_α of the matrix element H_α is zero, $V_\alpha = 0$, and A_α can be written in the form $A_\alpha = f_\pi \varphi_\pi p_\alpha$, where φ_π is the pion wave function, p_α is the pion 4-momentum, and f_π is a parameter of the dimension of mass (the constant $F_\pi = f_\pi / \sqrt{2}$ is sometimes used instead of f_π). The amplitude for this process is

$$M = \sqrt{\tfrac{1}{2}} \, G A_\alpha L^\alpha \cos\theta = \sqrt{\tfrac{1}{2}} \, G f_\pi \varphi_\pi p_\alpha \bar{u}_\nu \gamma_\alpha (1 + \gamma_5) u_\ell \cos\theta.$$

By using momentum conservation,

$$p_\pi = p_\nu + p_\ell,$$

and the Dirac equation ($\hat{p} - m)u = 0$, let us rewrite M in the form

$$M = \sqrt{\tfrac{1}{2}} \, G f_\pi m_\ell \varphi_\pi \bar{u}_\nu (1 - \gamma_5) u_\ell \cos\theta.$$

We find that the amplitude is proportional to m_ℓ and vanishes for $m_\ell \to 0$. The reason for this is as follows. It has been mentioned above that helicity is conserved in weak interaction. This fact, and the conservation of angular momentum forbid the spin-zero meson to decay into a left-handed neutrino

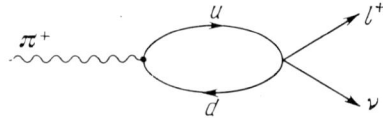

Fig. 5.1.

and a right-handed antilepton. We now find in a straightforward manner

$$\overline{|M|^2} = \tfrac{1}{2}G^2 f_\pi^2 m_\ell^2 8(p_\nu p_\ell)\cos^2\theta = 4G^2 f_\pi^2 m_\ell^2 m_\pi E_\nu \cos^2\theta,$$

where m_π is the pion mass, and E_ν is the neutrino energy in the pion rest frame ($p_\nu p_\ell = (p_\nu, p_\pi - p_\nu) = p_\nu p_\pi = m_\pi E_\nu$). Writing the decay probability in the form

$$\Gamma = \frac{\overline{|M|^2}}{2m}\Phi,$$

where Φ is the phase-space volume

$$\Phi = \int \frac{dk_\nu}{2E_\nu(2\pi)^3} \frac{dk_\ell}{2E_\ell(2\pi)^3} (2\pi)^4 \delta^4(k_\nu + k_\ell - p_\pi)$$

$$= \frac{1}{4\pi}\int \delta\left(E_\nu + \sqrt{E_\nu^2 + m_\ell^2} - m_\pi\right)\frac{E_\nu dE_\nu}{E_\ell} = \frac{E_\nu}{4\pi m_\pi},$$

we finally obtain

$$\Gamma = \frac{G^2 f_\pi^2 m_\ell^2 E_\nu^2 \cos^2\theta}{2\pi m_\pi} = \frac{G^2 f_\pi^2 m_\ell^2 m_\pi \cos^2\theta}{8\pi}\left(1 - \frac{m_\ell^2}{m_\pi^2}\right)^2.$$

A comparison of this expression with the experimentally measured $\pi \to \mu\nu$ decay rate yields $f_\pi \cong 130$ MeV. We shall see later that this quantity enters all relations derived on the basis of PCAC. The ratio of probabilities of $\pi \to e\nu$ and $\pi \to \mu\nu$ is independent of f_π and equals

$$\frac{\Gamma(\pi \to e\nu)}{\Gamma(\pi \to \mu\nu)} = \left(\frac{m_e}{m_\mu}\right)^2 \left(\frac{1 - m_e^2/m_\pi^2}{1 - m_\mu^2/m_\pi^2}\right)^2 \cong 1.3 \cdot 10^{-4}.$$

It must be emphasized that f_π is exactly the same for both decays, because the matrix element of the *ud* current is a function of only the total 4-momentum of the leptons, and is independent of the 4-momenta of each individual lepton.

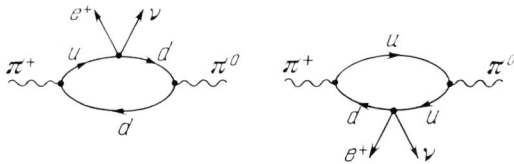

Fig. 5.2.

5.2. $\pi^+ \to \pi^0 e^+ \nu$ decay

The decay $\pi^+ \to \pi^0 e^+ \nu$ is represented by a sum of quark graphs (see fig. 5.2). Its amplitude is

$$M = \sqrt{\tfrac{1}{2}} \, G V_\alpha L^\alpha \cos\theta,$$

where

$$L_\alpha = \bar{\nu}\gamma_\alpha(1+\gamma_5)e,$$
$$V_\alpha = (f_+ p_\alpha + f_- q_\alpha)\varphi_2 \varphi_1.$$

Here φ_1 and φ_2 are wave functions of the π^+ and π^0 mesons, p_1 and p_2 are their 4-momenta, $p = p_1 + p_2$, $q = p_1 - p_2$. In the general case, the scalar dimensionless parameters f_+ and f_- are functions of q^2. We can assume, nevertheless, that the form factors f_+ and f_- are constant in the whole physical range of q^2,

$$f_+(q^2) \cong f_+(0), \qquad f_-(q^2) \cong f_-(0),$$

because of the negligibly small release of energy in the π_{e3} decay.

Because of the conservation of the vector current,

$$q^\alpha V_\alpha = 0.$$

This equality must hold in the limit of absolute isotopic invariance, with the electromagnetic interaction switched off. In this limit, $qp = m_{\pi^+}^2 - m_{\pi^0}^2 = 0$ and the transversality condition becomes

$$f_- q^2 = 0.$$

Hence, $f_- = 0$. As for $f_+(0)$, its magnitude is fixed because the vector current ud enters the same triplet as the isovector electromagnetic current. We have demonstrated in chapter 4 that $f_+(0)$ is the weak vector charge whose value in the transition $\pi^+ \to \pi^0 e^+ \nu$ is $\sqrt{T(T+1) - T_3(T_3+1)} = \sqrt{2}$. This value of $f_+(0)$ can easily be estimated from the graphs of fig. 5.2, if we take into account that weak charges in the transitions of quark ($u \to d e^+ \nu$)

and antiquark ($\bar{d} \to \bar{u}e^+\nu$) are $+1$ and -1, respectively. The matrix element in question is

$$\left\langle \pi^0 = \sqrt{\tfrac{1}{2}}(u\bar{u} - d\bar{d}) \,\middle|\, \bar{d}\gamma_0 u \,\middle|\, \pi^+ = u\bar{d} \right\rangle = \sqrt{2}.$$

Summarizing, the amplitude of the π_{e3} decay is given by

$$M = \sqrt{\tfrac{1}{2}}\, G\varphi_1\varphi_2\sqrt{2}\, p^\alpha \bar{u}_\nu \gamma_\alpha (1+\gamma_5) u_\ell \cos\theta.$$

Problem: Using this amplitude, calculate the width of the π_{e3} decay.
Answer:

$$\Gamma = \frac{G^2 \Delta^5}{30\pi^3}\left(1 - \frac{5m_e^2}{\Delta^2} - \frac{3}{2}\frac{\Delta}{m_\pi}\right)\cos^2\theta.$$

Here $\Delta = m_{\pi^+} - m_{\pi^0} \cong 4.5$ MeV; higher powers of the ratios m_e/Δ and Δ/m_π were neglected. The calculated probability of the π_{e3} decay comes to approximately 10^{-8} of the total width of the π-meson. The theoretical prediction for $\Gamma(\pi^+ \to \pi^0 e^+ \nu)$ is confirmed by experiment.

5.3. β-decay of the neutron

Both V_α and A_α are distinct from zero in the decay $n \to p e^- \bar{\nu}_e$. In the most general form,

$$V_\alpha = \bar{u}_p\!\left(f_1\gamma_\alpha + f_2\sigma_{\alpha\beta}q^\beta + f_3 q_\alpha\right)u_n,$$

$$A_\alpha = \bar{u}_p\!\left(g_1\gamma_\alpha + g_2\sigma_{\alpha\beta}q^\beta + g_3 q_\alpha\right)\gamma_5 u_n.$$

All six form factors f_i and g_i are functions of q^2, where $q = p_n - p_p = p_e + p_{\bar{\nu}}$. The characteristic scale of q^2 for these functions is of the order of 0.5 GeV^2. The only exception is $g_3(q^2)$, whose characteristic scale is much smaller (it is of the order of $m_\pi^2 \cong 0.02$ GeV^2). The energy released in β-decay is in all cases so small that all six form factors can be considered as constants. Experimentally, f_i and g_i as functions of q^2 are found from neutrino experiments

$$\nu_\mu(\nu_e) + n \to p + \mu^-(e^-),$$

$$\bar{\nu}_\mu(\bar{\nu}_e) + p \to n + \mu^+(e^+).$$

These experiments are discussed in Chapter 17.

5.4. Vector form factors

Let us consider first the vector form factors f_1, f_2, f_3. The weak charge is found from conservation of the vector current: $f_1(0) = 1$. This can be obtained, of course, by adding quark amplitudes of the graphs of fig. 5.3. In the case under discussion, however, it is simpler to resort to the general formula which gives the weak charge as $\sqrt{T(T+1) - T_3(T_3+1)}$. In the transition n → p we have $T = \frac{1}{2}$, $T_3 = -\frac{1}{2}$; consequently, $f_1(0) = 1$, and the weak charge of nucleons is equal to that of the d, u quarks. The data on the neutron β-decay and on superallowed transitions in nuclei (induced precisely by the weak vector charge, see Chapter 4) then yield the value of $\cos\theta$.

Usually the quantity $f_2 \sigma_{\alpha\beta} q_\beta$ is called the weak magnetism. The value of f_2 can be predicted by using the similarity of the weak and electromagnetic currents. Indeed, the electromagnetic nucleon vertices are

$$\gamma_\alpha + \frac{1.79}{2m_p} \sigma_{\alpha\beta} q^\beta \quad \text{for the proton,}$$

and

$$-\frac{1.91}{2m_p} \sigma_{\alpha\beta} q^\beta \quad \text{for the neutron.}$$

This means that the isoscalar vertex is

$$\gamma_\alpha - \frac{0.12}{2m_p} \sigma_{\alpha\beta} q^\beta \quad (\bar{p}p + \bar{n}n),$$

and the isovector vertex is

$$\gamma_\alpha + \frac{3.7}{2m_p} \sigma_{\alpha\beta} q^\beta \quad (\bar{p}p - \bar{n}n).$$

The isovector electromagnetic current enters the same triplet as the weak vector current; hence, the last expression is also valid for V_α. Since $f_1(0) = 1$,

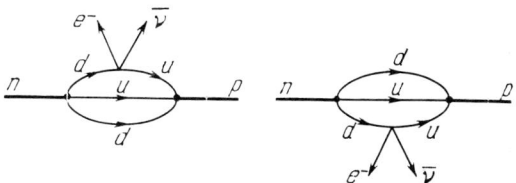

Fig. 5.3.

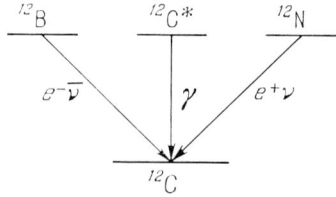

Fig. 5.4.

we obtain $f_2(0) = 3.7/2m_p$. At the same time this gives $f_3 = 0$. The term $f_3 q_\alpha$ is usually quoted as the effective scalar. This originates in the possibility of expressing this term as a scalar:

$$f_3 q_\alpha \bar{u}_\nu \gamma_\alpha (1 + \gamma_5) u_e = m_e f_3 \bar{u}_\nu (1 - \gamma_5) u_e.$$

Two facts are responsible for $f_3 = 0$. First of all, f_3 vanishes because of the transversality of V_α ($q^\alpha V_\alpha = 0$ yields $q^2 f_3 = 0$), and secondly, because V_α is G even and the isovector scalar must be G odd. The best experimental restrictions on the value of f_3 are found in μ-capture, $\mu^- p \to n\nu$.

Let us return to the weak magnetism. In the β-decay of the neutron, this term is unfortunately very small and therefore very difficult to observe. Its contribution is considerably higher in β-transitions of nuclei with higher energy release. Isotopically symmetric β-transitions of boron and nitrogen nuclei give a good example (see fig. 5.4). In these transitions, weak magnetism is calculated from the matrix element in the radiative decay of the excited level $^{12}C^*$; this level is in the same isotopic triplet with the ground states of ^{12}B and ^{12}N (all three levels have $J^P = 1^+$).

5.5. Axial form factors

The names given to the three axial form factors are: g_1 axial charge, g_2 weak "electrism", g_3 effective pseudo-scalar. We begin with the term $g_2 \sigma_{\alpha\beta} q_\beta \gamma_5$. This term is similar to the electric dipole moment of the neutron, which is known to be forbidden by the CP invariance of the electromagnetic interaction. Likewise, the weak electric term is also forbidden, being G even while the axial current is G odd. The terms $g_1 \gamma_\alpha \gamma_5$ and $g_3 q_\alpha \gamma_5$ are G odd and therefore allowed. They are related to each other owing to partial conservation of the axial current (PCAC). In the limit of massless pions, the axial current is conserved and matrix elements are transverse:

$$q^\alpha A_\alpha = 0.$$

The term $g_1 \gamma_\alpha \gamma_5$ is in itself not transverse because of the non-vanishing nucleon masses:

$$q_\alpha \bar{u}_p \gamma_\alpha \gamma_5 u_n = (m_p + m_n) \bar{u}_p \gamma_5 u_n.$$

In order to make it transverse, we multiply it by a transverse expression $(\delta_\beta^\alpha - q^\alpha q_\beta / q^2)$. Multiplication of the expression obtained, $g_1 \bar{u}_p \gamma_\beta \gamma_5 (\delta_\alpha^\beta - q^\beta q_\alpha / q^2) u_n$, by q^α yields zero. But in addition to the term $g_1 \bar{u}_p \gamma_\alpha \gamma_5 u_n$, this expression contains one more term which is easily recognizable as the effective pseudoscalar:

$$- g_1 \bar{u}_p \gamma_\beta \gamma_5 \frac{q^\beta q_\alpha}{q^2} u_n = \frac{2m g_1}{q^2} \bar{u}_p \gamma_5 u_n q_\alpha.$$

(We have used the equalities $\hat{p}_n u_n = m_n u_n$, $\bar{u}_p \hat{p}_p = m_p u_p$, and the approximation $m_p = m_n = m$). We find that the effective pseudoscalar form factor contains a π-meson pole at $q^2 = 0$ (we recall that so far the π-meson is assumed to be massless). Phenomenologically, this pole corresponds to the graph of fig. 5.5. The constants f_π and g which characterize this graph are known. The product $f_\pi q_\alpha$ characterizes the decay $\pi \to e\nu$, and we have found at the beginning of this chapter that $f_\pi \cong 130$ MeV. The constant g characterizes the π-meson-nucleon strong interaction. If the $\pi \bar{N} N$ vertex is written in the form

$$g \varphi_\pi \bar{u}_N \gamma_5 \tau u_N = g \Big[\sqrt{2} \varphi_{\pi^-} \bar{u}_p \gamma_5 u_n + \sqrt{2} \varphi_{\pi^+} \bar{u}_n \gamma_5 u_p + \varphi_{\pi^0} \bar{u}_p \gamma_5 u_p - \varphi_{\pi^0} \bar{u}_n \gamma_5 u_n \Big],$$

then the experimentally obtained value of g is such that $g^2/4\pi \cong 14$. In order to specify the normalization, let us remark that a similar quantity in electrodynamics is $e^2/4\pi = \alpha = \frac{1}{137}$. Some time ago the constant $g^2/4\pi$ was considered to be as fundamental for the strong interaction as $e^2/4\pi$ for the electromagnetic one. Now it is clear that $g^2/4\pi$ is a purely phenomenological quantity and has nothing to do with the fundamental lagrangian of the strong interaction. Let us recast the π-nucleon vertex in the gradient form:

$$\sqrt{2} g \varphi_\pi \bar{u}_p \gamma_5 u_n = - \frac{\sqrt{2} g}{2m} q^\beta \varphi_\pi \bar{u}_p \gamma_\beta \gamma_5 u_n = - \sqrt{2} \frac{q^\beta}{f} \varphi_\pi \bar{u}_p \gamma_\beta \gamma_5 u_n.$$

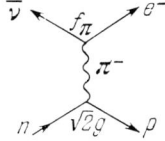

Fig. 5.5.

We have $f = 2m/g \cong 140$ MeV. (The quantity f introduced here must not be confused with f_1, f_2, f_3 and f_π introduced earlier, or with $f = g/2m$ often encountered in the literature). It is clear that the pole term in the amplitude as given by the graph of fig. 5.5 equals

$$-\sqrt{2}\,\frac{f_\pi}{f}\,\frac{q_\alpha q^\beta}{q^2}\,\bar{u}_p \gamma_\beta \gamma_5 u_n.$$

On the other hand, multiplication of the matrix element $g_1 \bar{u}_p \gamma_\beta \gamma_5 u_n$ by $\delta_\alpha^\beta - q_\alpha q^\beta/q^2$ yielded

$$g_1 \bar{u}_p \gamma_\beta \gamma_5 u_n (\delta_\alpha^\beta - q_\alpha q^\beta/q^2).$$

Consequently,

$$g_1 = \sqrt{2}\,\frac{f_\pi}{f}.$$

This is the so-called Goldberger-Treiman relation. Using numerical values of f_π and f, we obtain $g_1 \cong 1.3$ which is in good agreement with the experimental ratio $g_A/g_V = 1.25$. (The symbols g_V and g_A are often used in the literature instead of our f_1 and g_1; sometimes g_A stands for $-g_1$). It should be kept in mind that experimental values of f_π and f are found for $q^2 = m_\pi^2$, and g_1 is obtained for $q^2 = 0$. We assume that a shift of $m_\pi^2 \cong 0.02$ GeV^2 has a negligible effect on these quantities.

The pion pole in the matrix element of the axial current appears because of the spontaneous breaking of the chiral symmetry. As has been mentioned in chapter 4, chiral symmetry is realized in a non-linear manner (owing to the non-zero masses of the nucleons), by invoking an isotopic triplet of massless pseudoscalar particles. Such spin-zero massless particles accompanying the spontaneous breaking of symmetry, are called the Goldstone bosons. β-decay gives an example of how the conservation of the axial current (transversality of A_α) is provided by the pole term corresponding to a massless Goldstone meson:

$$-g_1 \bar{u}_p \gamma_\beta \gamma_5 u_n \frac{q^\beta q_\alpha}{q^2}.$$

Multiplication of this term by the leptonic matrix element yields an expression which is the effective pseudoscalar:

$$-g_1 \bar{u}_p \gamma_\beta \gamma_5 u_n \frac{q^\beta q_\alpha}{q^2} L_\alpha = -g_1 \frac{2m \cdot m_\ell}{q^2} \bar{u}_p \gamma_5 u_n \bar{u}_\ell (1 + \gamma_5) u_\nu.$$

The effective pseudoscalar is very small in β-decay because of the small electron mass. It is much larger in the case of μ-capture.

The finite mass of the π-meson cannot be ignored any more if the numerical value of the effective pseudoscalar is discussed. Normally one takes account of the pion mass in a most naive manner, by inserting m_π^2 into the propagator "by hand", so that the "constant" in the effective pseudoscalar, $g_P = m_\ell g_3$, becomes

$$g_P = \frac{2mm_\ell g_1}{q^2 - m_\pi^2}.$$

We see that g_P is a function of q^2. Practically all energy (of the order of m_μ) in the process $\mu^- p \to n\nu$ is carried away by the neutrino, with the nucleon gaining a momentum of the order of m_μ, so that $q^2 \cong -m_\mu^2$. It follows then that $g_P/g_1 \cong -6$. This high value of the effective pseudoscalar is in agreement with experiments on μ-capture.

5.6. The probability of β-decay. Angular correlations

It has been mentioned above that the lagrangian of the interaction between leptonic currents and currents of light quarks is CP invariant. This means, by virtue of the CPT theorem, that the lagrangian is also invariant under time reversal T. The T-invariance of the lagrangian can be shown to result in real form factors f_1, f_2, f_3 and g_1, g_3 entering the matrix element of the β-decay of the neutron. The same is true for the form factors in pion decays.

We have thus completed an analysis of all six form factors in the β-decay of the neutron. The vector and axial charges are the main contributors to this process, so that the decay amplitude can be written in the form

$$M = \sqrt{\tfrac{1}{2}}\, G \bar{u}_p \gamma_\alpha (1 + \alpha\gamma_5) u_n \bar{u}_e \gamma^\alpha (1 + \gamma_5) u_\nu \cos\theta,$$

where

$$\alpha = g_1/f_1 = 1.253 \pm 0.007.$$

Problem 1: Calculate the $e\nu$ correlation in the decay of the non-polarized neutron.

Answer: The decay probability is proportional to $1 + \lambda v_e \cdot n_\nu$ where v_e is the electron velocity vector (as before, in units where $c = 1$), n_ν is the neutrino velocity vector ($|n_\nu|^2 = 1$), and $\lambda = (1 - \alpha^2)/(1 + 3\alpha^2)$. For $\alpha \cong 1.25$, $\lambda \cong -0.1$.

Problem 2: Calculate the electron spectrum in the neutron decay.

Answer:

$$d\Gamma = \frac{G^2(1+3\alpha^2)\cos^2\theta}{2\pi^3}(E_e^2 - m_e^2)^{1/2}(\Delta - E_e)^2 E_e dE_e,$$

where

$$\Delta = m_n - m_p \cong 1.29 \text{ MeV}.$$

Problem 3: Calculate the total probability of the neutron decay.
Answer: Integration of the preceding equation yields

$$\Gamma = \frac{G^2\Delta^5\cos^2\theta}{60\pi^3}(1+3\alpha^2)0.47.$$

Problem 4: Calculate the longitudinal polarization of electrons in the neutron decay.
Answer: $P_e = -v_e$, where v_e is the electron velocity.

Problem 5: Calculate the angular distribution of electrons in the decay of the polarized neutron.
Answer: $1 + \rho_e v_e \eta$, where η is a unit vector in the direction of the neutron polarization, and

$$\rho_e = -\frac{2(\alpha^2 - \alpha)}{1+3\alpha^2} \cong -0.1 \quad \text{for } \alpha \cong 1.25.$$

Problem 6: Calculate the angular distribution of neutrinos in the decay of the polarized neutron.
Answer: $1 + \rho_\nu n_\nu \eta$, where

$$\rho_\nu = \frac{2(\alpha^2 + \alpha)}{1+3\alpha^2} \cong 1.$$

Note that $\eta[p_e \times p_\nu]$-type correlation in the neutron decay must be absent in the Born approximation (by virtue of T-invariance). Indeed, the substitution $t \to -t$ changes sign in each factor of this product; consequently, the whole product is T odd. The coefficient of this correlation is proportional to $\sin\delta$, where $\delta = \arg\alpha$. The experimental value is $\delta = (0.20 \pm 0.19)°$.

5.7. $\Sigma^\pm \to \Lambda e^\pm \nu$ decays

Being strangeness-conserving, $\Sigma^\pm \to \Lambda e^\pm \nu$ decays are therefore induced by the ud current. The general form of V_α and A_α for these decays is the same as for the β-decay of the neutron, but $f_1(0) = 0$ because Σ- and Λ-hyperons are not in the same isotopic multiplet. This is found most easily if we use the condition $q^\alpha V_\alpha = 0$, which gives

$$(m_\Sigma - m_\Lambda) f_1(q^2) + q^2 f_3(q^2) = 0.$$

Hence, $f_1(q^2) \to 0$ for $q^2 \to 0$. As a result, the term predominant in V_α is $f_2 \sigma_{\alpha\beta} q^\beta$, where $f_2 = \mu_{\Sigma\Lambda}/e$, and the quantity $\mu_{\Sigma\Lambda}$ determines the width of the Σ^0 hyperon with respect to the $\Sigma^0 \to \Lambda^0 \gamma$ decay. ($\Sigma^\pm \to \Lambda e^\pm \nu$ finds its total electromagnetic analog in $\Sigma^0 \to \Lambda e^+ e^-$.) The contribution of V_α to the total probability of $\Sigma^\pm \to \Lambda e^\pm \nu$ decays is negligible since $f_2 |q|$ is small ($f_2 \sim 1/m_\Lambda$, $|q|_{max} \sim m_\Sigma - m_\Lambda$). As for the axial matrix element A_α, it contains $g_1(0) \neq 0$ so that the probability of $\Sigma^\pm \to \Lambda e^\pm \nu$ decays is determined by this quantity, identical for both decays. Since $m_{\Sigma^-} - m_\Lambda \cong 82$ MeV and $m_{\Sigma^+} - m_\Lambda \cong 74$ MeV, we obtain

$$\frac{\Gamma(\Sigma^+ \to \Lambda e^+ \nu)}{\Gamma(\Sigma^- \to \Lambda e^- \bar{\nu})} = \left(\frac{74}{82}\right)^5 \cong 0.6.$$

Experimental data are in agreement with this prediction.

CHAPTER 6

Leptonic decays of K-mesons and hyperons

Most leptonic decays of strange particles are strangeness changing. The interaction that gives rise to these decays is the coupling of the *us* current to the leptonic currents $e\nu$ and $\mu\nu$. The following notation is used in the literature for muonic decays of K^+ and K^0 mesons:

$$K^+ \to \mu^+ \nu \qquad (K^+_{\mu 2}), \qquad K^+ \to \mu^+ \nu \pi^+ \pi^- \qquad (K^+_{\mu 4}),$$

$$K^+ \to \mu^+ \nu \pi^0 \qquad (K^+_{\mu 3}), \qquad K^+ \to \mu^+ \nu \pi^0 \pi^0 \qquad (K^{+\prime}_{\mu 4}),$$

$$K^0 \to \mu^+ \nu \pi^- \qquad (K^0_{\mu 3}), \qquad K^0 \to \mu^+ \nu \pi^0 \pi^- \qquad (K^0_{\mu 4}).$$

Similar notation is used for the decays of K^- and $\overline{K}^0$ mesons. If a decay produces a pair $e\nu$ instead of $\mu\nu$, the symbols change to K^+_{e2}, K^+_{e3}, and so on. In cases where it is immaterial whether an electron or a muon is emitted, we use the symbols $K_{\ell 2}$, $K_{\ell 3}$, $K_{\ell 4}$ where ℓ denotes a charged lepton ($\ell =$ e or μ).

No special contracted notation was introduced for leptonic decays of hyperons. The following hyperon decays will be considered in this chapter:

$$\Lambda \to p\ell^- \bar{\nu}, \qquad \Xi^- \to \Sigma^0 \ell^- \bar{\nu},$$

$$\Sigma^- \to n\ell^- \bar{\nu}, \qquad \Xi^0 \to \Sigma^+ \ell^- \bar{\nu},$$

$$\Xi^- \to \Lambda \ell^- \bar{\nu}, \qquad \Omega^- \to \Xi^0 \ell^- \bar{\nu}.$$

6.1. $|\Delta S| = 1$ and $\Delta Q = \Delta S$ rules

All strangeness-changing leptonic decays occur through the *us* current which changes strangeness by unity; therefore, all these decays must be subject to the selection rule $|\Delta S| = 1$. This rule holds in all the decays listed above.

Decays violating this rule,

$$\Xi^- \to n\ell^-\bar{\nu}, \qquad \Xi^0 \to p\ell^-\bar{\nu},$$
$$\Omega^- \to \Lambda^0\ell^-\bar{\nu}, \qquad \Omega^- \to \Sigma^0\ell^-\bar{\nu}, \qquad \Omega^- \to n\ell^-\bar{\nu}$$

have never been observed experimentally.

Another typical prediction of the theory is the selection rule $\Delta Q = \Delta S$ for the change of the charge and strangeness of hadrons. In the $s \to u$ transition both the strangeness S and electric charge Q increase by unity; hence, the selection rule is $\Delta Q = \Delta S$. Decays violating this rule,

$$K^0 \to \ell^-\bar{\nu}\pi^+, \qquad \overline{K}^0 \to \ell^+\nu\pi^-, \qquad K^+ \to \ell^-\bar{\nu}\pi^+\pi^+, \qquad \Sigma^+ \to n\ell^+\nu$$

have never been observed. It has to be underlined, though, that both these decays and the above-mentioned decays with $\Delta S > 1$ are forbidden theoretically to an extent much higher than the achieved experimental accuracy. Indeed, the selection rules are violated only in second and higher orders of perturbation theory, so that the anticipated accuracy must be of the order of 10^{-14}.

6.2. SU(3) and SU(2) properties of the *us* current

The coupling of the *us* current to the $e\nu$ or $\mu\nu$ current is proportional to $\sin\theta$, where θ is the Cabibbo angle ($\theta \simeq 13°$). We shall therefore present the decay amplitudes in the form

$$M = \sqrt{\tfrac{1}{2}}\, GH_\alpha L^\alpha \sin\theta,$$

where L^α is the leptonic matrix element, $L_\alpha = \bar{u}_\ell \gamma^\alpha(1+\gamma_5)u_\nu$, and H_α is the hadronic matrix element

$$H_\alpha = \langle f | \bar{u} O_\alpha s | i \rangle,$$

where $O_\alpha = \gamma_\alpha(1+\gamma_5)$.

We readily find that the currents $\bar{u}O_\alpha d$, $\bar{d}O_\alpha u$, $\bar{u}O_\alpha s$, $\bar{s}O_\alpha u$ are in the same octet

$$\bar{q}^i O_\alpha q_k - \tfrac{1}{3}\delta^i_k \bar{q}^m O_\alpha q_m,$$

where $q_1 = u$, $q_2 = d$, $q_3 = s$. Two more diagonal left-handed currents are in this octet: the isovector current $\sqrt{\tfrac{1}{2}}(\bar{u}O_\alpha u - \bar{d}O_\alpha d)$ and the isoscalar one $\sqrt{\tfrac{1}{6}}(\bar{u}O_\alpha u + \bar{d}O_\alpha d - 2\bar{s}O_\alpha s)$, participating in the weak interaction of the neutral currents. Also included are strangeness-changing neutral currents $\bar{d}O_\alpha s$ and $\bar{s}O_\alpha d$ which do not take part in the weak interaction. These unequal roles of different components of the current octet should not be

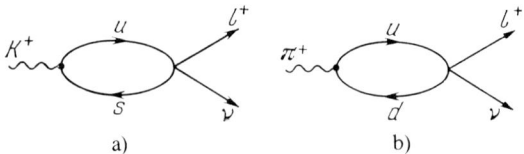

Fig. 6.1.

surprising since the weak interaction is not SU(3) invariant. What we do use is the SU(3) invariance of the strong interaction which enables us to find the relationships between distinct matrix elements of the currents us and ud.

As for the isotopic properties of the us current, it is obviously an isotopic spinor. Indeed, the isospins of the u- and s-quark are $\frac{1}{2}$ and 0, respectively, so that the us current must change the hadron isospin by $\frac{1}{2}$. We shall discuss this selection rule in more detail when considering $K_{\ell 3}$ decays.

6.3. $K_{\ell 2}$ decays

The $K^+ \to \ell^+ \nu$ decays (fig. 6.1a) are exact analogs of the $\pi^+ \to \ell^+ \nu$ decays (fig. 6.1b) in the unbroken SU(3) symmetry approximation. In this approximation, $m_u = m_d = m_s$, $m_K = m_\pi$, and the amplitudes of $K_{\ell 2}$ and $\pi_{\ell 2}$ decays differ only in the factors $\sin \theta$ and $\cos \theta$, respectively. Let us write the $K_{\ell 2}$ decay amplitude in the form

$$M = \sqrt{\tfrac{1}{2}}\, G f_K\, p^\alpha \bar{u}_\nu \gamma_\alpha (1 + \gamma_5) u_\ell \sin \theta,$$

where p is the 4-momentum of the K-meson. The SU(3) symmetry then yields $f_K = f_\pi$, and although SU(3) is violated, experimentally f_K is close to f_π. Calculation of the decay probability gives

$$\Gamma(K \to \ell \nu) = \frac{G^2}{8\pi} f_K^2 m_K m_\ell^2 \left(1 - \frac{m_\ell^2}{m_K^2}\right)^2 \sin^2 \theta.$$

Assuming $\sin \theta = 0.21$ (see the subsection below on $K_{\ell 3}$ decays), we obtain from the data on the width $\Gamma(K^+ \to \mu^+ \nu)$ that $f_K/f_\pi = 1.27$. The agreement with SU(3) is sufficiently good, especially if one takes into account that SU(3) is violated much more strongly in the masses of the K- and π-mesons.

Note that the ratio of the widths of K_{e2} and $K_{\mu 2}$ decays is much closer to its asymptotic limit, m_e^2/m_μ^2, than in the case of $\pi_{\ell 2}$ decays; this is caused by the phase-space volume being larger in the former case.

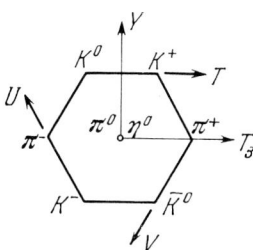

Fig. 6.2.

6.4. $K_{\ell 3}$ decays

Let us consider first $K_{\ell 3}^0$ decay. Its amplitude can be written in the form

$$M = \sqrt{\tfrac{1}{2}}\, G \sin\theta \left[f_+(q^2) p_\alpha + f_-(q^2) q_\alpha \right] \varphi_\pi \varphi_K \bar{u}_\nu \gamma^\alpha (1+\gamma_5) u_\ell,$$

where $p = p_K + p_\pi$, and $q = p_K - p_\pi$ is the 4-momentum carried by leptons. According to the Dirac equation, the f_- term gives a contribution proportional to m_ℓ, and thus can be neglected for K_{e3} decay. In the approximation of unbroken SU(3) symmetry, and for $q^2 = 0$, the term proportional to f_+ is readily related to a similar term in the π_{e3} decay, which yields that in $K_{\ell 3}^0$ decay, $f_+^0(0) = 1$. It can be demonstrated that the correction to this must be quadratic in the SU(3) symmetry breaking parameter, and therefore be small (this is the so-called Ademollo-Gatto theorem). The equality $f_+^0(0) = 1$ could be obtained in several ways. For instance, a glance at the $T_3 Y$ diagram of the octet of pseudoscalar mesons (fig. 6.2) clearly shows that $\bar{K}^0$ and π^+ are the upper and lower components, respectively, of the V-doublet. The $\int u^+(x) s(x) d^3x$ operator being the generator in V-space raising the projection of V-spin, the "weak charge" for the $\bar{K}^0 \to \pi^+$ transition is given by

$$f_+^0(0) = \sqrt{V(V+1) - V_3(V_3+1)} = 1.$$

By virtue of the $\Delta T = \tfrac{1}{2}$ selection rule, $f_+(0)$ for the $K_{\ell 3}^+$ decay must be smaller by a factor of $\sqrt{2}$:

$$f_+^+(0)/f_+^0(0) = 1/\sqrt{2}.$$

This relation must hold to the same accuracy as violation of isotopic invariance of the strong interaction, that is to about one percent. This relation can be derived by using, for example, the well-known Clebsch-Gordan coefficients and assuming formally that lepton pairs in the decays $K^+ \to \pi^0 e^+ \nu$ and $K^0 \to \pi^- e^+ \nu$ behave as a "spurious particle" with $T = \tfrac{1}{2}$

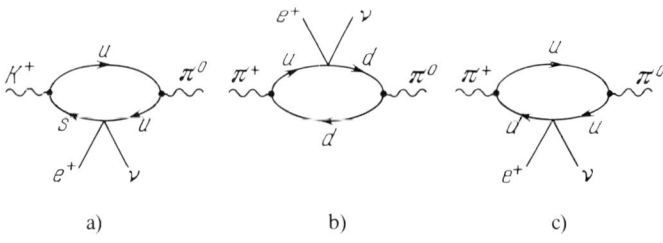

Fig. 6.3.

and $T_3 = +\frac{1}{2}$. Consequently, $f_+^+(0) = 1/\sqrt{2}$ in the approximation of unbroken SU(3) symmetry. This is also easily derived if one compares the quark graphs for K_{e3}^+ decay (fig. 6.3a) and π_{e3} decay (figs. 6.3b and c). Only one graph contributes to K_{e3} decay, while two graphs must be taken into consideration for π_{e3} decay. This occurs because the isoscalar pair $s\bar{s}$ does not enter the isovector π^0 meson. If we recall now that, in π_{e3} decay, $f_+(0) = \sqrt{2}$, we immediately obtain $f_+^+(0) = 1/\sqrt{2}$ for K_{e3}^+ decay. Experimentally, $f_+^+(q^2)$ and $f_+^0(q^2)$ are smooth functions of q^2 which are normally written in the linear approximation as

$$f_+^+(q^2) = f_+^+(0)\left(1 + \lambda_+^+ \frac{q^2}{m_\pi^2}\right) \simeq \frac{1}{\sqrt{2}}\left(1 + \lambda_+^+ \frac{q^2}{m^2}\right),$$

$$f_+^0(q^2) = f_+^0(0)\left(1 + \lambda_+^0 \frac{q^2}{m_\pi^2}\right) \simeq 1 + \lambda_+^0 \frac{q^2}{m^2},$$

where $\lambda_0^0 = \lambda_+^0 \simeq 0.030$. A standard calculation which we suggest as an exercise for the reader, yields

$$\Gamma(K_{e3}^+) = \frac{G^2 \sin^2 \theta}{768 \pi^3}\left|f_+^+(0)\right|^2 m_K^5 (0.57 + 0.14 \lambda_+^+).$$

A comparison of this expression with the experimental width of K_{e3}^+ decay (under the assumption $\sqrt{2} f_+^+(0) = 1$) yields that $\sin \theta \simeq 0.21$.

6.5. $K_{\mu 3}$ decays

$K_{\mu 3}$ decay is of interest because it could furnish information on the value of $f_-(q_2)$, provided we take for granted the $V - A$ character of the coupling and the μ-e universality which are fundamental to the standard theory of weak interactions. In the limit of SU(3) symmetry, $f_-(0) = 0$. The Ademollo

-Gatto theorem is not valid, however, with respect to $f_-(0)$, so that SU(3) can be expected to be broken strongly in this case. Thus, it is not impossible that $|\xi(0)| \sim 1$, where

$$\xi(q^2) = f_-(q^2)/f_+(q^2).$$

PCAC serves to derive the relation between the form factors f_K of $K_{\mu 2}$ decay and $f_+(m_K^2)$ and $f_-(m_K^2)$ of $K_{\mu 3}$ decay [the last two are taken at a non-physical point in which the pion 4-momentum is zero; hence, $q^2 = (p_K - p_\pi)^2 = p_K^2 = m_K^2$]. The expression in question is called the Callan-Treiman formula:

$$f_+^0(m_K^2) + f_-^0(m_K^2) = f_K/f_\pi.$$

Using the experimental value of the ratio $f_K/f_\pi = 1.27$, and assuming $f_+^0(m_K^2) = 1 + 0.03 \, m_K^2/m_\pi^2$, we find $f_-(m_K^2) \cong -0.1$. In principle, the data on f_- can be extracted from the results on muon polarization in $K_{\mu 3}$ decay, from the angular and energy distributions of muons and pions and from accurate measurements of the ratio $\Gamma(K_{\mu 3})/\Gamma(K_{e3})$. So far we have no generally accepted value of $\xi(0)$ owing to the small scale of the effect, and the quantity λ_- in the formula $f_-(q^2) = f_-(0)[1 + \lambda_- q^2/m^2]$ is still unknown.

6.6. K_{e4} decays

The decay $K^+ \to \pi^+\pi^- e^+ \nu_e$ is the best investigated among the K_{e4} decays. Its amplitude can be written in the form

$$M = \sqrt{\tfrac{1}{2}} \, G(V_\alpha + A_\alpha)L^\alpha \sin\theta,$$

where

$$A_\alpha = f_1(p_1 + p_2)_\alpha + f_2(p_1 - p_2)_\alpha + f_3(p - p_1 - p_2)_\alpha,$$

$$V_\alpha = f_4 m_K^{-2} \varepsilon_{\alpha\mu\nu\rho} p_\mu p_{1\nu} p_{2\rho},$$

p_1, p_2, p being the momenta of the π^-, π^+, and K mesons, respectively. The form factors f_1, f_2, f_3, f_4 are functions of three scalar variables, $pp_1, pp_2, p_1 p_2$. The contribution of the term f_3 is negligibly small because of the smallness of the electron mass. The contribution of the f_4 term is also small owing to high powers of momentum. An estimate that can be obtained for the form factors f_1 and f_2, assuming them constant and using PCAC, is

$$f_1 \approx f_2 \approx 1/f_\pi.$$

The K_{e4} decay is especially interesting because it enables us to study in pure form the scattering of pions on pions.

6.7. Leptonic decays of hyperons

Let us derive relations between the amplitudes for leptonic decays of baryons, which follow from the SU(3) symmetry of strong interactions and from the fact that the currents $\bar{u}d$ and $\bar{u}s$ belong to the same octet. Consider the baryon octet with $J^P = \frac{1}{2}^+$ (see Chapter 29, sect. 2):

$$B^i_k = \begin{pmatrix} \sqrt{\frac{1}{6}}\Lambda + \sqrt{\frac{1}{2}}\Sigma^0 & \Sigma^+ & p \\ \Sigma^- & \sqrt{\frac{1}{6}}\Lambda - \sqrt{\frac{1}{2}}\Sigma^0 & n \\ \Xi^- & \Xi^0 & -\sqrt{\frac{2}{3}}\Lambda \end{pmatrix}.$$

Here i is the number of the row, and k the number of the column, $i, k = 1, 2, 3$. The trace of B^i_k is zero: $B^i_i = 0$. We shall study the amplitudes

$$\langle B^m_i | j^i_k | B^k_m \rangle \quad \text{and} \quad \langle B^k_m | j^i_k | B^m_i \rangle$$

for the case of

$$j^1_2 = \bar{q}^1 q_2 \equiv \bar{u}d \quad \text{and} \quad j^1_3 = \bar{q}^1 q_3 \equiv \bar{u}s.$$

The current j^1_2 turns the d-quark into the u-quark, and j^1_3 turns s into u. In each amplitude, three octets are to be multiplied, and a scalar thereby obtained. Note that $8 \times 8 = 1 + 8 + 8 + 10 + \overline{10} + 27$ (see Chapter 29), so that two distinct scalars are present in the product $8 \times 8 \times 8$: the symmetrical one, D, and the antisymmetrical one, F. The amplitude in the general form is

$$(D + F)\langle B^m_i | j^i_k | B^k_m \rangle + (D - F)\langle B^k_m | j^i_k | B^m_i \rangle,$$

where D and F are scalar parameters which in the general case are different for different matrix elements of currents, that is for form factors $f_1, f_2, f_3, g_1, g_2, g_3$ (see Chapter 5 where f_i and g_i are defined for β-decay of the neutron). If the initial baryons are described by the matrix B^i_k, the final ones are given by $\bar{B}^k_i$ which is obtained from B^k_i by conjugation of each element and transposition. It is convenient to multiply B by $\bar{B}$ without writing out the explicit expression of the latter. The matrix elements of the current $j^1_2 \cos\theta$ are

$$\cos\theta\Big\{(D+F)\Big[\Big(\sqrt{\tfrac{1}{6}}\bar{\Lambda}^0 + \sqrt{\tfrac{1}{2}}\bar{\Sigma}^0, \Sigma^-\Big) + \Big(\bar{\Sigma}^+, \sqrt{\tfrac{1}{6}}\Lambda - \sqrt{\tfrac{1}{2}}\Sigma^0\Big) + (\bar{p}n)\Big]$$
$$+ (D-F)\Big[\Big(\bar{\Sigma}^+, \sqrt{\tfrac{1}{6}}\Lambda^0 + \sqrt{\tfrac{1}{2}}\Sigma^0\Big)$$
$$+ \Big(\sqrt{\tfrac{1}{6}}\Lambda^0 - \sqrt{\tfrac{1}{2}}\Sigma^0, \Sigma^-\Big) + (\bar{\Xi}^0\Xi^-)\Big]\Big\}.$$

Here the expression proportional to $(D + F)$ is obtained by multiplying the first row of B (with bars over the particle symbols) by the second row of the same matrix (this corresponds to the product $\bar{B}_1^m B_m^2$ since the first column $\bar{B}_1^m$ in matrix $\bar{B}$ is obtained by transposition (and conjugation) of the first row B_m^1 in matrix B). The term proportional to $(D - F)$ is obtained by multiplying the second column of matrix B (with bars over the particle symbols) by the first column of the same matrix (this corresponds to the product $\bar{B}_m^2 B_1^m$).

Likewise, we obtain the matrix elements of the current $j_3^1 \sin\theta$ by writing out the products of the first row by the third row, and of the third column by the first one:

$$\sin\theta \left\{ (D+F)\left[\left(\sqrt{\tfrac{1}{6}}\,\Lambda^0 + \sqrt{\tfrac{1}{2}}\,\Sigma^0,\, \Xi^-\right) + (\bar\Sigma^+, \Xi^0) + \left(\bar p, -\sqrt{\tfrac{2}{3}}\,\Lambda\right)\right] \right.$$
$$\left. + (D-F)\left[\left(\bar p, \sqrt{\tfrac{1}{6}}\,\Lambda^0 + \sqrt{\tfrac{1}{2}}\,\Sigma^0\right) + (\bar n, \Sigma^-) + \left(-\sqrt{\tfrac{2}{3}}\,\bar\Lambda, \Xi^-\right)\right] \right\}.$$

The expressions obtained immediately yield six amplitudes for strangeness-conserving transitions and six amplitudes for transitions changing strangeness by unity:

n → p:	$(D+F)\cos\theta$,	$\Lambda \to$ p:	$-\sqrt{\tfrac{1}{6}}(D+3F)\sin\theta$
$\Xi^- \to \Xi^0$:	$(D-F)\cos\theta$,	$\Xi^- \to \Lambda$:	$-\sqrt{\tfrac{1}{6}}(D-3F)\sin\theta$
$\Lambda \to \Sigma^+$:	$\sqrt{\tfrac{2}{3}}\,D\cos\theta$,	$\Sigma^0 \to$ p:	$\sqrt{\tfrac{1}{2}}(D-F)\sin\theta$
$\Sigma^- \to \Lambda$:	$\sqrt{\tfrac{2}{3}}\,D\cos\theta$,	$\Xi^- \to \Sigma^0$:	$\sqrt{\tfrac{1}{2}}(D+F)\sin\theta$
$\Sigma^0 \to \Sigma^+$:	$-\sqrt{2}\,F\cos\theta$,	$\Sigma^- \to$ n:	$(D-F)\sin\theta$
$\Sigma^- \to \Sigma^0$:	$+\sqrt{2}\,F\cos\theta$,	$\Xi^0 \to \Sigma^+$:	$(D+F)\sin\theta$.

Three strangeness-conserving transitions (n → p, $\Sigma^\pm \to \Lambda$) and all the strangeness-changing ones were observed experimentally. (The decay $\Sigma^0 \to$ pe$^-\bar\nu$ cannot be observed since the electromagnetic decay $\Sigma^0 \to \Lambda^0\gamma$ is predominant. Nevertheless, it is possible to observe the reaction $\tilde\nu_\mu + $ p $\to \Sigma^0 + \mu^+$ which has the same matrix element).

We have thus established that the twelve amplitudes are functions of three parameters, D, F, and θ. This is true for each of the six form factors describing baryon decays: f_1, f_2, f_3 and g_1, g_2, g_3. The predominant role in baryon decays is played by the vector form factor $f_1(q^2)$ and the axial form factor $g_1(q^2)$, where q is the 4-momentum carried by the leptons. The static limit is a fairly good approximation since $|q| \ll M$, where M is the baryon mass; consequently, we shall consider only the vector and axial charges, $f_1(0)\gamma_\alpha$ and $g_1(0)\gamma_\alpha\gamma_5$. The contributions of other form factors will be neglected.

Let us now discuss the values of the parameters D_V and F_V giving $f_1(0)$, and D_A and F_A giving $g_1(0)$. The vector charges $\int \bar{q}\gamma_0 \frac{1}{2}\lambda q \mathrm{d}^3 x$ are generators of SU(3), so that their matrix elements for transitions between octet states j and k are equal to the structural constants f_{ijk} of SU(3) (see appendix). This means, in our notation, that $F_V = 1$, $D_V = 0$. This conclusion agrees with that drawn in chapter 5 where the isotopic subgroup SU(2) of SU(3) was considered. Indeed, we had concluded that $f_1(0) = 0$ for the $\Sigma \to \Lambda e\nu$ decays (and this implies $D_V = 0$), because Σ and Λ belong to different isotopic multiplets. We have also established there that $f_1(0) = 1$ for the β-decay of the neutron (the $T = \frac{1}{2} \to T = \frac{1}{2}$ transition). In the notation of this chapter, this gives $F_V = 1$.

Both D_A and F_A are distinct from zero and have to be determined by experiment (hereafter the subscript A is omitted). The matrix elements of the fundamental decays are:

$\mathrm{n} \to \mathrm{pe}^-\nu$: $\quad \left[\gamma_\alpha + (D + F)\gamma_\alpha\gamma_5\right]\cos\theta$,

$\Sigma^\pm \to \Lambda \mathrm{e}^\pm \nu$: $\quad \gamma_\alpha\gamma_5\sqrt{\frac{2}{3}}\, D\cos\theta$,

$\Lambda \to \mathrm{pe}^-\nu$: $\quad -\sqrt{\frac{3}{2}}\left[\gamma_\alpha + (F + \frac{1}{3}D)\gamma_\alpha\gamma_5\right]\sin\theta$,

$\Sigma^- \to \mathrm{ne}^-\nu$: $\quad -\left[\gamma_\alpha + (F - D)\gamma_\alpha\gamma_5\right]\sin\theta$,

$\Xi^- \to \Lambda \mathrm{e}^-\nu$: $\quad \sqrt{\frac{3}{2}}\left[\gamma_\alpha + (F - \frac{1}{3}D)\gamma_\alpha\gamma_5\right]\sin\theta$,

$\Xi^- \to \Sigma^0 \mathrm{e}^-\nu$: $\quad \sqrt{\frac{1}{2}}\left[\gamma_\alpha + (F + D)\gamma_\alpha\gamma_5\right]\sin\theta$,

$\Xi^0 \to \Sigma^+ \mathrm{e}^-\nu$: $\quad \left[\gamma_\alpha + (F + D)\gamma_\alpha\gamma_5\right]\sin\theta$.

The best fit values of $\sin\theta$, D, and F for all available decay rates and the values of the ratio $g_A/g_V (\equiv g_1(0)/f_1(0))$ for the leptonic decays of baryons are: $\sin\theta \simeq 0.23$, $D \simeq 0.80$, $F \simeq 0.45$ (note that g_A/g_V measured in neutron decay fixes the sum $F + D = 1.25 \pm 0.01$). Calculation of D/F in the framework of SU(6) symmetry yields $D/F = 3/2$; hence, $D = 0.75$, $F = 0.5$.

Unfortunately, the value of g_A/g_V has been measured so far only for three decays: $\mathrm{n} \to \mathrm{pe}^-\nu$, $\Lambda \to \mathrm{pe}^-\nu$, and $\Sigma^- \to \mathrm{ne}^-\nu$. Moreover, there is a discrepancy between two most accurate experimental results concerning the decay of Σ^-. One of them gives $|g_A/g_V| = 0.435 \pm 0.035$, while the other gives $0.17^{+0.07}_{-0.09}$. In both experiments the ratio $|g_A/g_V|$ is obtained from the energy spectra of the neutrons. It seems highly improbable that improved precision of the experiments will reveal serious deviations from the SU(3) picture of semileptonic decays, which by now is well established. Nevertheless, precise measurements of g_A/g_V for a number of decays would be very welcome. It would be interesting, for instance, to see whether $D \cong F$. In this case we must have $g_A \cong 0$ in the $\Sigma^- \to$ ne$^-\nu$ decay.

Problem: Derive the expression for the neutron spectrum in the $\Sigma^- \to$ ne$^-\nu$ decay. Assume the neutron to be non-relativistic (i.e. its kinetic energy is negligible compared to that carried away by the electron and neutrino). If we neglect the electron mass, and integrate over the lepton phase space, we obtain

$$d\Gamma = c\left[g_V^2 \delta_0^\alpha \delta_0^\beta + g_A^2(-g^{\alpha\beta} + \delta_0^\alpha \delta_0^\beta)\right]\left(-g_{\alpha\beta}q^2 + q_\alpha q_\beta\right) d\mathbf{q}$$
$$= c\left[g_V^2(-q^2 + q_0^2) + g_A^2(2q^2 + q_0^2)\right] d\mathbf{q}$$
$$= c\left[g_V^2 \mathbf{q}^2 + g_A^2(3\Delta^2 - 2\mathbf{q}^2)\right] d\mathbf{q}.$$

Here $\Delta = m_\Sigma - m_n = q_0$, $q = (q_0, \mathbf{q})$ is the total 4-momentum of the leptons, and $-\mathbf{q}$ is the neutron momentum. By integrating over $d\mathbf{q}$ ($0 \leq |\mathbf{q}| \leq \Delta$) and comparing the expression obtained $\frac{4}{5}\pi c(g_V^2 + 3g_A^2)$ to that for the total probability of the decay,

$$\Gamma = \frac{G^2(g_V^2 + 3g_A^2)\Delta^5}{60\pi^3}$$

(see Chapter 5), we obtain $c = G^2/48\pi^4$.

CHAPTER 7

Strangeness-changing non-leptonic interactions

In this chapter we shall analyse non-leptonic decays of strange particles: K $\to$ 2π, K $\to$ 3π, $\Lambda \to$ Nπ, $\Sigma \to$ Nπ, $\Xi \to \Lambda\pi$, $\Omega \to \Lambda\overline{K}$, and $\Omega \to \Xi\pi$. All these decays are produced by the coupling of *us*- and *du*-currents (this coupling transforms the s-quark into a d-quark). For example, the decay K$^+ \to \pi^+\pi^0$ is mostly due to the quark graphs of fig. 7.1. Another example is the decay $\Sigma^+ \to$ nπ^+, which is due mostly to quark graphs of the type illustrated in fig. 7.2. Virtual strong interactions are more essential in non-leptonic than in semi-leptonic interactions discussed in the preceding chapters. The gluon exchanges play a particularly important role.

7.1. Properties of the bare non-leptonic lagrangian

The initial product of currents, sometimes termed the bare non-leptonic lagrangian (a hint at the subsequent "dressing" with gluons), takes the form

$$2\sqrt{2}\, G \cos\theta \sin\theta \left[(\bar{d}_L \gamma_\alpha u_L)(\bar{u}_L \gamma^\alpha s_L) + (\bar{u}_L \gamma_\alpha d_L)(\bar{s}_L \gamma^\alpha u_L) \right].$$

(For the sake of brevity, we discuss hereafter mostly the processes which are caused by the first term of this expression, i.e. the term transforming the s- to a d-quark.) Let us discuss the properties of this lagrangian. It is *CP* invariant, but not invariant with respect to *P* and *C* separately. It satisfies the condition

$$|\Delta S| = 1,$$

where ΔS is the difference in strangeness of the initial and final states. (Recall that by definition, the strangeness of the s-quark is -1.) The decays with $|\Delta S| > 1$ are forbidden. These decays (such as $\Xi \to$ Nπ, $\Omega \to \Lambda\pi$, or $\Omega \to$ NK) are allowed only in higher orders of perturbation theory in weak interactions, and their expected decay rates are smaller by approximately fourteen orders of magnitude than the decay rates of the processes allowed

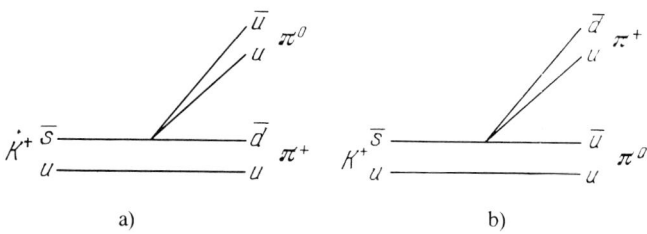

Fig. 7.1.

by the selection rule $|\Delta S| = 1$. None of the forbidden decays are observed, but the experimental upper bounds are too rough compared to the accuracy of the theoretical selection rules.

As for the isospin selection rules, it is easily seen that the $(\bar{d}u)(\bar{u}s)$ interaction, comprising three isospinors and one isoscalar, satisfies the condition

$$\Delta T = \tfrac{1}{2}, \tfrac{3}{2}.$$

We shall demonstrate later that experimentally observed transition amplitudes with $\Delta T = \tfrac{1}{2}$ are greater by approximately an order of magnitude than the amplitudes with $\Delta T = \tfrac{3}{2}$. No quantitative calculation "to the end" is available so far on the mechanisms of enhancement of amplitudes with $\Delta T = \tfrac{1}{2}$, and suppression of those with $\Delta T = \tfrac{3}{2}$. Nevertheless, it is qualitatively clear, that the approximate rule $\Delta T = \tfrac{1}{2}$ for non-leptonic decays is dynamical in nature. A significant role is played here by virtual gluons. We cannot yet calculate their contribution exactly, so we shall try to obtain an estimate. We shall assume that the contribution of soft gluons (gluons with low q^2) is already covered by the quark wave function of a hadron, and take account of the hard gluons. By virtue of the asymptotic freedom of quantum chromodynamics, this calculation can be realized in the framework of perturbation theory. "Dressing" of the bare non-leptonic lagrangian with hard gluons yields the so-called effective non-leptonic lagrangian (see the end of this chapter) which differs from the bare one both in the value of the

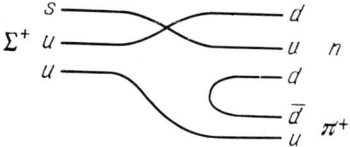

Fig. 7.2.

effective constant in the initial operator expression, and in the appearance of additional operator terms (the so-called Wilson operator expansion).

7.2. The inclusion of hard gluons

Before discussing the contribution of virtual gluons, let us separate the bare lagrangian into two terms,

$$2(\bar{d}_L \gamma_\alpha u_L)(\bar{u}_L \gamma^\alpha s_L) = I_3 + I_6,$$

where

$$I_3 = (\bar{d}_L \gamma_\alpha u_L)(\bar{u}_L \gamma^\alpha s_L) - (\bar{u}_L \gamma_\alpha u_L)(\bar{d}_L \gamma^\alpha s_L),$$

$$I_6 = (\bar{d}_L \gamma_\alpha u_L)(\bar{u}_L \gamma^\alpha s_L) + (\bar{u}_L \gamma_\alpha u_L)(\bar{d}_L \gamma^\alpha s_L).$$

The first of these terms is antisymmetric, while the second one is symmetric with respect to the permutation $\bar{d} \leftrightarrow \bar{u}$ or $u \leftrightarrow s$. The subscripts 3 and 6 indicate the dimension in color space; their meaning will be clear shortly. Until this section, color variables were omitted because weak currents are color singlets. Here we want to discuss the contribution of color gluons, so that color indices of the quark creation and annihilation operators must be written explicitly. Then,

$$I_{3,6} = (\bar{d}_L^i \gamma_\alpha u_{Li})(\bar{u}_L^k \gamma^\alpha s_{Lk}) \mp (\bar{u}_L^i \gamma_\alpha u_{Li})(\bar{d}_L^k \gamma^\alpha s_{Lk}).$$

Let us consider the transition $us \leftrightarrow ud$ (fig. 7.3) realized by I_3 and I_6. In the first case the initial and final states are antisymmetric with respect to the permutation of color indices. Hence, they belong to a color triplet (more precisely, to the anti-triplet $\bar{3}$). In the second case they are symmetric and belong to a sextet (we recall that $3 \times 3 = 6 + \bar{3}$). It is easy to see that the interaction I_3 satisfies the $\Delta T = \frac{1}{2}$ selection rule. Indeed, owing to antisymmetrization, the final state ud (fig. 7.3) is isoscalar ($ud - du$ has $T = 0$), but then $\Delta T = \frac{1}{2}$ since the initial state is an isospinor. In the case of the interaction I_6, the final state $ud + du$ has $T = 1$. Consequently, I_6 includes transitions both with $\Delta T = \frac{1}{2}$ and with $\Delta T = \frac{3}{2}$.

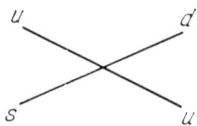

Fig. 7.3.

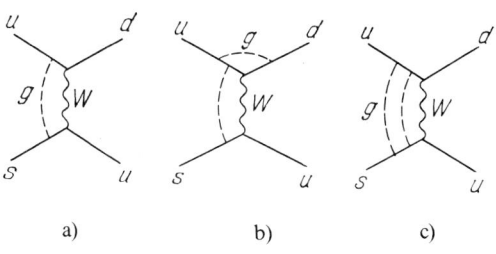

Fig. 7.4.

We shall begin with just those graphs in which a W-boson connects two different quark lines. We mean the graphs of fig. 7.4 and more complex graphs with an infinite number of hard virtual gluons (soft gluons are not shown here and in subsequent graphs). In this case the inclusion of hard virtual gluons does not modify the form of the operators I_3 and I_6 but only changes their coefficients. The result is

$$\sqrt{2}\, G(a_3 I_3 + a_6 I_6) \sin\theta \cos\theta.$$

In the leading logarithm approximation,

$$a_i = \left(\frac{\alpha_s(\mu)}{\alpha_s(m_W)} \right)^{\gamma_i}, \quad i = 3, 6.$$

Here α_s is the so-called running coupling constant of the strong interaction, m_W is the W-boson mass, μ is the characteristic momentum of quarks in a hadron ($1/\mu$ is the confinement radius), and

$$\gamma_3 = \frac{4}{b}, \quad \gamma_6 = -\frac{2}{b}.$$

We recall that according to quantum chromodynamics, the strong interaction "constant" $\alpha_s = g^2/4\pi$ (g is the color charge) is a logarithmic function of the momentum transfer $Q \equiv \sqrt{-q^2}$, and contains the parameter b:

$$\alpha_s(Q) = \frac{2\pi}{b \ln(Q/\Lambda)}.$$

Here $\Lambda \sim 100$ MeV is a universal constant defining the scale of hadronic dimensions, and the dimensionless constant b is a function of $N_f(Q)$, that is of the number of flavors of quarks with masses below Q:

$$b = 11 - \tfrac{2}{3} N_f(Q).$$

If $Q \lesssim m_c$ (m_c is the mass of the c-quark), then u-, d-, and s-quarks are operative, $N_f(Q) = 3$ and $b = 9$. If $Q \gg m_c$, then u-, d-, s-, and c-quarks are operative, $N_f(Q) = 4$ and $b = \frac{25}{3}$. If $Q \gg m_b$, then u-, d-, s-, c-, and b-quarks are operative, $N_f(Q) = 5$ and $b = \frac{23}{3}$. The leading logarithm approximation mentioned above assumes that $\alpha_s(Q) \ll 1$, but that $\alpha_s(Q) \ln(Q/\Lambda)$ is not small. This means that terms of the order of $[\alpha_s(Q)]^{n+1}(\ln(Q/\Lambda))^n$ are neglected, and that the terms of the order of $(\alpha_s(Q) \ln(Q/\Lambda))^n$ are summed over to all orders of perturbation theory.

For the estimates in the above expression for a_i we shall assume that $\alpha_s(\mu) = 1$. This yields $\mu \cong 140$ MeV for $\Lambda \cong 70$ MeV. (In this case, $\alpha_s(m_\rho) \cong 0.3$, and $\alpha_s(2m_c) \cong 2.5$ GeV) $\cong 0.2$, which is in agreement with the charmonium data. Some authors derive $\Lambda \cong 350$ GeV from the data on deep-inelastic scattering, which yields $\mu \cong 700$ MeV.) For $m_W \cong 80$ GeV and $\Lambda \cong 70$ MeV we obtain $\alpha_s(\mu)/\alpha_s(m_W) \cong 10$; hence

$$a_3 \cong 3 \quad \text{and} \quad a_6 = 0.6.$$

This demonstrates that hard gluons enhance the $\Delta T = \frac{1}{2}$ transitions and suppress those with $\Delta T = \frac{3}{2}$. We shall find afterwards, however, that these theoretically predicted enhancement and suppression effects are not sufficient to match the experimentally observed large amplitudes with $\Delta T = \frac{1}{2}$, and small amplitudes with $\Delta T = \frac{3}{2}$.

7.3. The gluonic monopole

One additional class of graphs is possible when virtual gluons are taken into account. This class was not discussed in the preceding pages. We mean the graphs in which a virtual W-boson is emitted and absorbed by the same quark which is coupled to other quarks only by gluons. Elementary graphs of this type are shown in fig. 7.5. Let us now calculate their contribution; note that the result will not be significantly changed when these graphs are

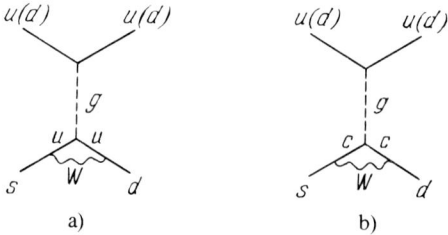

Fig. 7.5.

dressed, in the leading logarithm approximation, by an infinite number of gluons. If we neglect the mixing of d- and s-quarks with b-quarks, the charged currents take the form

$$\bar{u}_L \gamma_\alpha d'_L = \bar{u}_L \gamma_\alpha (d_L \cos\theta + s_L \sin\theta),$$

$$\bar{c}_L \gamma_\alpha s'_L = \bar{c}_L \gamma_\alpha (-d_L \sin\theta + s_L \cos\theta).$$

The graphs 5a and 5b would cancel out, therefore, if the masses of u- and c-quarks were identical. Actually $m_c \gg m_u$, and the compensation is realized only for momenta of virtual particles exceeding m_c. For lower momenta, the graph 7.5a is predominant. We can then conclude that the sum of the two graphs of fig. 7.5 is approximately equal to the contribution of the first of them, but with the momenta of the virtual particles in the loop cut off at m_c. Taking into account that $m_c \ll m_W$, we can reduce the graph in fig. 7.5a to the four-fermion point interaction (see fig. 7.6a). The gap in the four-fermion vertex in fig. 7.6a separates two white $V - A$ currents found in parentheses in the expression

$$\sqrt{\tfrac{1}{2}}\, G \sin\theta \cos\theta \left(\bar{d}^{\alpha i} (O_\mu)_\alpha^\beta \delta_i^k u_{\beta k} \right)\left(\bar{u}^{\gamma l}(O^\mu)_\gamma^\delta \delta_l^m s_{\delta m} \right),$$

where $O_\mu = \gamma_\mu(1 + \gamma_5)$, Greek letters denote Dirac indices $(0, 1, 2, 3)$, and Latin letters denote color indices $(1, 2, 3)$. The calculation is conveniently simplified by the Fierz transformation: the operators $\bar{u}$ and u are now in one bracket, and $\bar{d}$ and s in the other one (see fig. 7.6b). By using the relations

$$(O_\mu)_\alpha^\beta (O^\mu)_\gamma^\delta = -(O_\mu)_\alpha^\delta (O^\mu)_\gamma^\beta,$$

$$\delta_i^k \delta_l^m = \tfrac{1}{3}\delta_i^m \delta_l^k + \tfrac{1}{2}\lambda_i^m \lambda_l^k,$$

(see appendix), and taking into account that fermion operators anticommute, we obtain

$$(\bar{d}O_\mu u)(\bar{u}O^\mu s) = \tfrac{1}{3}(\bar{u}O_\mu u)(\bar{d}O^\mu s) + \tfrac{1}{2}(\bar{u}O_\mu \lambda^a u)(\bar{d}O^\mu \lambda^a s).$$

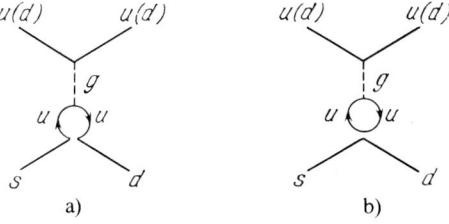

Fig. 7.6.

It is necessary now to calculate the u-quark loop in fig. 7.6b. The upper vertex (emission of a gluon) of this loop is

$$g\bar{u}\frac{\lambda^b}{2}\gamma_\nu u.$$

In loop calculations, we have to find the trace over Dirac and color suffices. Hence, the term $\frac{1}{3}(\bar{u}O_\mu u)(\bar{d}O^\mu s)$ of the lower vertex gives zero, and only the part $\frac{1}{2}(\bar{u}\gamma_\mu \lambda^a u)(\bar{d}O^\mu \lambda^a s)$ of the term $\frac{1}{2}(\bar{u}O_\mu \lambda^a u)(\bar{d}O^\mu \lambda^a s)$ will give a non-zero contribution.

The loop in question is exactly the same (with the exception of additional color matrices) as that found in quantum electrodynamics in the case of vacuum polarization calculations:

$$-\operatorname{Tr}\frac{\lambda^a}{2}\frac{\lambda^b}{2}\int \operatorname{Tr}\gamma_\mu \frac{1}{\hat{p}+\hat{q}}\gamma_\nu \frac{1}{\hat{p}}\frac{d^4p}{(2\pi)^4} = \frac{\delta^{ab}}{2}\left(q^2 g_{\mu\nu} - q_\mu q_\nu\right)\mathcal{P},$$

where $a, b = 1, \ldots, 8$, and q is the 4-momentum carried by the gluon (see fig. 7.6b). We have ignored the u-quark mass since $m_u \ll \mu$. The minus sign in front of the above expression comes from the following standard factors (they are discussed in all textbooks on quantum field theory): (-1) for each closed fermion loop, $(+i)$ for each fermion propagator, $(-i)$ for each propagator of the vector particle, (i) for each vertex, and $(+i)$ for the transition in d^4p from dp_0 to idp_4. In addition, the relation between S- and T-matrices, $S = 1 + iT$, and that between S and the lagrangian, $S = \exp(i\mathcal{L}d^4x)$, must be taken into account.

Up to a dimensionless factor $\mathcal{P}$, the form of the integral is determined by the transversality condition (which demands, in particular, that quadratic divergence be absent). In order to find $\mathcal{P}$, multiply the left-hand and right-hand parts by $g^{\mu\nu}$. This gives

$$3q^2\mathcal{P} = -\int \operatorname{Tr}\gamma_\mu \frac{1}{\hat{p}+\hat{q}}\gamma^\mu \frac{1}{\hat{p}}\frac{d^4p}{(2\pi)^4} = 2\int \operatorname{Tr}\frac{1}{\hat{p}}\hat{q}\frac{1}{\hat{p}}\hat{q}\frac{1}{\hat{p}}\frac{1}{\hat{p}}\frac{d^4p}{(2\pi)^4}$$

$$= \frac{1}{(2\pi)^4}\int \frac{d^4p}{p^6}\left[2(pq)(pq) - q^2 p^2\right]$$

$$= -\frac{q^2}{4\pi^2}\int \frac{dp^2}{p^2},$$

since

$$\int \frac{d^4p}{p^6} 2p_\mu p_\nu = \tfrac{1}{2}g_{\mu\nu}\int \frac{d^4p}{p^4}.$$

Finally,

$$\mathcal{P} = -\frac{1}{12\pi^2}\int \frac{dp^2}{p^2} = -\frac{1}{12\pi^2}\ln\frac{m_c^2}{\mu^2},$$

where the upper integration limit is chosen to be m_c^2 (since the graphs of figs. 7.5a and 7.5b cancel out for $p^2 > m_c^2$). The lower integration limit is chosen to be μ^2 because the free quark propagator must be modified at $p^2 < \mu^2$ due to confinement. As a result, the contribution of the graph of fig. 7.6a becomes

$$-\sqrt{2}\, G\sin\theta\cos\theta \frac{g^2}{12\pi^2}\ln\frac{m_c^2}{\mu^2}\left(\bar{d}_L\gamma_\mu\lambda^a s_L\right)\frac{\delta^{ab}}{2}\left(\bar{u}\gamma^\mu\frac{\lambda^b}{2}u + \bar{d}\gamma^\mu\frac{\lambda^b}{2}d\right)$$

$$= -\sqrt{2}\, G\sin\theta\cos\theta\frac{\alpha_s(\mu)}{6\pi}\ln\frac{m_c}{\mu}\left(\bar{d}_L\gamma_\mu\lambda^a s_L\right)\left(\bar{u}\gamma^\mu\lambda^a u + \bar{d}\gamma^\mu\lambda^a d\right).$$

Note that the factor $1/q^2$ in the gluon propagator is cancelled out by the factor q^2 produced by the quark loop. The effective four-fermion interaction described by fig. 7.6a thus looks local from large distances ($r \gg 1/2m_c$). The effective vertex of gluon emission in the s ↔ d transition becomes zero for $q^2 = 0$ (q is the gluon 4-momentum). So we can refer to it as the monopole vertex. Indeed, it is well-known that in the nuclear monopole transitions, the emission of a real photon with $q^2 = 0$ is forbidden, and thus an e^+e^- pair with $q^2 > 0$ is emitted. (Strictly speaking, the gds vertex is a sum of a monopole, $q^2\gamma_\alpha$, and an anapole, $q^2\gamma_\alpha\gamma_5$).

We must explain now why $\alpha_s(\mu)$, and not $\alpha_s(m_c)$, was substituted into the above expression. The cut-off of the loop is carried out at the point m_c, so that on the face of it the graphs of fig. 7.6 correspond to $\alpha_s(m_c)$. It can be demonstrated, however, that if we dress these graphs with additional gluons, a renormalization factor appears: $(\alpha_s(\mu)/\alpha_s(m_c))^{2\gamma_3}$ where $2\gamma_3 = 8/b = 0.96 \simeq 1$. Consequently, $\alpha_s(m_c)$ in the expression under consideration is replaced by $\alpha_s(\mu)$ (recall that by our convention, $\alpha_s(\mu) = 1$). Denoting

$$a_R = \frac{1}{6\pi}\ln\frac{m_c}{\mu},$$

we obtain $a_R \simeq 0.12$ for $m_c/\mu \simeq 10$. Hence, a_R is quite small:

$$a_R \sim \tfrac{1}{5}a_6 \sim \tfrac{1}{25}a_3.$$

The contribution of the graphs of the type of fig. 7.6 is nevertheless very substantial because of the specific structure of the operator

$$-\left(\bar{d}_L\gamma_\mu\lambda^a s_L\right)\left(\bar{u}\gamma^\mu\lambda^a u + \bar{d}\gamma^\mu\lambda^a d\right).$$

In fact, this operator is a sum of two terms,

$$-(\bar{d}_L\gamma_\mu\lambda^a s_L)(\bar{u}_L\gamma^\mu\lambda^a u_L + \bar{d}_L\gamma^\mu\lambda^a d_L),$$

$$-(\bar{d}_L\gamma_\mu\lambda^a s_L)(\bar{u}_R\gamma^\mu\lambda^a u_R + \bar{d}_R\gamma^\mu\lambda^a d_R).$$

The first of these terms involves only left-handed spinors and, because of the smallness of the coefficient a_R, is much less important than I_3 and I_6 which have the same helicity structure. Therefore, as a crude approximation, the contribution of this operator can be neglected. The second operator (we denote it by I_R) contains both the left-handed and right-handed spinors. In a number of cases this may considerably enhance the contribution of I_R to the amplitude of non-leptonic decays. This phenomenon will be demonstrated in Chapter 9.

7.4. Effective non-leptonic lagrangian

Now let us collect all the terms of the simplified effective non-leptonic quark interaction with $\Delta S = 1$:

$$\tilde{\mathcal{L}}_{\text{eff}}(\Delta S = 1) = \sqrt{2}\, G \sin\theta \cos\theta [a_3 I_3 + a_6 I_6 + a_R I_R],$$

where

$$I_{3,6} = (\bar{d}_L\gamma_\alpha u_L)(\bar{u}_L\gamma^\alpha s_L) \mp (\bar{u}_L\gamma_\alpha u_L)(\bar{d}_L\gamma^\alpha s_L),$$

$$I_R = -(\bar{d}_L\gamma_\alpha \lambda s_L)(\bar{u}_R\gamma^\alpha \lambda u_R + \bar{d}_R\gamma^\alpha \lambda d_R),$$

$$a_3 \cong 3, \qquad a_6 \cong 0.6, \qquad a_R \cong 0.12.$$

Similarly to I_3, the operator I_R is an isotopic spinor and therefore results only in transitions with $\Delta T = \tfrac{1}{2}$.

If in addition to the u- and d-quarks the upper lines in figs. 7.5 and 7.6 symbolize s-quarks as well, and if the dressing of these graphs with an infinite number of gluons were also included, then the total effective non-leptonic lagrangian would have the form given by Shifman, Vainshtein and Zakharov

$$\mathcal{L}_{\text{eff}}(\Delta S = 1) = \sqrt{2}\, G \sin\theta \cos\theta \sum_{i=1}^{6} c_i O_i,$$

where O_i are four-fermion operators which transform as irreducible representations of the isospin group SU(2) and the flavor group SU(3)$_f$:

$$O_1 = \bar{d}_L\gamma_\mu s_L \bar{u}_L\gamma^\mu u_L - \bar{u}_L\gamma_\mu s_L \bar{d}_L\gamma^\mu u_L \qquad (\{8f\}, \Delta T = \tfrac{1}{2}),$$

$$O_2 = \bar{d}_L \gamma_\mu s_L \bar{u}_L \gamma^\mu u_L + \bar{u}_L \gamma_\mu s_L \bar{d}_L \gamma^\mu u_L + 2\bar{d}_L \gamma_\mu s_L \bar{d}_L \gamma^\mu d_L$$
$$+ 2\bar{d}_L \gamma_\mu s_L \bar{s}_L \gamma^\mu s_L \qquad (\{8d\}, \Delta T = \tfrac{1}{2}),$$

$$O_3 = \bar{d}_L \gamma_\mu s_L \bar{u}_L \gamma^\mu u_L + \bar{u}_L \gamma_\mu s_L \bar{d}_L \gamma^\mu u_L + 2\bar{d}_L \gamma_\mu s_L \bar{d}_L \gamma^\mu d_L$$
$$- 3\bar{d}_L \gamma_\mu s_L \bar{s}_L \gamma^\mu s_L \qquad (\{27\}, \Delta T = \tfrac{1}{2}),$$

$$O_4 = \bar{d}_L \gamma_\mu s_L \bar{u}_L \gamma^\mu u_L + \bar{u}_L \gamma_\mu s_L \bar{d}_L \gamma^\mu u_L$$
$$- \bar{d}_L \gamma_\mu s_L \bar{d}_L \gamma^\mu d_L \qquad (\{27\}, \Delta T = \tfrac{3}{2}),$$

$$O_5 = \bar{d}_L \gamma^\mu \lambda^a s_L (\bar{u}_R \gamma_\mu \lambda^a u_R + \bar{d}_R \gamma^\mu \lambda^a d_R + \bar{s}_R \gamma^\mu \lambda^a s_R) \qquad (\{8\}, \Delta T = \tfrac{1}{2}),$$

$$O_6 = \bar{d}_L \gamma_\mu s_L (\bar{u}_R \gamma^\mu u_R + \bar{d}_R \gamma^\mu d_R + \bar{s}_R \gamma^\mu s_R) \qquad (\{8\}, \Delta T = \tfrac{1}{2}).$$

The characteristic numerical values of the coefficients c_i (for parameter values of $\mu = 0.14$ GeV, $m_W = 100$ GeV, $m_c = 2$ GeV, and $\alpha_s(\mu) = 1$) are:

$c_1 = -2.75, \quad c_2 = 0.06, \quad c_3 = 0.08, \quad c_4 = 0.39,$
$c_5 = -0.14, \quad c_6 = -0.05.$

Clearly, $I_3 = -O_1$, $15I_6 = 3O_2 + 2O_3 + 10O_4$. The operator I_R would be identical to O_5 if the upper lines in the graphs of figs. 7.5 and 7.6 represented not only u- and d- but s-quarks as well. (Terms of the type $(\bar{d}s)(\bar{s}s)$ were ignored so far, since they cannot contribute to simple quark diagrams representing decays of strange particles. This will be shown later.) The operator O_6 corresponds to the as yet undiscussed graph of fig. 7.7, involving the exchange of two gluons, and to more complicated graphs in which the gluons connecting the quark lines form a white system. The result is almost unaltered when O_6 is taken into account. Namely, O_5 is replaced by the combination $O_5 + \tfrac{3}{16} O_6$ (the factor $\tfrac{3}{16}$ is easily obtained by means of the Fierz transformation).

The effective lagrangian $\mathcal{L}_{\text{eff}}(\Delta S = 1)$ will be applied in chapters 9 and 10 to a dynamic analysis of non-leptonic decays of hyperons and K-mesons. Before doing that, however, we devote chapter 8 to kinematic (independent of dynamics) properties of hyperon-decay amplitudes.

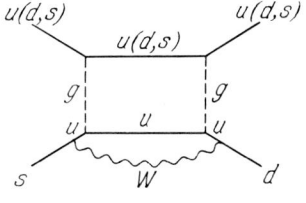

Fig. 7.7.

CHAPTER 8

Phenomenology of non-leptonic decays of hyperons

8.1. Relativistically invariant amplitude

The non-leptonic decay $B_1 \to B_2 + \pi$ of a baryon with $J^P = \frac{1}{2}^+$ is characterized by its amplitude M which is given, in the most general form, as

$$M = Gm_\pi^2 \bar{u}_2 (A + B\gamma_5) u_1.$$

Here u_1 and u_2 are spinor wave functions of the initial and final baryons, and φ_π is the π-meson wave function. The multiplier m_π^2, where m_π is the charged pion mass, is inserted "by hand" in order to obtain a correct dimension and reasonable order of magnitude of the amplitude. Indeed, if we divide the nuclear time of the order of 10^{-24}s by $G^2 m_\pi^4 \sim 10^{-14}$, the result is a time of the order of 10^{-10}s, that is the characteristic lifetime of hyperons. The dimensionless coefficients A and B are numbers (complex numbers in the general case; see below). Making use of the Dirac equation one can easily show that the expression $\bar{u}_2 (A + B\gamma_5) u_1$ is the most general Lorentz-invariant characterizing the vertex $\bar{B}_2 B_1 \pi$. By taking into account that φ_π is a pseudoscalar, we conclude that P-parity is not conserved by amplitude A and conserved by B. Obviously, A corresponds to the S-wave, that is to zero orbital momentum ($l = 0$) of the system $B_2 \pi$, and B corresponds to the P-wave ($l = 1$).

A standard calculation gives for the decay width

$$\Gamma = \frac{G^2 m_\pi^4}{8\pi} q \left\{ \left[\frac{(M+m)^2 - \mu^2}{M^2} \right] |A|^2 + \left[\frac{(M-m)^2 - \mu^2}{M^2} \right] |B|^2 \right\},$$

where q is the momentum of the decay products in the rest frame of the decaying hyperon ($q = |\mathbf{q}|$), $M(m)$ is the mass of the decaying (produced) baryon, μ is the mass of the pion (π^0 or $\pi^\pm$) which is produced in the decay in question, and m_π is the mass of the charged pion.

8.2. Non-relativistic form of the amplitude

Let us express the four-component Dirac spinors u_1 and u_2 in terms of the two-component Pauli spinors ψ_1 and ψ_2:

$$u_1 = \begin{pmatrix} \psi_1 \\ \chi_1 \end{pmatrix} = \begin{pmatrix} \psi_1 \\ 0 \end{pmatrix},$$

$$u_2 = \begin{pmatrix} \psi_2 \\ \chi_2 \end{pmatrix} = \sqrt{\frac{E+m}{2E}} \begin{pmatrix} \psi_2 \\ \boldsymbol{\sigma}\cdot\boldsymbol{q}\,\psi_2/\sqrt{E+m} \end{pmatrix}.$$

We obtain

$$M \sim \psi_2^+ (S + P\boldsymbol{\sigma}\cdot\boldsymbol{n})\psi_1,$$

where $\boldsymbol{n} = \boldsymbol{q}/|\boldsymbol{q}|$, and S and P are amplitudes of the S- and P-waves, respectively. It is easy to see that

$$\frac{P}{S} = \sqrt{\frac{E-m}{E+m}} \frac{B}{A}.$$

(The normalization of the amplitudes S and P will not be taken into consideration.)

8.3. Spin correlations in hyperon decays

Consider the decay

$$\Lambda \to p + \pi^-.$$

Let $\boldsymbol{\eta}$ and $\boldsymbol{\zeta}$ be unit vectors characterizing the polarization of the Λ-hyperon and the proton, respectively, in the rest frames of each of these particles:

$$\psi_1\psi_1^+ = \tfrac{1}{2}(1 + \boldsymbol{\eta}\cdot\boldsymbol{\sigma}),$$

$$\psi_2\psi_2^+ = \tfrac{1}{2}(1 + \boldsymbol{\zeta}\cdot\boldsymbol{\sigma}).$$

Let $\boldsymbol{n}$ be a unit vector in the direction of the proton momentum in the Λ-hyperon rest frame. Let us find the decay probability as a function of $\boldsymbol{\eta}$,

ζ, and n:

$$W(\eta, \zeta, n) \sim |M|^2 \sim \mathrm{Tr}(1 + \zeta \cdot \sigma)(S + P\sigma \cdot n)(1 + \eta \cdot \sigma)$$
$$\times (S^* + P^*\sigma \cdot n)$$
$$\sim \{|S|^2(1 + \eta \cdot \zeta) + |P|^2(1 + 2(\eta \cdot n)(\zeta \cdot n) - \eta \cdot \zeta)$$
$$+ (SP^* + S^*P)(\eta \cdot n + \zeta \cdot n) + i(SP^* - S^*P)\zeta[\eta \cdot n]\}$$
$$\sim \{1 + \alpha(\eta \cdot n + \zeta \cdot n) + \beta\zeta[\eta \cdot n] + \gamma\eta \cdot \zeta$$
$$+ (1 - \gamma)(\eta \cdot n)(\zeta \cdot n)\}.$$

Here

$$\alpha = \frac{SP^* + S^*P}{|S|^2 + |P|^2}, \quad \beta = i\frac{SP^* - S^*P}{|S|^2 + |P|^2}, \quad \gamma = \frac{|S|^2 - |P|^2}{|S|^2 + |P|^2}.$$

(It is easily found that $\alpha^2 + \beta^2 + \gamma^2 = 1$). The following relations were used to calculate the trace (see appendix):

$$\sigma_i \sigma_k = \delta_{ik} + i\varepsilon_{ikl}\sigma_l,$$
$$\mathrm{Tr}\,\sigma_i \sigma_k = 2\delta_{ik},$$
$$\mathrm{Tr}\,\sigma_i \sigma_k \sigma_l = 2i\varepsilon_{ikl},$$
$$\mathrm{Tr}\,\sigma_i \sigma_k \sigma_l \sigma_m = 2(\delta_{ik}\delta_{lm} + \delta_{im}\delta_{kl} - \delta_{il}\delta_{km}).$$

Let us analyze the expression for $W(\eta, \zeta, n)$. The decay probability in the S-wave is zero if η and ζ are antiparallel. This result is quite natural: with zero orbital momentum, the proton spin must be in the same direction as that of the Λ-hyperon. This is not true for the P-wave: the probability is maximum when the proton spin is directed along the vector $2n(\eta \cdot n) - \eta$.

If the proton polarization is not measured, we put $\zeta = 0$. In this case the angular distribution of protons takes the form $1 + \alpha\eta \cdot n$. The P-odd angular asymmetry is a result of interference between the S- and P-waves. If the decaying hyperon is not polarized, then $\eta = 0$ and the decay probability is proportional to $1 + \alpha\zeta \cdot n$. This means that the proton is polarized longitudinally (its spin is directed along its momentum), and that the degree of this polarization is α. As can be readily found from the expression for $W(\eta, \zeta, n)$, the proton polarization is given by

$$P = \frac{n(\alpha + \eta \cdot n) + \beta[\eta \cdot n] + \gamma[n[\eta \cdot n]]}{1 + \alpha\eta \cdot n}.$$

(The decay probability $W(P, \zeta)$ is proportional to $1 + \zeta \cdot P$ and reaches a maximum for $\zeta \parallel P$.)

The nucleon polarization along the normal to the plane containing the vectors η and n, is proportional to Im SP^* and is non-vanishing only for the non-zero relative phases, Δ, of the S- and P-wave amplitudes:

$$\alpha = \frac{2|S||P|\cos\Delta}{|S|^2 + |P|^2}, \qquad \beta = -\frac{2|S||P|\sin\Delta}{|S|^2 + |P|^2}.$$

We show below that Δ can be expressed as a function of pion-proton scattering phases, provided time-reversal invariance holds.

8.4. Isotopic amplitudes and the $\Delta T = \frac{1}{2}$ rule

The initial state in the $\Lambda \to p + \pi^-$ decay has $T = 0$, while the final state is a superposition of $T = \frac{1}{2}$ and $T = \frac{3}{2}$:

$$p\pi^- = \sqrt{\tfrac{1}{3}}\,\psi_{3/2} - \sqrt{\tfrac{2}{3}}\,\psi_{1/2}.$$

Similarly, we have for the $\Lambda \to n + \pi^0$ decay:

$$n\pi^0 = \sqrt{\tfrac{2}{3}}\,\psi_{3/2} + \sqrt{\tfrac{1}{3}}\,\psi_{1/2}.$$

If the $\Delta T = \frac{1}{2}$ rule were absolute, the transitions would be only to the state $\psi_{1/2}$, so that the ratio of the amplitudes Λ^0_- and Λ^0_0 describing the decays $\Lambda^0 \to p\pi^-$ and $\Lambda^0 \to n\pi^0$, respectively, would be

$$\frac{\Lambda^0_-}{\Lambda^0_0} = -\sqrt{2},$$

valid both for A and for B amplitudes. As a result, the correlation coefficients α, β, γ for the two decays would be identical, with the following ratio for the decay rates:

$$\Gamma(\Lambda \to p\pi^-) : \Gamma(\Lambda \to n\pi^0) = 2.$$

All these predictions are well confirmed experimentally (see appendix).

In order to derive the corollaries of the $\Delta T = \frac{1}{2}$ rule for decays of Σ-hyperons, it will be convenient to introduce the concept of the spurion, S, and consider the reactions

$$S\Sigma^+ \to p\pi^0 \quad \Sigma_0^+ = \left\langle \sqrt{\tfrac{2}{3}}\psi_{3/2} - \sqrt{\tfrac{1}{3}}\psi_{1/2} \,\middle|\, H \,\middle|\, \sqrt{\tfrac{1}{3}}\varphi_{3/2} + \sqrt{\tfrac{2}{3}}\varphi_{1/2} \right\rangle$$

$$= \tfrac{1}{3}\sqrt{2}\,(\Sigma_{3/2} - \Sigma_{1/2}),$$

$$S\Sigma^+ \to n\pi^+ \quad \Sigma_+^+ = \left\langle \sqrt{\tfrac{1}{3}}\psi_{3/2} + \sqrt{\tfrac{2}{3}}\psi_{1/2} \,\middle|\, H \,\middle|\, \sqrt{\tfrac{1}{3}}\varphi_{3/2} + \sqrt{\tfrac{2}{3}}\varphi_{1/2} \right\rangle$$

$$= \tfrac{1}{3}(\Sigma_{3/2} + 2\Sigma_{1/2}),$$

$$S\Sigma^- \to n\pi^- \quad \Sigma_-^- = \langle \psi_{3/2} | H | \varphi_{3/2} \rangle = \Sigma_{3/2}.$$

Here ψ and φ are the isotopic wave functions of the pion-nucleon and spurion-Σ systems, respectively. (The isotopic properties of the spurion are identical to those of the K^0 meson; but the spurion is a fictitious particle which carries neither energy nor momentum. Introduction of the spurion allows one to operate with the isotopically invariant interaction hamiltonian H.) Consequently, the three observed amplitudes are given in terms of two isotopic amplitudes $\Sigma_{3/2}$ and $\Sigma_{1/2}$, which yields the triangle relation,

$$\Sigma_+^+ + \sqrt{2}\,\Sigma_0^+ = \Sigma_-^-,$$

both for the S- and P-wave amplitudes of Σ decays. The amplitudes as given by experiment are schematically presented in fig. 8.1. The triangle in the figure is not closed because of the transitions with $\Delta T = \tfrac{3}{2}$. It is of great interest that the triangle is almost right-angled, and that its legs are almost on the coordinate axes. This will be taken up again in the next chapter when the dynamics of hyperon decays is discussed.

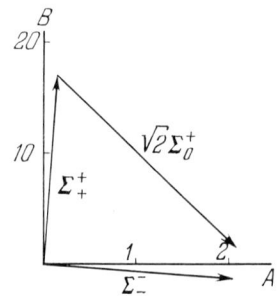

Fig. 8.1.

By using the spurion we easily establish that the $\Delta T = \frac{1}{2}$ rule leads to the relations

$$M(\Xi^- \to \pi^- \Lambda^0) : M(\Xi^0 \to \pi^0 \Lambda^0) = \sqrt{2},$$
$$M(\Omega^- \to \Xi^0 \pi^-) : M(\Omega^- \to \Xi^- \pi^0) = \sqrt{2}.$$

We shall see in chapter 9 that in the latter case, theory predicts, and experiment confirms, that the admixture of $\Delta T = \frac{3}{2}$ causes a considerable deviation from $\sqrt{2}$.

8.5. Phases of the S- and P-wave amplitudes

We shall demonstrate that the phases of S- and P-wave amplitudes for hyperon decay to a given isotopic state are equal to the phases of the scattering of decay products in this isotopic state. This statement is based on the unitarity of the S-matrix and on its symmetry. This last property follows from the invariance of the lagrangian with respect to time reversal T. (Here we neglect very weak violation of CP invariance (or T-invariance, by virtue of the CPT theorem) which results in small phases of S- and P-wave amplitudes, even in the case of no interaction in the final state).

The proof will be obtained in the most general form, so that it may be used in other cases as well: decays of K-mesons, neutrino reactions, and so on. First of all we want to find a relation between the matrix elements of the direct and inverse reactions. This relation is a result of the unitarity of the S-matrix,

$$SS^+ = 1,$$

provided the amplitudes of these reactions connecting the initial states i and final states f are small. Let us separate S into two terms,

$$S = S^0 + S^1,$$

where $|S^0| \gg |S^1|$. The terms S^0 and S^1 are chosen to satisfy the conditions:

$$S^0_{fi} = 0, \quad S^0_{ii'} \neq 0, \quad S^0_{ff'} \neq 0,$$
$$S^1_{fi} \neq 0, \quad S^1_{ii'} = 0, \quad S^1_{ff'} = 0,$$

where i, i' (f, f') are states from a group of initial (final) states. Therefore,

$$S_{fi} = S^1_{fi}, \quad S_{ii'} = S^0_{ii'}, \quad S_{ff'} = S^0_{ff'}.$$

The unitarity condition is now expressed by

$$(S^0 + S^1)^+ (S^0 + S^1) = 1.$$

Neglecting the terms $S^1 S^{1+}$, we obtain

$$S^{0+} S^0 = 1,$$
$$S^{0+} S^1 + S^{1+} S^0 = 0,$$

whence

$$S^1 = -S^0 S^{1+} S^0$$

or

$$S^1_{fi} = -S^0_{ff'} S^{1+}_{f'i'} S^0_{i'i}.$$

Using now the definition of hermitian conjugation and the equality $S^1_{fi} = S_{fi}$, we derive

$$S_{fi} = -S^0_{ff'} S^*_{i'f'} S^0_{i'i}.$$

If the states i and f are diagonal (i.e. $i' = i, f' = f$), then

$$S^0_{i'i} = \delta_{i'i} e^{2i\delta_i}, \qquad S^0_{ff'} = \delta_{ff'} e^{2i\delta_f},$$

where δ_i and δ_f are scattering phases in the initial and final states, respectively. In this case

$$S_{fi} = -e^{+2i(\delta_i + \delta_f)} S^*_{if}.$$

Note that so far we have used only the unitarity of the S-matrix and the weakness of the interaction inducing the transitions between i and f (this last property was used when $S^1 S^{1+}$ was ignored).

Let us take into account now that in the case of T-reversibility the S-matrix must be symmetric,

$$S_{fi} = S_{if}.$$

Substitution of this condition into the preceding expression yields

$$S_{fi} = -e^{+2i(\delta_i + \delta_f)} S^*_{fi},$$

or

$$\arg S_{fi} = \delta_f + \delta_i + \tfrac{1}{2}\pi.$$

We now take into consideration that the amplitudes M differ from the corresponding elements of S by a factor i. Moreover, $\delta_i = 0$ in the decay of a particle. This immediately gives

$$\arg M_{fi} = \delta_f.$$

It then follows that for the amplitudes of hyperon decay into channels with given l and T

$$S_T = \pm |S_T| e^{i\delta_S^T},$$

$$P_T = \pm |P_T| e^{i\delta_P^T}.$$

The relative phase of S- and P-wave amplitudes, Δ, equals: $\Delta = \delta_S - \delta_P + n\pi$. (We recall that the correlation $\zeta[\boldsymbol{\eta} \cdot \boldsymbol{n}]$ is proportional to $\sin \Delta$.)

8.6. SU(3) relation between amplitudes of hyperon decays

Let us derive a relation between the S-wave amplitudes of hyperon decays. We assume that the predominant part of the non-leptonic interaction, satisfying the $\Delta T = \frac{1}{2}$ rule, is a component of an octet. In accordance with this hypothesis we consider the spurion to be the sixth component of the octet ($\propto \lambda_6$), like the K^0 meson. The SU(3)-invariant amplitude, describing processes

$$S + B \to B + P,$$

in which both the fictitious particle S and the real particles (baryons B and pseudoscalar mesons P) are components of octets, can be written in the most general form as

$$A = \sum_{i=1}^{9} A_i J_i,$$

where A_i are numbers and J_i are SU(3) scalars constructed from octet wave functions. Nine such scalars can be formed in the general case:

$$J_1 = (S\bar{B}\tilde{P}B), \quad J_4 = (SB\bar{B}\tilde{P}), \quad J_7 = (S\bar{B})(B\tilde{P}),$$

$$J_2 = (SB\tilde{P}\bar{B}), \quad J_5 = (S\tilde{P}\bar{B}B), \quad J_8 = (SB)(\bar{B}\tilde{P}),$$

$$J_3 = (S\bar{B}B\tilde{P}), \quad J_6 = (S\tilde{P}B\bar{B}), \quad J_9 = (S\tilde{P})(\bar{B}B),$$

where parentheses stand for "trace"; for instance, $(SB) = S_k^i B_i^k$, $(S\bar{B}\tilde{P}B) = S_k^i \bar{B}_l^k \tilde{P}_m^l B_i^m$. The nine scalars are related by a linear constraint:

$$\sum_{i=1}^{6} J_i = \sum_{i=7}^{9} J_i,$$

which will not be used here. We now easily obtain from the explicit form of the meson and baryon octets the formulas

$$A(\Lambda^0_-) = \sqrt{\tfrac{1}{6}}\,(A_1 + A_3 - 2A_4),$$
$$A(\Xi^-_-) = \sqrt{\tfrac{1}{6}}\,(A_2 - 2A_3 + A_4),$$
$$A(\Sigma^+_0) = \sqrt{\tfrac{1}{2}}\,(A_1 - A_3),$$
$$A(\Sigma^+_+) = A_1 + A_7,$$
$$A(\Sigma^-_-) = A_3 + A_7.$$

Later we shall need the transformation properties of octet matrices and invariants J_i under charge conjugation. These are

$$B = B^i_k \xrightarrow{C} B_C = \bar{B}^k_i = \tilde{\bar{B}},$$

$$P \xrightarrow{C} \tilde{P},$$

$$S \xrightarrow{C} S,$$

so that

$$J_1 \to J_1, \qquad J_3 \leftrightarrow J_5,$$
$$J_2 \to J_2, \qquad J_4 \leftrightarrow J_6,$$
$$J_9 \to J_9, \qquad J_7 \leftrightarrow J_8.$$

The amplitude $A = \Sigma A_i J_i$ corresponds to transitions with $P = -1$ and $CP = +1$, and hence with $C = -1$. Consequently,

$$A \xrightarrow{C} -A.$$

This yields

$$A_1 = A_2 = A_9 = 0, \qquad A_3 = -A_5, \qquad A_4 = -A_6, \qquad A_7 = -A_8.$$

By using the relation $A_1 = A_2 = 0$, we obtain

$$A(\Lambda^0_-) + 2A(\Xi^-_-) = \sqrt{3}\,A(\Sigma^+_0).$$

This is the so-called Lee-Sugawara equation, supported rather well by experiment. It should also be mentioned that a similar relation between P-wave amplitudes was also found experimentally, although it is not predicted by SU(3) symmetry (under C-conjugation

$$B = \sum_{i=1}^{9} B_i J_i \to +B,$$

and none of the nine amplitudes B_i vanish).

CHAPTER 9

Dynamics of non-leptonic decays of hyperons

In this chapter we calculate amplitudes for several non-leptonic decays of hyperons. Calculations are based on the quark model of hadrons and on the total effective non-leptonic lagrangian

$$\mathcal{L}_{\text{eff}} = \sqrt{2}\, G \sin\theta \cos\theta \sum_{i=1}^{6} c_i O_i.$$

Explicit expressions for the operators O_i and the values of the coefficients c_i in the above expression were given at the end of Chapter 7. The present chapter deals mostly with decays of the Λ-hyperon. Decays of the Ω-hyperon are discussed at the end of the chapter.

9.1. Quark graphs

The weak interaction of quarks responsible for the non-leptonic decays of hyperons can be of two types: scattering (fig. 9.1) or decay (fig. 9.2). Let us consider, for instance, decays of the Λ-hyperon. Quark graphs describing these decays can be divided into two classes: external (fig. 9.3) and internal graphs (figs. 9.4 and 9.5). The s-quark decay in external graphs occurs as if the decay were isolated, i.e. the quark emits a free π-meson made up of a quark and an antiquark produced in the weak vertex. Creation of the π-meson in internal graphs is more complex and requires participation of other quarks. So far we cannot calculate these more complicated graphs. It is readily apparent, however, that they possess a spectacular property: they allow only transitions with $\Delta T = \frac{1}{2}$. To understand the mechanism forbidding the transitions with $\Delta T = \frac{3}{2}$, just look at the simplified effective lagrangian $\tilde{\mathcal{L}}$ derived in chapter 7. We recall that this lagrangian consists of three terms:

$$\tilde{\mathcal{L}}_{\text{eff}} = \sqrt{2}\, G(a_3 I_3 + a_6 I_6 + a_R I_R) \sin\theta \cos\theta,$$

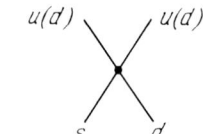

Fig. 9.1.

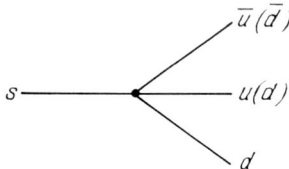

Fig. 9.2.

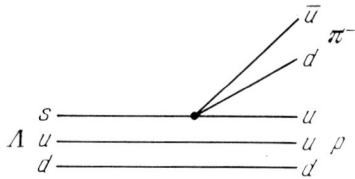

Fig. 9.3.

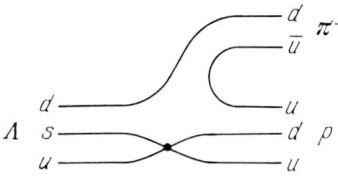

Fig. 9.4.

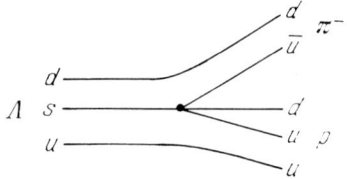

Fig. 9.5.

where

$$I_{3,6} = \bar{u}_L\gamma_\alpha s_L \bar{d}_L\gamma^\alpha u_L \mp \bar{d}_L\gamma_\alpha s_L \bar{u}_L\gamma^\alpha u_L,$$
$$I_R = -\bar{d}_L\gamma_\alpha \lambda^a s_L(\bar{u}_R\gamma^\alpha \lambda^a u_R + \bar{d}_R\gamma^\alpha \lambda^a d_R),$$
$$a_3 \approx 3, \quad a_6 \approx 0.6, \quad a_R \approx 0.12.$$

Among the three terms, only I_6 allows transitions with $\Delta T = \frac{3}{2}$, while I_3 and I_R are pure isotopic spinors and therefore allow transitions only with $\Delta T = \frac{1}{2}$. But it is precisely I_6 which cannot contribute to the internal graphs. The point is, the operator I_6 is symmetrized over quark pairs. (Both initial and final quarks are symmetrized in the case of scattering (figs. 9.1 and 9.4), and only final quarks are symmetrized in the case of decay (figs. 9.2 and 9.5)). At the same time, any pair of quarks in a white baryon is in the color-antisymmetric state and forms an antitriplet $\bar{3}$. Note that owing to this reason, three "symmetric" operators O_2, O_3 (the only O-operator with $\Delta T = \frac{3}{2}$) and O_4, of the total effective lagrangian do not contribute to internal graphs. Among the remaining operators, O_1 is predominant in these graphs (because of a large value of c_1).

9.2. Factorization of external diagrams for the decay $\Lambda^0 \to p\pi^-$

Let us turn now to the calculation of external graphs. We consider the decay $\Lambda^0 \to p\pi^-$ and determine the contribution of each term of the operators O_i to the amplitudes. If we assume that quarks forming the π-meson in fig. 9.3 do not interact with quarks in the initial and final baryons, then the decay amplitude is factorized into a product of two matrix elements. This becomes especially lucid if we take the term $\bar{d}_L\gamma_\alpha u_L \bar{u}_L\gamma^\alpha s_L$:

$$\langle \pi^- p | \bar{d}_L\gamma_\alpha u_L \bar{u}_L\gamma^\alpha s_L | \Lambda \rangle = \langle \pi^- | \bar{d}_L\gamma_\alpha u_L | 0 \rangle \langle p | \bar{u}_L\gamma^\alpha s_L | \Lambda \rangle.$$

Both factors are already familiar:

$$\langle \pi^- | \bar{d}_L\gamma_\alpha u_L | 0 \rangle = \tfrac{1}{2}\langle \pi | \bar{d}\gamma_\alpha\gamma_5 u | 0 \rangle = \tfrac{1}{2} f_\pi \varphi_\pi k_\alpha,$$

where φ_π is the π-meson wave function, k_α is its 4-momentum, and $f_\pi \approx 0.95 m_\pi$ is a well-known constant characterizing the decays $\pi \to e\nu$ and $\pi \to \mu\nu$. As for the matrix element $\langle \bar{p} | \bar{u}_L\gamma_\alpha s_L | \Lambda \rangle$, it determines the amplitude of Λ-hyperon β-decay, and is written in the form

$$\langle p | \bar{u}_L\gamma_\alpha s_L | \Lambda \rangle = \tfrac{1}{2}\langle p | \bar{u}\gamma_\alpha(1+\gamma_5)s | \Lambda \rangle = +\tfrac{1}{2}\bar{u}_p(g_V\gamma_\alpha + g_A)\gamma_\alpha\gamma_5)u_\Lambda.$$

By virtue of SU(3) symmetry, $g_V = -3/\sqrt{6}$, $g_A/g_V = F + \tfrac{1}{3}D$ (see chapter 6). The experimental value of g_V in the decays $\Lambda \to p\ell\nu$ does not contradict the theoretical value; as for the ratio g_A/g_V, experiment gives 0.62 ± 0.05. We thus obtain

$$\langle \pi^- p | \bar{d}_L \gamma_\alpha u_L \bar{u}_L \gamma^\alpha s_L | \Lambda \rangle = -\tfrac{1}{8}\sqrt{6} f_\pi k_\alpha \bar{u}_p \left(\gamma_\alpha + \frac{g_A}{g_V} \gamma_\alpha \gamma_5 \right) u_\Lambda \varphi_\pi$$

$$= -\tfrac{1}{8}\sqrt{6} f_\pi \bar{u}_p \Bigg[(m_\Lambda - m_p)$$

$$- (m_\Lambda + m_p) \frac{g_A}{g_V} \gamma_5 \Bigg] u_\Lambda \varphi_\pi.$$

Let us recall now the general form of the amplitude for non-leptonic hyperon decay (it was given at the beginning of chapter 8):

$$M = G m_\pi^2 \bar{u}_2 (A + B \gamma_5) u_1 \varphi_\pi.$$

We see that the contributions of the term $\sqrt{2}\, G \bar{d}_L \gamma_\alpha u_L \bar{u}_L \gamma^\alpha s_L \sin\theta \cos\theta$ to the amplitudes A and B of the decay $\Lambda^0 \to p\pi^-$ are, respectively,

$$A(\Lambda^0_-): \quad -\tfrac{1}{4}\sqrt{3}\, \sin\theta \cos\theta \, \frac{f_\pi(m_\Lambda - m_p)}{m_\pi^2},$$

$$B(\Lambda^0_-): \quad +\tfrac{1}{4}\sqrt{3}\, \sin\theta \cos\theta \, \frac{f_\pi(m_\Lambda - m_p)}{m_\pi^2} \frac{g_A}{g_V}.$$

In order to find the amplitude corresponding to the term $\bar{u}_L \gamma_\alpha u_L \bar{d}_L \gamma^\alpha s_L$ we must permute the operators $\bar{u}_L$ and d_L. This is conveniently carried out by means of the Fierz transformation (see Chapter 29):

$$\delta^i_k [\gamma_\mu(1+\gamma_5)]^\alpha_\beta \delta^l_m [\gamma^\mu(1+\gamma_5)]^\gamma_\delta$$

$$= +\tfrac{1}{3} \delta^i_m [\gamma_\mu(1+\gamma_5)]^\alpha_\delta \delta^l_k [\gamma^\mu(1+\gamma_5)]^\gamma_\beta$$

$$+ \tfrac{1}{2} (\lambda^a)^i_m [\gamma_\mu(1+\gamma_5)]^\alpha_\delta (\lambda^a)^l_k [\gamma^\mu(1+\gamma_5)]^\gamma_\beta.$$

Here Greek letters denote Dirac indices, Latin letters denote colors, and the plus signs result from anticommutation of spinor operators. According to color conservation, the contribution of the second term vanishes, ($\langle \pi | \bar{d}_L \gamma_\alpha \lambda^a u_L | 0 \rangle = 0$), and we obtain

$$\langle \pi^- p | \bar{u}_L \gamma_\alpha u_L \bar{d}_L \gamma^\alpha s_L | \Lambda \rangle = \tfrac{1}{3} \langle \pi^- p | \bar{d}_L \gamma_\alpha u_L \bar{u}_L \gamma^\alpha s_L | \Lambda \rangle.$$

It is apparent that the terms $\bar{d}_L \gamma_\alpha d_L \bar{d}_L \gamma^\alpha s_L$, $\bar{d}_R \gamma_\alpha d_R \bar{d}_L \gamma^\alpha s_L$, and $\bar{d}_R \gamma_\alpha \lambda^a d_R \bar{s}_L \gamma^\alpha \lambda^a s_L$ do not contribute to the external graph of the decay $\Lambda^0 \to p\pi^-$ since they do not contain the $\bar{u}$-quark required to form the π^- meson. In order to calculate the contributions of the terms $\bar{u}_R \gamma_\alpha u_R \bar{d}_L \gamma^\alpha s_L$

and $\bar{u}_R\gamma_\alpha\lambda^a u_R \bar{d}_L\gamma^\alpha\lambda^a s_L$, we shall use the Fierz transformations

$$\delta^i_k[\gamma_\mu(1+\gamma_5)]^\alpha_\beta \delta^l_m[\gamma^\mu(1-\gamma_5)]^\gamma_\delta$$
$$= -2\left[\tfrac{1}{3}\delta^i_m(1+\gamma_5)^\alpha_\delta \delta^l_k(1-\gamma_5)^\gamma_\beta + \tfrac{1}{2}(\lambda^a)^i_m(1+\gamma_5)^\alpha_\delta(\lambda^a)^l_k(1-\gamma_5)^\gamma_\beta\right],$$

$$(\lambda^a)^i_k[\gamma_\mu(1+\gamma_5)]^\alpha_\beta(\lambda^a)^l_m[\gamma^\mu(1-\gamma_5)]^\gamma_\delta$$
$$= -2\left[\tfrac{16}{9}\delta^i_m(1+\gamma_5)^\alpha_\delta \delta^l_k(1-\gamma_5)^\gamma_\beta - \tfrac{1}{3}(\lambda^a)^i_m(1+\gamma_5)^\alpha_\delta(\lambda^a)^l_k(1-\gamma_5)^\gamma_\beta\right].$$

(The minus signs in front of these two expressions take account of the anticommutation of spinor operators.) Carrying out factorization and taking into account conservation of color, we obtain

$$\langle \pi^- p | \bar{u}_R\gamma^\alpha u_R \bar{d}_L\gamma^\alpha s_L | \Lambda \rangle = -\tfrac{2}{3}\langle \pi^- | \bar{d}_L u_R | 0\rangle \langle p | \bar{u}_R s_L | \Lambda \rangle,$$
$$\langle \pi^- p | \bar{u}_R\gamma^\alpha\lambda^a u_R \bar{d}_L\gamma^\alpha\lambda^a s_L | \Lambda \rangle = -\tfrac{32}{9}\langle \pi^- | \bar{d}_L u_R | 0\rangle \langle p | \bar{u}_R s_L | \Lambda \rangle.$$

9.3. Enhancement of the contribution of right-handed quarks

Now compare the "right-handed" amplitude

$$\langle \pi^- | \bar{d}_L u_R | 0\rangle \langle p | \bar{u}_R s_L | \Lambda \rangle$$

comprising creation and annihilation operators of the right-handed quarks (which are present in O_5 and O_6) with the "left-handed" amplitude

$$\langle \pi^- | \bar{d}_L\gamma_\alpha u_L | 0\rangle \langle p | \bar{u}_L\gamma^\alpha s_L | \Lambda \rangle,$$

which does not comprise the right-handed quarks (this type of amplitude is generated by the operators O_1, O_2, O_3, O_4). The former is a product of scalars while the latter is a product of vectors. In order to reduce them to identical form, we shall use the relation

$$\langle \pi^- | \bar{d}\gamma_\alpha\gamma_5 u | 0\rangle = \frac{m_u + m_d}{m_\pi^2}\langle \pi^- | \bar{d}\gamma_5 u | 0\rangle k_\alpha.$$

In order to verify this equality, it is sufficient to multiply it by k_α and use the Dirac equation for quarks. We recall that $k = k_\Lambda - k_p$, $k^2 = m_\pi^2$; at the same time, $k = k_d + k_{\bar{u}} = k_s - k_u$. The last equality enables us to write

$$k_\alpha \langle p | \bar{u}(\gamma^\alpha + \gamma^\alpha\gamma_5)s | \Lambda\rangle = \langle p | \bar{u}[(m_s - m_u) - (m_s + m_u)\gamma_5]s | \Lambda\rangle$$
$$= m_s \langle p | \bar{u}(1-\gamma_5)s | \Lambda\rangle.$$

(Here the u-quark mass is neglected compared to that of the s-quark. Recall that $m_s \approx 150$ MeV, $m_u \approx 4$ MeV, $m_d \approx 7$ MeV). We thus find that the "right-handed" amplitudes contain a large multiplier

$$\chi = \frac{m_\pi^2}{(m_u + m_d)m_s} \approx 12.$$

One can easily find that

$A(\text{"right-handed"}) = -\chi A (\text{"left-handed"}),$

$B(\text{"right-handed"}) = \chi B (\text{"left-handed"}).$

It is now possible to give the final result for the contribution of the external graph to the amplitude for the decay $\Lambda^0 \to p\pi^-$ taking into account all terms of the effective non-leptonic lagrangian $\mathcal{L}$:

$$A(\Lambda^0_-) = \tfrac{1}{4}\sqrt{3}\, \sin\theta \cos\theta \frac{f_\pi(m_\Lambda - m_p)}{m_\pi^2} [\tfrac{2}{3}c_1 - \tfrac{4}{3}c_2 - \tfrac{4}{3}c_3 - \tfrac{4}{3}c_4$$

$$- \tfrac{32}{9}\chi c_5 - \tfrac{2}{3}\chi c_6],$$

$$B(\Lambda^0_-) = -\tfrac{1}{4}\sqrt{3}\, \sin\theta \cos\theta \frac{f_\pi(m_\Lambda + m_p)}{m_\pi^2} \frac{g_A}{g_V}$$

$$\times [\tfrac{2}{3}c_1 - \tfrac{4}{3}c_2 - \tfrac{4}{3}c_3 - \tfrac{4}{3}c_4 + \tfrac{32}{9}\chi c_5 + \tfrac{2}{3}\chi c_6].$$

Substitution of the values $c_1 = -2.75$, $c_2 = 0.06$, $c_3 = 0.08$, $c_4 = 0.39$, $c_5 = -0.14$, $c_6 = -0.05$ yields

$A(\Lambda^0_-) \approx 0.44,$

$B(\Lambda^0_-) \approx 7.4.$

Experimental data give

$A(\Lambda^0_-) = 1.47 \pm 0.01,$

$B(\Lambda^0_-) = 9.98 \pm 0.24.$

Agreement proves better for amplitude B than for A. It is possible that internal graphs neglected in the above calculations are important in the case of A.

9.4. $\Lambda^0 \to n\pi^0$ decay

Similar calculations for the decay $\Lambda^0 \to n\pi^0$ readily yield

$$\langle \pi^0 n | u_L \gamma_\alpha u_L \bar{d}_L \gamma^\alpha s_L | \Lambda \rangle = \langle \pi^0 | \bar{u}_L \gamma_\alpha u_L | 0 \rangle \langle n | \bar{d}_L \gamma^\alpha s_L | \Lambda \rangle,$$

$$\langle \pi^0 n | \bar{d}_L \gamma_\alpha u_L \bar{u}_L \gamma^\alpha s_L | \Lambda \rangle = \tfrac{1}{3} \langle \pi^0 | \bar{u}_L \gamma_\alpha u_L | 0 \rangle \langle n | \bar{d}_L \gamma^\alpha s_L | \Lambda \rangle,$$

§4] $\Lambda^0 \to n\pi^0$ decay

$$\langle \pi^0 n | \bar{d}_L \gamma_\alpha d_L \bar{d}_L \gamma^\alpha s_L | \Lambda \rangle = \tfrac{4}{3} \langle \pi^0 | \bar{d}_L \gamma_\alpha d_L | 0 \rangle \langle n | \bar{d}_L \gamma^\alpha s_L | \Lambda \rangle,$$

$$\langle \pi^0 n | \bar{d}_R \gamma_\alpha \lambda^a d_R \bar{d}_L \gamma^\alpha \lambda^a s_L | \Lambda \rangle = -\tfrac{32}{9} \langle \pi^0 | \bar{d}_L d_R | 0 \rangle \langle n | \bar{d}_R s_L | \Lambda \rangle,$$

$$\langle \pi^0 n | (\bar{d}_R \gamma_\alpha d_R + \bar{u}_R \gamma_\alpha u_R) \bar{d}_L \gamma^\alpha s_L | \Lambda \rangle = -\tfrac{2}{3} \langle \pi^0 | \bar{d}_L d_R | 0 \rangle \langle n | \bar{d}_R s_L | \Lambda \rangle.$$

These relations make it possible to write out the final expression for the contribution of the external graphs of fig. 9.6 to the amplitudes for $\Lambda^0 \to n\pi^0$ decay:

$$A(\Lambda_0^0) = \frac{\sqrt{3}}{4\sqrt{2}} \sin\theta \cos\theta \frac{f_\pi(m_\Lambda - m_p)}{m_\pi^2} \left[-\tfrac{2}{3}c_1 + \tfrac{4}{3}c_2 + \tfrac{4}{3}c_3 - \tfrac{8}{3}c_4 \right.$$

$$\left. + \tfrac{32}{9}\chi c_5 + \tfrac{2}{3}\chi c_6 \right],$$

$$B(\Lambda_0^0) = -\frac{\sqrt{3}}{4\sqrt{2}} \sin\theta \cos\theta \frac{f_\pi(m_\Lambda + m_p)}{m_\pi^2} \frac{g_A}{g_V}$$

$$\times \left[-\tfrac{2}{3}c_1 + \tfrac{4}{3}c_2 + \tfrac{4}{3}c_3 - \tfrac{8}{3}c_4 - \tfrac{32}{9}\chi c_5 - \tfrac{2}{3}\chi c_6 \right].$$

Let us find now the contribution of operator O_4 giving rise to transitions with $\Delta T = \tfrac{3}{2}$:

$$A(\Lambda_-^0) + \sqrt{2} A(\Lambda_0^0) = -\sqrt{3} \sin\theta \cos\theta \frac{f_\pi(m_\Lambda - m_p)}{m_\pi^2} c_4 = -0.17,$$

$$B(\Lambda_-^0) + \sqrt{2} B(\Lambda_0^0) = +\sqrt{3} \sin\theta \cos\theta \frac{f_\pi(m_\Lambda + m_p)}{m_\pi^2} \frac{g_A}{g_V} c_4$$

$$= +1.32.$$

(The right-hand sides of these expressions would be exactly zero if the $\Delta T = \tfrac{1}{2}$ rule were strict.) The experimental values are

$$A(\Lambda_-^0) + \sqrt{2} A(\Lambda_0^0) = -0.09 \pm 0.03,$$

$$B(\Lambda_-^0) + \sqrt{2} B(\Lambda_0^0) = -0.66 \pm 0.81.$$

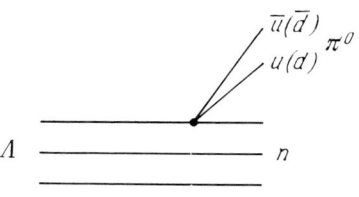

Fig. 9.6.

We see that the theory gives for the S-wave a somewhat milder suppression of the amplitude with $\Delta T = \frac{3}{2}$ than is found experimentally. Although the sign in the case of the P-wave is different from the predicted one, this is not really a contradiction because experimental error is very large. It would be very interesting to obtain more accurate experimental data since the theoretical prediction for the amplitudes with $\Delta T = \frac{3}{2}$ is very reliable because of the absence of a contribution from the internal diagrams. Indeed, the predicted amplitude may vary by a factor of 2 but its sign is unambiguous.

9.5. Ω-hyperon decays

Let us consider now the decays

$$\Omega^- \to \Xi^0 \pi^-,$$
$$\Omega^- \to \Xi^- \pi^0.$$

The ratio of the rates for these two decays would be exactly 2 if the $\Delta T = \frac{1}{2}$ rule were strict. Let us calculate this ratio taking into account the contribution of the operator O_4 giving rise to transitions with $\Delta T = \frac{3}{2}$ in external graphs (figs. 9.7 and 9.8). Two waves participate in the decay of the Ω-hyperon which has $J^P = \frac{3}{2}^+$: P($l = 1$) and D($l = 2$). The scale of the P-wave amplitude is approximately identical to that in decays of hyperons with $J^P = \frac{1}{2}^+$. As for the D-wave amplitude, it must be small because it contains a factor $(kR)^2 \ll 1$ due to the centrifugal barrier. Consequently, we take into account only the P-wave amplitude. By retracing the line of arguments followed in the above case of the Λ-hyperon, we easily obtain

$$\frac{\Gamma(\Omega^- \to \Xi^0 \pi^-)}{\Gamma(\Omega^- \to \Xi^- \pi^0)} \simeq \frac{|P(\Omega_-^-)|^2}{|P(\Omega_0^-)|^2}$$

$$\simeq 2 \frac{|c_1 + 2c_2 + 2c_3 + 2c_4 - \frac{16}{3}\chi c_5 - \chi c_6|^2}{|-c_1 + 2c_2 + 2c_3 - 4c_4 - \frac{16}{3}\chi c_5 - \chi c_6|^2}$$

$$\simeq 2.94.$$

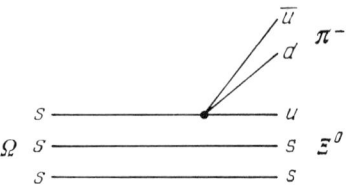

Fig. 9.7.

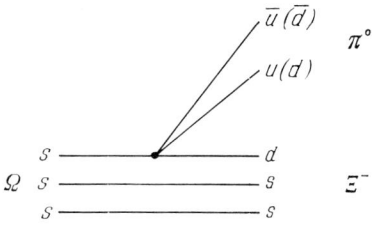

Fig. 9.8.

A preliminary experimental result for this ratio is 2.93 ± 0.45. An analysis of Λ-hyperon decays has demonstrated that the theoretical value of $\chi(c_5 + c_6)$ is perhaps underestimated, while that of c_4 overestimated. It is reasonable to expect therefore, that the experimental value for the ratio $\Gamma(\Omega^- \to \Xi^0 \pi^-) : \Gamma(\Omega^- \to \Xi^- \pi^0)$ will be close to 2.5 when accuracy is improved. Note that χ would increase by a factor of 1.5 if quark masses were smaller, by the same factor, than the currently accepted values (namely, if they were $m_u \approx 3$ MeV, $m_d \approx 5$ MeV, $m_s \approx 100$ MeV). Incidentally, with a lower mass of the s-quark it would be clearer why nature demonstrates such a good $SU(3) \times SU(3)$ chiral symmetry.

CHAPTER 10

Non-leptonic decays of K-mesons

In this chapter we discuss decays of charged and neutral K-mesons into two and three pions: $K \to 2\pi$ and $K \to 3\pi$. Before going into the problems connected with the isotopic selection rules and quark dynamics, we have to analyze restrictions imposed on these decays by conservation of CP parity. (Small effects in which CP invariance is broken will be discussed in Chapter 12).

10.1. K_1^0 and K_2^0 mesons

Consider the decay of a neutral kaon into two pions, $\pi^0\pi^0$ or $\pi^+\pi^-$. The $\pi^0\pi^0$ system is CP even: $CP(\pi^0\pi^0) = (CP(\pi^0))^2 = (-1)^2 = +1$. The same is true for the $\pi^+\pi^-$ system:

$$CP(\pi^+\pi^-) = C(\pi^+\pi^-)P(\pi^+\pi^-) = (-1)^l(-1)^l = +1.$$

(Here l is the orbital momentum of the system $\pi^+\pi^-$; $l = 0$ in K-meson decays). Because of this, and by virtue of the conservation of CP parity, only a CP-even system can decay into two pions. However, neither K^0 nor $\overline{K}^0$ have definite CP parity:

$$CP|K^0\rangle = |\overline{K}^0\rangle,$$

$$CP|\overline{K}^0\rangle = |K^0\rangle.$$

As a result, two pions are produced in the decay of a linear superposition of K^0 and $\overline{K}^0$:

$$K_1^0 = \frac{K^0 + \overline{K}^0}{\sqrt{2}}.$$

Its CP-parity is $+1$:

$$CP|K_1^0\rangle = +|K_1^0\rangle.$$

An orthogonal superposition

$$K_2^0 = \frac{K^0 - \bar{K}^0}{\sqrt{2}}$$

is CP odd ($CP|K_2^0\rangle = -|K_2^0\rangle$), and will decay into two pions only in interactions violating CP invariance.

Consider now decays into three pions: $CP|\pi^0\pi^0\pi^0\rangle = -|\pi^0\pi^0\pi^0\rangle$ so that the decay of K_1^0 into three neutral pions is forbidden while the decay $K_2^0 \to 3\pi^0$ is allowed. The CP parity of the system $\pi^+\pi^-\pi^0$ is a function of the orbital state of the pions. The total angular momentum J of the three pions in the decay $K \to 3\pi$ is zero. We can write J as a sum,

$$J = l + L,$$

where l is the orbital momentum of the pair $\pi^+\pi^-$, and L is the orbital momentum of the π^0 meson with respect to the center of mass of the pair $\pi^+\pi^-$. From $J = 0$ we obtain $L = l$. The pair $\pi^+\pi^-$ is CP even irrespective of l, and the π^0 meson is CP odd, so that

$$CP|\pi^+\pi^-\pi^0\rangle = (-1)^{l+1}|\pi^+\pi^-\pi^0\rangle.$$

The system $\pi^+\pi^-\pi^0$ can therefore be produced with even l in the decay of the K_2^0 meson, and with odd l in the decay of the K_1^0 meson. Non-zero orbital momenta suppress the decay amplitude; hence,

$$|M(K_1^0 \to \pi^+\pi^-\pi^0)| \ll |M(K_2^0 \to \pi^+\pi^-\pi^0)|.$$

Consequently, predominant decays must be $K_1^0 \to 2\pi$ and $K_2^0 \to 3\pi$. The former violate and the latter conserve P-parity. The corresponding interactions in the effective non-leptonic lagrangian are of the same order of magnitude (see chapters 7 and 9). Thus, the ratio of decay rates for $K_1^0 \to 2\pi$ and $K_2^0 \to 3\pi$ is determined by the ratio of phase-space volumes in these decays, and therefore by the energy release in the two cases. As a result, the decay rate of $K_1^0 \to 2\pi$ is larger by approximately three orders of magnitude than the decay rate of $K_2^0 \to 3\pi$.

Let us turn now to isotopic relations between different charged channels of non-leptonic decays.

10.2. Isotopic relations for $K \to 2\pi$ decays

Two pions can be found in the states with $T = 0$, 1, and 2. Let one pion be described by an isotopic vector $\boldsymbol{a}$ and another by an isotopic vector $\boldsymbol{b}$. The state with $T = 0$ is described by an isoscalar $\boldsymbol{a} \cdot \boldsymbol{b} = a_i b_i$. The state with $T = 1$ is represented by an isovector $a_i b_j \varepsilon_{ijk}$, that is by $[\boldsymbol{a} \times \boldsymbol{b}]$. The states with $T = 2$ are represented by a symmetric tensor of rank two with zero trace:

$$a_i b_j + a_j b_i - \tfrac{2}{3} \delta_{ij} (\mathbf{a} \cdot \mathbf{b}).$$

(This tensor has five independent components.) The tensor and the scalar are symmetrical with respect to permutation of the vectors $\mathbf{a}$ and $\mathbf{b}$, and the vector is anti-symmetrical. Two pions produced in the K-meson decay have $l = 0$ and are in a spatially symmetrical state; consequently, they cannot be in the isospin-antisymmetric state by virtue of the generalized Bose principle and therefore cannot have $T = 1$. This means that the two pions can have $T = 0, 2$ in the $K_1^0 \to 2\pi$ decay, and only $T = 2$ in the decay $K^+ \to \pi^+ \pi^0$ ($T = 0$ is impossible in this last case since $T_3 = 1$).

Let us now apply the $\Delta T = \tfrac{1}{2}$ rule. Were it an exact rule, two pions could not be in a state with $T = 2$ in the decay of the K-meson having isospin $\tfrac{1}{2}$. Consequently, the decay $K^+ \to \pi^+ \pi^0$ would be forbidden, and the decays $K_1^0 \to 2\pi$ would be to the state of two pions with $T = 0$. From the wave function of this state

$$a_i b_i = a_+ b_- + a_- b_+ + a_0 b_0,$$

where subscripts indicate charge states of pions, we readily conclude that

$$\frac{\Gamma(K_1^0 \to \pi^+ \pi^-)}{\Gamma(K_1^0 \to \pi^0 \pi^0)} = 2.$$

Experimentally this ratio is close to 2.2, and the decay $K^+ \to \pi^+ \pi^0$ is not completely forbidden even if suppressed by a factor of several hundred as compared to the decays $K_1^0 \to 2\pi$. The deviation from the $\Delta T = \tfrac{1}{2}$ rule is caused in both cases by transitions with $\Delta T = \tfrac{3}{2}$ produced owing to the term $I_6(O_4)$ in the effective non-leptonic lagrangian.

By using "spurions" with $T = \tfrac{1}{2}$ and $T = \tfrac{3}{2}$ and the table of the Clebsch-Gordan coefficients (see appendix, Chapter 29, sect. 2.5), we easily derive the following relations (it is essential that the wave function of two π-mesons is symmetrized over the relevant isotopic variables):

$$M(K_1^0 \to \pi^+ \pi^-) = \sqrt{\tfrac{2}{3}} A_0 e^{i\delta_0} + \sqrt{\tfrac{1}{3}} A_2 e^{i\delta_2},$$

$$M(K_1^0 \to \pi^0 \pi^0) = \sqrt{\tfrac{1}{3}} A_0 e^{i\delta_0} - \sqrt{\tfrac{2}{3}} A_2 e^{i\delta_2},$$

$$M(K^+ \to \pi^+ \pi^0) = \tfrac{1}{2} \sqrt{3} A_2 e^{i\delta_2},$$

where A_0 and A_2 are real amplitudes of transitions to states with $T = 0$ and $T = 2$, respectively, and δ_0 and δ_2 are phases of the $\pi\pi$ scattering in these states. (These phases are taken into account in correspondence with the relations derived in chapter 8.) The ratio for widths becomes

$$\frac{\Gamma(K_1^0 \to \pi^+\pi^-)}{\Gamma(K_1^0 \to \pi^0\pi^0)} = \frac{p_\pm}{p_0}\left[2 + 6\sqrt{2}\,\frac{A_2}{A_0}\cos(\delta_2 - \delta_0)\right],$$

$$\frac{\Gamma(K^+ \to \pi^+\pi^0)}{\Gamma(K_1^0 \to 2\pi)} = \frac{3}{4}\left(\frac{A_2}{A_0}\right)^2.$$

Using the ratio of phase-space volumes equal to the ratio of momenta $p_\pm/p_0 = 0.986$, and using also the data on S-wave $\pi\pi$ scattering, $\delta_0 - \delta_2 = 53° \pm 6°$, we find that the two ratios are in agreement with experiment if $A_2/A_0 = 4.5 - 4.6\%$.

10.3. Quark graphs for $K^\pm \to \pi^\pm\pi^0$ decays

As an example, let us calculate the amplitude of the $K^\pm \to \pi^\pm\pi^0$ decays on the basis of the effective non-leptonic interaction O_4 (we recall that the remaining terms of the effective non-leptonic lagrangian discussed in chapter 7 are subject to the $\Delta T = \frac{1}{2}$ selection rule and therefore make no contribution to the $K^+ \to \pi^+\pi^0$ decay). In order to avoid dealing with the $\bar{s}$-quark and to operate only with the s-quark, let us consider only the $K^- \to \pi^-\pi^0$ decay whose amplitude is equal to that of $K^+ \to \pi^+\pi^0$.

It is quite apparent that annihilation-type graphs shown in fig. 10.1 make no contribution to the amplitude. Indeed, the π^0 mesons in graphs 10.1a and 10.1b are produced with opposite signs (from $u\bar{u}$ in fig. 10.1a and from $d\bar{d}$ in fig. 10.1b). We now take into account that two pions produced in the decay are in a symmetric state (this corresponds to the up-down symmetry of the graphs). It is then immediately apparent that the two graphs 10.1a and 10.1b cancel each other out.

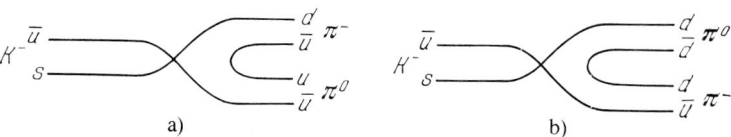

Fig. 10.1.

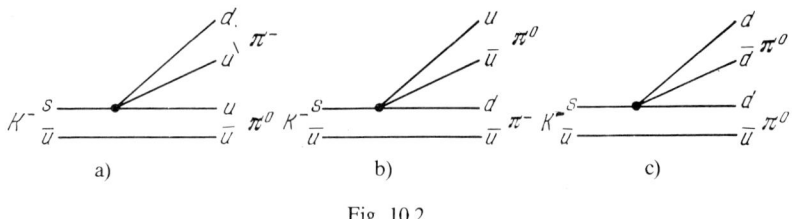

Fig. 10.2.

Let us consider now the external graphs in which an s-quark emits a pion (see fig. 10.2). We recall (see the end of chapter 7) that

$$O_4 = \bar{d}_L \gamma_\alpha u_L \bar{u}_L \gamma^\alpha s_L + \bar{u}_L \gamma_\alpha u_L \bar{d}_L \gamma^\alpha s_L - \bar{d}_L \gamma_\alpha d_L \bar{d}_L \gamma^\alpha s_L.$$

First factorize the amplitude corresponding to the graph of fig. 10.2a. The contribution of the first term of O_4 to the amplitude is

$$\langle \pi^- \pi^0 | \bar{d}_L \gamma_\alpha u_L \bar{u}_L \gamma^\alpha s_L | K^- \rangle_a = \langle \pi^- | \bar{d}_L \gamma_\alpha u_L | 0 \rangle \langle \pi^0 | \bar{u}_L \gamma^\alpha s_L | K^- \rangle.$$

The contribution of the second term of O_4 to the amplitude of fig. 10.2a can be found by means of the Fierz transformation (see appendix and chapters 7 and 9):

$$\langle \pi^- \pi^0 | \bar{u}_L \gamma_\alpha u_L \bar{d}_L \gamma^\alpha s_L | K^- \rangle_a = \tfrac{1}{3} \langle \pi^- \pi^0 | \bar{d}_L \gamma_\alpha u_L \bar{u}_L \gamma^\alpha s_L | K^- \rangle.$$

Obviously, the contribution of the third term of O_4 to the amplitude of fig. 10.2a is zero: this term contains no u-quark operators. The total contribution of operator O_4 to the amplitude corresponding to graph 10.2a is

$$\langle \pi^- \pi^0 | O_4 | K^- \rangle_a = \tfrac{4}{3} \langle \pi^- | \bar{d}_L \gamma_\alpha u_L | 0 \rangle \langle \pi^0 | \bar{u}_L \gamma^\alpha s_L | K^- \rangle$$

$$= \tfrac{1}{3} \langle \pi^- | \bar{d} \gamma_\alpha \gamma_5 u | 0 \rangle \langle \pi^0 | \bar{u} \gamma^\alpha s | K^- \rangle.$$

A similar analysis in the case of the graph of fig. 10.2b yields

$$\langle \pi^- \pi^0 | \bar{d}_L \gamma_\alpha u_L \bar{u}_L \gamma^\alpha s_L | K^- \rangle_b = \tfrac{1}{3} \langle \pi^- \pi^0 | \bar{u}_L \gamma_\alpha u_L \bar{d}_L \gamma^\alpha s_L | K^- \rangle_b$$

$$= \tfrac{1}{3} \langle \pi^0 | \bar{u}_L \gamma_\alpha u_L | 0 \rangle \langle \pi^- | \bar{d}_L \gamma^\alpha s_L | K^- \rangle.$$

Hence,

$$\langle \pi^- \pi^0 | O_4 | K^- \rangle_b = \tfrac{1}{3} \langle \pi^0 | \bar{u} \gamma_\alpha \gamma_5 u | 0 \rangle \langle \pi^- | \bar{d} \gamma^\alpha s | K^- \rangle.$$

Consider finally the amplitude corresponding to the graph of fig. 10.2c.

Only the last term of O_4 contributes to it:

$$\langle \pi^-\pi^0|O_4|K^-\rangle_c = -\langle \pi^-\pi^0|\bar{d}_L\gamma_\alpha d_L \bar{d}_L\gamma^\alpha s_L|K^-\rangle_c$$

$$= -\tfrac{4}{3}\langle \pi^0|\bar{d}_L\gamma_\alpha d_L|0\rangle\langle \pi^-|\bar{d}_L\gamma^\alpha s_L|K^-\rangle$$

$$= -\tfrac{1}{3}\langle \pi^0|\bar{d}\gamma_\alpha\gamma_5 d|0\rangle\langle \pi^-|\bar{d}\gamma^\alpha s|K^-\rangle.$$

Taking into account now that

$$\langle \pi^0|\bar{u}\gamma_\alpha\gamma_5 u|0\rangle = -\langle \pi^0|\bar{d}\gamma_\alpha\gamma_5 d|0\rangle = \tfrac{1}{2}\sqrt{2}\,\langle \pi^-|\bar{d}\gamma_\alpha\gamma_5 u|0\rangle,$$

$$\langle \pi^-|\bar{d}\gamma_\alpha s|K^-\rangle = \sqrt{2}\,\langle \pi^0|\bar{u}\gamma_\alpha s|K^-\rangle,$$

we notice that graphs 10.2a, 2b, and 2c give identical contributions. Finally,

$$\langle \pi^-\pi^0|O_4|K^-\rangle = \langle \pi^-\pi^0|O_4|K^-\rangle_a + \langle \pi^-\pi^0|O_4|K^-\rangle_b$$

$$+ \langle \pi^-\pi^0|O_4|K^-\rangle_c$$

$$= \langle \pi^-|\bar{d}\gamma_\alpha\gamma_5 u|0\rangle\langle \pi^0|\bar{u}\gamma^\alpha s|K^-\rangle$$

$$= \sqrt{\tfrac{1}{2}}\,f_\pi k_{1\alpha}\bigl[f_+(k+k_2)^\alpha + f_-(k-k_2)^\alpha\bigr]$$

$$= \sqrt{\tfrac{1}{2}}\,f_\pi\bigl[f_+(m_K^2 - m_\pi^2) + f_-m_\pi^2\bigr].$$

These expressions take into account the relations (see chapters 5 and 6)

$$\langle \pi^-|\bar{d}\gamma_\alpha\gamma_5 u|0\rangle = f_\pi k_{1\alpha},$$

$$\langle \pi^0|\bar{u}\gamma^\alpha s|K^-\rangle = \sqrt{\tfrac{1}{2}}\bigl(f_+(k+k_2)^\alpha + f_-(k-k_2)^\alpha\bigr),$$

where k is the momentum of the K-meson, and k_1 and k_2 are momenta of the π^- and π^0 mesons, respectively. In the case of $K^- \to \pi^-\pi^0$ decay, $k = k_1 + k_2$. If we neglect m_π^2 in comparison to m_K^2 (moreover, we take into account that $f_- < f_+ = 1$), then

$$\langle \pi^-\pi^0|O_4|K^-\rangle \approx \sqrt{\tfrac{1}{2}}\,f_\pi m_K^2.$$

Recalling that

$$\mathcal{L}_{\text{eff}} = \sqrt{2}\,G\sin\theta\cos\theta\sum_{i=1}^{6} c_i O_i,$$

where $c_4 = 0.39$ (see the end of Chapter 7), we obtain for the decay amplitude

$$M(K^{\pm} \to \pi^{\pm}\pi^0) = \langle \pi^-\pi^0 | \mathcal{L}_{\text{eff}} | K^- \rangle$$
$$= \sqrt{2}\, Gc_4 \sin\theta \cos\theta \langle \pi^-\pi^0 | O_4 | K^- \rangle$$
$$= Gm_K^2 f_\pi c_4 \sin\theta \cos\theta.$$

Experimentally

$$|M(K^+ \to \pi^+\pi^0)| \approx 0.05 G m_K^2 m_\pi.$$

We find that the experimental value of c_4 is smaller by a factor of 1.5 than the value calculated in the framework of the theory taking account of hard virtual gluons (i.e. of corrections for strong interactions at small distances). Recall that the data on the $\Lambda \to N\pi$ decays also indicate that c_4 is smaller by the same factor of 1.5 than the result of theoretical calculations. These calculations are unsatisfactory possibly because only the main logarithmic terms are taken into account. Furthermore, the contribution of intermediate distances is neglected (large distances are covered by the meson wave functions).

We have thus discussed the $K^+ \to \pi^+\pi^0$ decay due to transitions with $\Delta T = \frac{3}{2}$. The $K_S \to 2\pi$ decays dominated by transitions with $\Delta T = \frac{1}{2}$ can be analyzed in a similar manner. It can be shown, in particular, that the main contribution to the $K_S^0 \to \pi^+\pi^-$ decay is made by the operators O_5 and O_6 and that the theoretical value of the amplitude obtained fits well the corresponding experimental data.

10.4. $K \to 3\pi$ decays

In a general case, three pions may form a state with $T = 0, 1, 2$, and 3:

$$3 \times 3 \times 3 = 1 + 3 + 3 + 3 + 5 + 5 + 7.$$

Let us denote isotopic functions of pions by a, b, c. A state with $T = 0$ is written as $a[bc]$; it is anti-symmetric and therefore can comprise only three different pions with odd orbital momenta l and L (see the beginning of this chapter). Such a state could be produced in a suppressed (as yet unobserved) decay $K_1^0 \to \pi^+\pi^-\pi^0$. The state with $T = 3$ is completely symmetrical. Two states with $T = 2$ have mixed symmetry. One of the three states with $T = 1$ is completely symmetrical, and the remaining two have mixed symmetry. The symmetrical state with $T = 1$ is predominant in the $K \to 3\pi$ decay because of the $\Delta T = \frac{1}{2}$ rule and low energy release:

$$A = a(bc) + b(ca) + c(ab).$$

Decays of the K^+ meson are represented by the component A_+:
$$A_+ = a_+(bc) + b_+(ca) + c_+(ab),$$
and decays of the K_2^0 meson by the component A_0:
$$A_0 = a_0(bc) + b_0(ca) + c_0(ab).$$
Recast A_+ in the form
$$\begin{aligned}A_+ = &+ a_+b_+c_- + a_+b_-c_+ + a_+b_0c_0\\ &+ b_+c_+a_- + b_+c_-a_+ + b_+c_0a_0\\ &+ c_+a_+b_- + c_+a_-b_+ + c_+a_0b_0.\end{aligned}$$
This yields the ratio of decay widths of the K^+ meson:
$$\frac{\Gamma(K^+ \to \pi^+\pi^-\pi^-)}{\Gamma(K^+ \to \pi^0\pi^0\pi^+)} = \frac{|2a_+b_+c_-|^2 + |2b_+c_+a_-|^2 + |2c_+a_+b_-|^2}{|a_+b_0c_0|^2 + |b_+c_0a_0|^2 + |c_+a_0b_0|^2}$$
$$= \frac{12}{3} = \frac{4}{1}.$$

In the explicit form A_0 is written as
$$\begin{aligned}A_0 = &+ a_0b_+c_- + a_0b_-c_+ + a_0b_0c_0\\ &+ b_0c_+a_- + b_0c_-a_+ + b_0c_0a_0\\ &+ c_0a_+b_- + c_0a_-b_+ + c_0a_0b_0.\end{aligned}$$
This yields the relation between the K_2^0 meson decay widths:
$$\Gamma(K_2^0 \to \pi^+\pi^-\pi^0) : \Gamma(K_2^0 \to \pi^0\pi^0\pi^0)$$
$$= \{|a_0b_+c_-|^2 + |a_0b_-c_+|^2 + |b_0c_+a_-|^2$$
$$+ |b_0c_-a_+|^2 + |c_0a_+b_-|^2 + |c_0a_-b_+|^2\} : |3a_0b_0c_0|^2$$
$$= 6 : 9 = 2 : 3.$$
Making use of the spurion, we easily obtain
$$M(K^0 \to A_0) : M(K^+ \to A_+) = 1 : \sqrt{2}.$$
If we now take into account that
$$M(K^0 \to A_0) = -M(\bar{K}^0 \to A_0)$$
and
$$K_2^0 = \sqrt{\tfrac{1}{2}}(K^0 - \bar{K}^0),$$

we conclude that

$$M(K_2^0 \to A_0) = M(K^+ \to A_+),$$

that is total widths of three-pion decays of the K^+ and K^0 mesons must be equal. Finally,

$$\Gamma(K^+ \to \pi^+\pi^+\pi^-) : \Gamma(K_2^0 \to \pi^0\pi^0\pi^0) : \Gamma(K_2^0 \to \pi^+\pi^-\pi^0) :$$
$$: \Gamma(K^+ \to \pi^0\pi^0\pi^+) = 4:3:2:1.$$

Before these ratios are compared to experimental values, we must take into account the differences in phase-space volumes due to differences in masses of K^+ and K^0 mesons and $\pi^\pm$ and π^0 mesons. Ratios of the phase-space volumes are

1 : 1.44 : 1.28 : 1.15.

This taken into account, the ratios $4:1$ and $3:2$ are satisfied with approximately 5% accuracy. The ratio

$$\Gamma(K^+ \to \pi^+\pi^+\pi^-) : \Gamma(K_2^0 \to \pi^+\pi^-\pi^0) = 2$$

is more sensitive to transitions with $\Delta T = \frac{3}{2}$. Experiment gives approximately 2.5.

CHAPTER 11

Neutral K-mesons in vacuum and in matter

Neutral K-mesons are a unique physical system which appears to have been created by nature specially to demonstrate, in the most impressive manner, a number of spectacular phenomena.

Two degenerate levels (K^0 and $\overline{K}^0$) are mixed by the strangeness-violating weak interaction. This mixing produces two levels K_1^0 and K_2^0 with small mass difference ($\Delta m \simeq 0.5 \cdot 10^{10}$ s^{-1}) but very different lifetimes ($\tau_1 \simeq 10^{-10}$ s, $\tau_2 \simeq 5 \cdot 10^{-8}$ s). The maximum breaking of P- and C-parity occurs in the decays of these two levels. Both the interaction and decays of neutral K-mesons are characterized by spectacular quantum-mechanical effects. If the K-mesons did not exist, they should have been invented "on purpose", in order to teach students the principles of quantum mechanics. In particular, K-mesons make it possible to realize in the laboratory, and over large distances, a number of gedanken experiments, such as, for example, the Einstein-Podolsky-Rosen experiment. And the last but not the least significant phenomenon is the breaking of CP parity in decays of K^0 mesons, which we shall discuss in the next chapter. The subject of the present chapter is the difference in masses of K_1^0 and K_2^0 mesons, and interference effects in decays of these particles.

11.1. $K^0 \leftrightarrow \overline{K}^0$ transitions and the K_1^0-K_2^0 mass difference

Most elementary particles have corresponding antiparticle counterparts, with the same mass, lifetime, and spin, but with opposite sign of charge (electric, baryonic, or leptonic). The electron-positron, proton-antiproton, and neutron-antineutron are examples of such pairs. The truly neutral particles which are identical to their antiparticles (the photon, π^0, η^0, ω^0, etc.) form a much smaller class of particles. The neutral K-mesons are at the junction of these two classes. On the one hand, K^0 and $\overline{K}^0$ are not identical: the former has positive strangeness and the latter negative, so that in strong

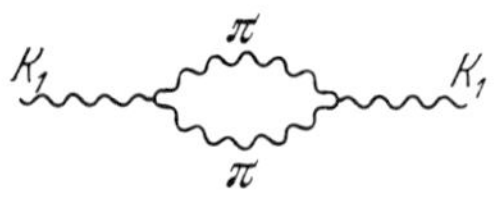

Fig. 11.1.

interactions with nuclei K^0 and $\overline{K}^0$ are just as different as the neutron n and antineutron $\bar{n}$. On the other hand, non-conservation of strangeness in weak interactions allows $K^0 \leftrightarrow \overline{K}^0$ transitions while the transition $n \leftrightarrow \bar{n}$ is ruled out because of conservation of baryonic charge.

The mixing of two degenerate levels in vacuum must result in level splitting. The levels with definite CP parity,

$$K_1^0 = \frac{K + \overline{K}^0}{\sqrt{2}} \quad \text{and} \quad K_2^0 = \frac{K - \overline{K}^0}{\sqrt{2}},$$

would have definite masses and lifetimes if CP invariance were strict. In fact, this invariance is broken but the corresponding corrections to mass and lifetime are negligibly small. The lifetime difference between K_1^0 and K_2^0 occurs because K_1^0 decays into two pions while only three-pion decays are allowed for K_2^0 (see chapter 10). Let us consider the graphs of figs. 11.1 and 11.2. The imaginary parts of these graphs are proportional to $\Gamma(K_1 \to 2\pi)$ and $\Gamma(K_2 \to 3\pi)$, respectively. The real parts contribute to the masses of the K_1 and K_2 mesons. If we assume that by order of magnitude the real parts are comparable to the imaginary ones, then $|\Delta m_{12}| \equiv |m_2 - m_1| \sim \Gamma_1 \sim G^2 m_\pi^5$. It is essential to emphasize that the mass difference Δm_{12} would not vanish even if, for instance, pions were twice as massive as they actually are and consequently the decays $K_1 \to 2\pi$ and $K_2 \to 3\pi$ were kinematically forbidden.

It should be instructive to regard the masses of the K_1^0 and K_2^0 mesons as mean values of the hamiltonian over the states K_1^0 and K_2^0:

$$m_1 = \langle K_1 | H | K_1 \rangle = \left\langle \frac{K + \overline{K}}{\sqrt{2}} \middle| H \middle| \frac{K + \overline{K}}{\sqrt{2}} \right\rangle$$

$$= \tfrac{1}{2} [\langle K|H|K \rangle + \langle \overline{K}|H|\overline{K} \rangle + \langle K|H|\overline{K} \rangle + \langle \overline{K}|H|K \rangle],$$

$$m_2 = \langle K_2 | H | K_2 \rangle = \left\langle \frac{K - \overline{K}}{\sqrt{2}} \middle| H \middle| \frac{K - \overline{K}}{\sqrt{2}} \right\rangle$$

$$= \tfrac{1}{2} [\langle K|H|K \rangle + \langle \overline{K}|H|\overline{K} \rangle - \langle K|H|\overline{K} \rangle - \langle \overline{K}|H|K \rangle].$$

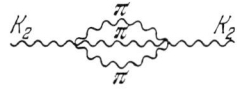

Fig. 11.2.

We see that

$$m_1 - m_2 = \langle K|H|\overline{K}\rangle + \langle \overline{K}|H|K\rangle,$$

that is the mass difference between K_1^0 and K_2^0 mesons is due to $K \leftrightarrow \overline{K}$ transitions which change strangeness by 2. Such transitions with $|\Delta S| = 2$ are absent in the weak lagrangian in the discussed quark theory of the weak interaction (see fig. 11.3). The mass difference Δm_{12} is very sensitive to such transitions. It would "respond" to them even if the constant G_2 (see fig. 11.3) were smaller by seven orders of magnitude than the Fermi constant G. Indeed, let us analyze the contribution of the graph in fig. 11.3 to the amplitude of the $K^0 \leftrightarrow \overline{K}^0$ transition (fig. 11.4). Factorization of the contributions of loops readily yields

$$\Delta m_{12} \sim G_2 f_K^2 m_K,$$

where $f_K \simeq 165$ MeV is a constant characterizing the decay amplitude of $K \to \mu\nu$, and m_K is the K-meson mass. Hence, G_2 is smaller by at least seven orders of magnitude than the Fermi constant G.

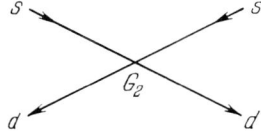

Fig. 11.3.

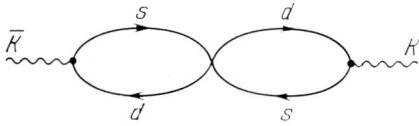

Fig. 11.4.

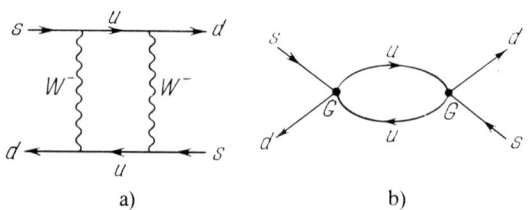

Fig. 11.5.

11.2. Glashow-Iliopoulos-Maiani mechanism

In the theory under discussion, ss ↔ dd transitions can, and must, arise in the second order of perturbation theory in G (fig. 11.5). Fig. 11.5a shows how this transition goes via exchange of W-bosons; the same process is shown in fig. 11.5b in the local four-fermion limit. Obviously, the graph of fig. 11.5b is quadratically divergent and yields the effective constant of the ss ↔ dd transition, of the order of $G^2\Lambda^2$, where Λ is the cutoff energy. As clearly follows from fig. 11.5a, this cutoff energy is of the order of m_W, that is of the mass of the intermediate boson. But as follows from the preceding section, the quantity $G_2 \sim G^2 m_W^2$ is by many orders of magnitude larger than is allowed by the experimental value of Δm_{12}.

It is very interesting that the theory of the weak interaction which is our subject here, copes with this difficulty by mutual compensation of the graphs containing virtual u- and c-quarks. The mechanism of this compensation is often abbreviated as GIM, by the first letters of the authors' names. It must be emphasized that Glashow, Iliopoulos and Maiani have published their paper several years before charmed particles were discovered, and greatly stimulated the experimental effort to detect these particles.

With the c-quarks taken into account, we must consider not just one square of fig. 11.5a, but four such squares (fig. 11.6). We recall that the

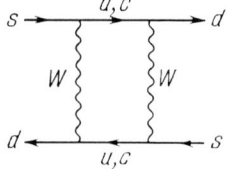

Fig. 11.6.

charged current emitting W-bosons is $\bar{u}d' + \bar{c}s'$ where
$$d' = d\cos\theta + s\sin\theta, \qquad s' = -d\sin\theta + s\cos\theta.$$
(We have neglected the contribution of the b- and t-quarks; its inclusion is straightforward (see Chapter 15)). Now it is easily verified that contributions of all four squares in fig. 11.6 are proportional to $\sin^2\theta \cos^2\theta$, with the plus sign in the case of $u\bar{u}$ and $c\bar{c}$, and with minus in the case of $u\bar{c}$ and $c\bar{u}$. The four graphs of fig. 11.6 therefore completely cancel one another for momenta of virtual quarks much greater than m_c. A non-zero contribution arises from the low-momentum range, because $m_c \neq m_u$:
$$\mathcal{L}_{\Delta S=2} = G_2 \bar{s}\gamma_\alpha(1+\gamma_5)d \cdot \bar{s}\gamma^\alpha(1+\gamma_5)d,$$
where
$$G_2 = \frac{1}{16\pi^2} G^2(m_c - m_u)^2 \sin^2\theta \cos^2\theta = \frac{1}{16\pi^2} G^2 m_c^2 \sin^2\theta \cos^2\theta.$$

It is now simple to calculate the mass difference between the K_1^0 and K_2^0 mesons:
$$\Delta m_{12} \equiv m_2 - m_1 \simeq \frac{4m_c^2 \cos^2\theta}{3\pi m_\mu^2} \Gamma(K^+ \to \mu^+\nu).$$

A comparison with the observed value of Δm_{12} yields that the c-quark mass is of the order of 1 GeV.

11.3. Oscillations of strangeness

Let us consider the behaviour of K^0 mesons in vacuum. Assume that at $t=0$ we have pure K^0 mesons produced, for example, in the reaction $\pi^- + p \to K^0 + \Lambda$. No $\bar{K}^0$ mesons are present at $t=0$. Take into account now that
$$K = \frac{K_1 + K_2}{\sqrt{2}}, \qquad K_1 = \frac{K + \bar{K}}{\sqrt{2}}, \qquad K_2 = \frac{K - \bar{K}}{\sqrt{2}}.$$
After a time t we then have
$$\tfrac{1}{2}\left[(K+\tilde{K})e^{-im_1 t - \Gamma_1 t/2} + (K-\tilde{K})e^{-im_2 t - \Gamma_2 t/2}\right].$$
If K_1^0 and K_2^0 mesons were stable ($\Gamma_1 = \Gamma_2 = 0$), then after $t = \pi/(m_2 - m_1)$ the beam would consist exclusively of $\bar{K}^0$ mesons. After another such interval the beam's strangeness would again become positive. Such oscillations of strangeness superposed over exponential damping due to $\Gamma_{1,2} \neq 0$

are indeed observed experimentally. For instance, they are observed in leptonic decays of neutral kaons.

We recall that the $\Delta Q = \Delta S$ rule forbids the decays $K^0 \to e^- \bar{\nu} \pi^+$ and $\overline{K}^0 \to e^+ \nu \pi^-$ and allows only

$$K^0 \to e^+ \nu \pi^- \quad \text{and} \quad \overline{K}^0 \to e^- \bar{\nu} \pi^+.$$

The oscillations of strangeness can therefore be observed by recording the number of electrons and positrons produced in K_{e3} decays. Indeed,

$$N_{e^+}(t) \propto N_K(t) = \tfrac{1}{4}\left[e^{-\Gamma_1 t} + e^{-\Gamma_2 t} + 2e^{-(\Gamma_1+\Gamma_2)/2} \cos(m_2 - m_1)t\right],$$

$$N_{e^-}(t) \propto N_{\overline{K}}(t) = \tfrac{1}{4}\left[e^{-\Gamma_1 t} + e^{-\Gamma_2 t} - 2e^{-(\Gamma_1+\Gamma_2)/2} \cos(m_2 - m_1)t\right].$$

Experiments of this type make it possible to determine Δm_{12}. Another possibility to observe oscillations of strangeness consists in recording interactions of K-mesons in plates of materials placed across the beam path. The point is, that the strong interactions of K^0 and $\overline{K}^0$ mesons are essentially different. For example, slow $\overline{K}^0$ mesons are effectively absorbed in collisions with protons, producing hyperons:

$$\overline{K}^0 + p \to \Lambda^0 + \pi^+.$$

At the same time, conservation of strangeness in strong interactions forbids absorption of K^0 mesons resulting in production of hyperons.

The "two-facedness" of neutral kaons is a purely quantum-mechanical property. The K^0 (or $\overline{K}^0$) meson is a superposition of K_1^0 and K_2^0, which, in their turn, are superpositions of K^0 and $\overline{K}^0$. The results of measurements depend on the character of the measurement process. If these are 2π or 3π decays, that is decays into channels with definite values of CP parity, then K_1^0 and K_2^0 are recorded. If these are decays of the type of $K^0 \to e^+ \nu \pi^-$ and $\overline{K}^0 \to e^- \bar{\nu} \pi^+$ which proceed differently for the particle and its antiparticle, K^0 and $\overline{K}^0$ are recorded.

In a sense, neutral kaons resemble the electron spin in mutually perpendicular magnetic fields. The strangeness and CP parity are similar to spin projections on the x and z axes. Once the spin projection on z is fixed, the projection on x is uncertain, and vice versa. The state with the spin projection on the z-axis equal to $+\tfrac{1}{2}$ is a linear superposition of states with projections on the x-axis equal to $+\tfrac{1}{2}$ and $-\tfrac{1}{2}$. Kaon decays into charge-symmetric ($2\pi, 3\pi$) and charge-asymmetric states ($e^+ \nu \pi^-, e^- \bar{\nu} \pi^+$) play the role of mutually perpendicular magnetic fields H_z and H_x.

11.4. Regeneration

Differences in nuclear properties of K^0 and $\overline{K}^0$ mesons result in an extremely interesting phenomenon called regeneration. Assume that at $t = 0$ we have a beam consisting of K^0 mesons only. Decays $K_1^0 \to 2\pi$ will take place in this beam during time $t \sim \tau_1 \equiv 1/\Gamma_1$. These decays will stop after $\tau_1 \ll t \ll \tau_2$ where $\tau_2 = 1/\Gamma_2$, since all K_1^0 mesons will decay and leave a pure beam of K_2^0 mesons. If the beam is now directed at a plate, say a copper plate, $K_1^0 \to 2\pi$ decays will reappear behind the plate. What happens is the regeneration of K_1^0 mesons in matter.

Let us find the regeneration amplitude f_{21}. Denote the amplitude of the K^0 ($\overline{K}^0$) meson scattering on a nucleus by f ($\bar{f}$). After scattering, a K_2^0 meson transforms into a linear superposition of K_2^0 and K_1^0:

$$K_2^0 = \sqrt{\tfrac{1}{2}}\,(K^0 - \overline{K}^0) \Rightarrow \sqrt{\tfrac{1}{2}}\,(fK^0 - \bar{f}\overline{K}^0)$$

$$= \tfrac{1}{2}\big[f(K_1^0 + K_2^0) - \bar{f}(K_1^0 - K_2^0)\big] = \tfrac{1}{2}(f + \bar{f})K_2^0 + \tfrac{1}{2}(f - \bar{f})K_1^0.$$

Hence,

$$f_{21} = \tfrac{1}{2}(f - \bar{f})$$

and we would have $f_{21} = 0$ if $f = \bar{f}$.

When K_1^0 mesons travel at a non-zero angle with respect to the incident beam of K_2^0 mesons, regeneration on different nuclei in the plate is incoherent (usually such incoherent regeneration is called diffractional). If, however, a K_1^0 meson travels forward, the amplitudes of regeneration on nuclei along the beam axis add up coherently. Consider a plate dx thick, and a K_2^0 beam incident on it from the left (fig. 11.7). The K_1^0 wave arriving at point A from a ring with radius ρ and volume $2\pi\rho\,d\rho\,dx$, is equal to

$$f_{21} N 2\pi\rho\,d\rho\,dx\,\frac{e^{ik_1 r}}{r},$$

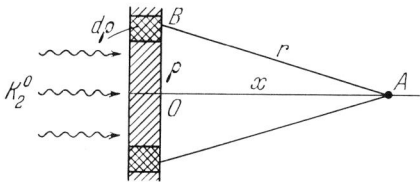

Fig. 11.7.

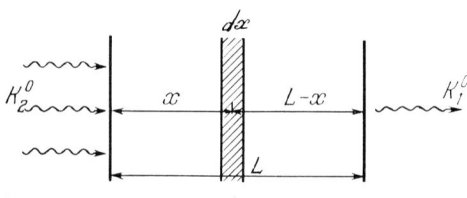

Fig. 11.8.

where N is the number of nuclei per 1 cm^3, and k_1 is the K_1^0 meson momentum. Making use of the equality $r\,dr = \rho\,d\rho$ and integrating over r, we obtain

$$\frac{2\pi i}{k_1} f_{21}(0) N\,dx\,e^{ik_1 x} = i\lambda f_{21}(0) N\,dx\,e^{ik_1 x},$$

where $\lambda = 2\pi/k_1$. (Only the lower limit of integration at $r = x$, corresponding to regeneration at zero angle, makes a non-zero contribution; hence, $f_{21}(0)$ appears in the above expression. At the upper limit the integral vanishes because of the oscillation in the integrand.) If we take into account that f_{21} is small and if we neglect the reverse transformation $K_1^0 \to K_2^0$, we easily derive a similar formula for a plate with finite thickness L (fig. 11.8). The K_1^0 wave generated in the dx thick layer at a distance x from the left-hand side of the plate, will reach the right-hand side of the plate with the amplitude da_1:

$$da_1 = iN\lambda f_{21}\,dx\,\exp\left[ik_2 x + ik_1(L - x) - \frac{L - x}{2\Lambda} - \frac{L}{2u}\right].$$

The first term in the exponent describes propagation of the K_2^0 wave till depth x, and the second term represents propagation of the K_1^0 wave from x to the right-hand side of the plate; the third term gives damping of the K_1^0 wave due to decays of K_1^0 mesons (Λ is the decay length, $\Lambda = \tau_1 \gamma v$, where τ_1 is the K_1^0 lifetime, v is its velocity, $\gamma = (1 - v^2)^{-1/2}$), and the fourth term describes absorption of mesons in the plate (u is the collision length, identical for K_1^0 and K_2^0).

The energy of the K-meson scattered forward remains practically unaltered so that

$$k_1^2 + m_1^2 = k_2^2 + m_2^2,$$

whence

$$k_1 - k_2 = (m_1 - m_2)\frac{m}{k}.$$

By integrating da_1 over x from 0 to L and taking the squared modulus of the expression obtained, we find the intensity of the coherent K_1^0 wave,

$$|a_1|^2 = \frac{4|f_{21}(0)|^2 N^2 \Lambda^2 \lambda^2}{1 + 4\delta^2} |e^{-i\delta l} - e^{-l/2}|^2 e^{-L/u},$$

where two dimensionless quantities are used:

$$l = \frac{L}{\Lambda} \quad \text{and} \quad \delta = (m_2 - m_1)\tau_1.$$

Consider now the diffraction regeneration of K_1^0 mesons. The differential cross section of regeneration is

$$\frac{d\sigma_{21}}{d\Omega} = |f_{21}|^2.$$

The intensity of K_1^0 mesons created within the layer of thickness dx and having travelled the distance $L - x$ is

$$d\left(\frac{dn_1}{d\Omega}\right) = |f_{21}|^2 N \, dx \exp\left[-\frac{L-x}{\Lambda} - \frac{L}{u}\right].$$

Integration over x yields

$$\frac{dn_1}{d\Omega} = |f_{21}|^2 N \Lambda (1 - e^{-l}) e^{-L/u}.$$

The ratio of coherent to diffraction intensity is

$$R = \frac{|a_1|^2}{dn_1/d\Omega} = \frac{4N\Lambda\lambda^2 |e^{-i\delta l} - e^{-l/2}|^2}{(1 - e^{-l})(1 + 4\delta^2)}$$

and is independent of f_{21}. The intensity of diffraction regeneration at zero angle is found by extrapolation. Measurement of R makes it possible to determine Δm_{12} with high accuracy.

The interference of K_1^0 waves which regenerated in two (or more) plates has been measured in a number of experiments. Among other things, these experiments yielded the sign of Δm_{12} and established that K_2^0 is heavier than K_1^0.

CHAPTER 12

Violation of *CP* invariance

12.1. $K_L^0 \to \pi^+\pi^-$ decay

Cristensen, Cronin, Fitch and Turlay have found in 1964 that long-lived neutral K-mesons decay into $\pi^+\pi^-$ with a small relative width ($B \simeq 2 \cdot 10^{-3}$). This discovery signified that decays of K_2^0 mesons violate *CP* invariance. Violation of *CP* invariance means that, if K_1^0 (K_2^0) denote, as before, the states with $CP = +1$ (-1), the long-lived neutral K-meson is not identical to K_2^0. It has no definite *CP* parity and is represented by a superposition of K_1^0 and K_2^0:

$$K_L^0 = \frac{1}{\sqrt{1+|\varepsilon|^2}} \left(K_2^0 + \varepsilon K_1^0 \right)$$

(subscript L for "long"). The short-lived meson is denoted by K_S^0 (S for "short"). Experiments have demonstrated that the complex parameter ε is small ($|\varepsilon| \simeq 2.3 \cdot 10^{-3}$, see below) so that $|\varepsilon|^2$ will be hereafter neglected. If the amplitudes of the $K_L \to \pi^+\pi^-$ and $K_S \to \pi^+\pi^-$ decays are denoted by $\langle \pi^+\pi^-|K_L\rangle$ and $\langle \pi^+\pi^-|K_S\rangle$, it is convenient to introduce a complex parameter

$$\eta_{+-} \equiv |\eta_{+-}|e^{i\phi_{+-}} = \frac{\langle \pi^+\pi^-|K_L\rangle}{\langle \pi^+\pi^-|K_S\rangle}.$$

The experimental values are $|\eta_{+-}| \simeq 2.3 \cdot 10^{-3}$, $\phi_{+-} \simeq 45°$.

Phenomenologically it is natural to distinguish between two possible mechanisms of *CP* violation in $K_L^0 \to \pi^+\pi^-$ decay. The first of these is a direct *CP*-odd decay $K_2^0 \to \pi^+\pi^-$, due to the *CP*-non-invariant interaction with $|\Delta S| = 1$ (S stands for strangeness). The strength of this interaction must be weaker by at least three orders of magnitude than that of the weak non-leptonic interaction. Hence, it was called the milliweak interaction. The second mechanism reduces the $K_L^0 \to \pi^+\pi^-$ decay to a *CP*-invariant decay of the εK_1^0 component contained in K_L^0. In this case violation of *CP* is

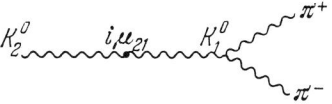

Fig. 12.1.

totally restricted to the mixing $K_2^0 \leftrightarrow K_1^0$ which forms states with definite masses and lifetimes K_L^0 and K_S^0. This mechanism is referred to as the superweak mixing (see below, p. 97). Obviously, in this case (neglecting $|\varepsilon|^2$ terms)

$$\langle \pi^+\pi^-|K_L^0\rangle = \varepsilon\langle \pi^+\pi^-|K_1^0\rangle = \varepsilon\langle \pi^+\pi^-|K_S\rangle.$$

Hence,

$$\eta_{+-} = \varepsilon.$$

The mixing $K_2^0 \leftrightarrow K_1^0$ and the subsequent decay $K_1^0 \to \pi^+\pi^-$ can be represented in perturbation theory by a graph of fig. 12.1. Here $i\mu_{21}$ is non-diagonal CP-odd "mass". If K_2^0 and K_1^0 have no common decay channels (this is assumed by the mechanism under discussion), then $i\mu_{21}$ is pure imaginary by virtue of CPT invariance (i.e. the hermiticity of the interaction hamiltonian transforming K_2 into K_1). As one easily finds from the graph of fig. 12.1,

$$\eta_{+-} = \frac{i\mu_{21}}{m_2 - m_1 - \tfrac{1}{2}i(\Gamma_2 - \Gamma_1)} = \frac{i\mu_{21}}{(m_L - m_S) - \tfrac{1}{2}i(\Gamma_L - \Gamma_S)},$$

whence

$$\tan\phi_{+-} = \frac{2(m_L - m_S)}{\Gamma_S - \Gamma_L} \simeq 0.96,$$

$$\phi_{+-} \simeq 44^0.$$

The mechanism of $K_2^0 \leftrightarrow K_1^0$ mixing thus yields the phase value, ϕ_{+-}, in agreement with experiment.

12.2. Other observed *CP* violating effects

We shall see that the mechanism of superweak mixing is quite adequate to describe two other *CP*-odd effects observed experimentally.

Let us denote the CP-odd decay amplitude for $K_L^0 \to \pi^0\pi^0$ by $\langle \pi^0\pi^0 | K_L \rangle$, and introduce a parameter

$$\eta_{00} = \frac{\langle \pi^0\pi^0 | K_L \rangle}{\langle \pi^0\pi^0 | K_S \rangle}.$$

As follows from the preceding section,

$$\eta_{00} \equiv |\eta_{00}| e^{i\phi_{00}} = \eta_{+-} = \varepsilon,$$

if the $K_2^0 \leftrightarrow K_1^0$ mixing mechanism is responsible for the CP breaking. The experimental accuracy of η_{00} measurements is poorer than that of η_{+-}:

$$|\eta_{00}| : |\eta_{+-}| = 1.03 \pm 0.04, \qquad \phi_{00} - \phi_{+-} = 10^0 \pm 6^0.$$

The equality $\eta_{00} = \eta_{+-}$ thus holds within experimental errors.

The $K_2^0 \leftrightarrow K_1^0$ mixing results in a charge asymmetry of leptonic decays of the K_L^0 meson. Denote the parameter of this asymmetry by δ:

$$\delta = \frac{\Gamma(K_L^0 \to \ell^+ \nu \pi^-) - \Gamma(K_L^0 \to \ell^- \tilde{\nu} \pi^+)}{\Gamma(K_L^0 \to \ell^+ \nu \pi^-) + \Gamma(K_L^0 \to \ell^- \tilde{\nu} \pi^+)}.$$

We shall take into account that $K_L = K_2 + \varepsilon K_1 = \sqrt{\frac{1}{2}}[K(1+\varepsilon) - \overline{K} \times (1-\varepsilon)]$, and apply the $\Delta Q = \Delta S$ rule which states that the $K \to \ell^+$ and $\overline{K} \to \ell^-$ decays are allowed while $K \to \ell^-$ and $\overline{K} \to \ell^+$ are forbidden. Also taking into account that $\Gamma(K \to \ell^+) = \Gamma(\overline{K} \to \ell^-)$, we obtain

$$\delta = \frac{|1+\varepsilon|^2 - |1-\varepsilon|^2}{|1+\varepsilon|^2 + |1-\varepsilon|^2} \simeq 2\,\mathrm{Re}\,\varepsilon.$$

Experimental values of charge asymmetry are available both for the K_{e3} and for $K_{\mu 3}$ decays. Within the experimental error, the values of δ coincide; the mean asymmetry parameter is $\delta = (3.3 \pm 0.12) \cdot 10^{-3}$ which must be compared with $\sqrt{2}|\eta_{+-}| \simeq 3.2 \cdot 10^{-3}$. The agreement is again good.

All the CP-odd decays listed above are among the decays of K_L^0 mesons. In spite of a thorough search, no violation of CP was detected in decays or interactions of other particles. This exceptional behavior of K_L^0 mesons is based on the enhancement of CP-odd effects in K_L^0 decays; this will be shown in the next section.

12.3. Superweak mixing

The parameter $i\mu_{21}$ was defined above as the non-diagonal mass,

$$i\mu_{21} = \langle K_1|H|K_2\rangle,$$

where H is the hamiltonian of the CP-non-invariant interaction giving rise to the $K_1 \leftrightarrow K_2$ transitions. Let us find the selection rules for this hamiltonian, and its effective strength. In doing this, we shall not decide in advance on whether the hamiltonian is a primary one or whether it appears as an effective hamiltonian in higher orders of the perturbation theory with respect to a primary CP-non-invariant interaction (see Chapter 15, sect. 4). It is apparent that H changes strangeness by 2:

$$|\Delta S| = 2.$$

Indeed,

$$i\mu_{21} = \langle K_1|H|K_2\rangle = \tfrac{1}{2}\langle K + \overline{K}|H|K - \overline{K}\rangle$$
$$= \tfrac{1}{2}[\langle K|H|K\rangle - \langle \overline{K}|H|\overline{K}\rangle + \langle \overline{K}|H|K\rangle - \langle K|H|\overline{K}\rangle]$$
$$= \tfrac{1}{2}[\langle \overline{K}|H|K\rangle - \langle K|H|\overline{K}\rangle].$$

The last equality appears because the particle and antiparticle have equal masses (which in the general case are complex if decays are taken into account):

$$\langle K|H|K\rangle = \langle \overline{K}|H|\overline{K}\rangle.$$

We recall that the mass difference $\Delta m_{LS} = m_L - m_S$ is also caused by transitions with $|\Delta S| = 2$ (see chapter 11):

$$\Delta m_{LS} = m_L - m_S = -\tfrac{1}{2}[\langle K|H|\overline{K}\rangle + \langle \overline{K}|H|K\rangle].$$

However, Δm_{LS} involves the sum and not the difference of amplitudes of the $K \to \overline{K}$ and $\overline{K} \to K$ transitions.

Assume that the $K \leftrightarrow \overline{K}$ transitions occur because of the effective four-fermion interaction turning a pair of s-quarks into a pair of d-quarks and vice versa (see figs. 12.2 and 12.3). Then the CP-invariant part of this interaction contributes to Δm_{LS}, and the CP-non-invariant part to $i\mu_{21}$. The former is proportional to $\mathrm{Re}\, G_2$, the real part of the constant G_2, while the latter is proportional to the imaginary part, $\mathrm{Im}\, G_2$. Taking into account that

$$|\mu_{21}| \sim 3 \cdot 10^{-3} |\Delta m_{LS}|,$$

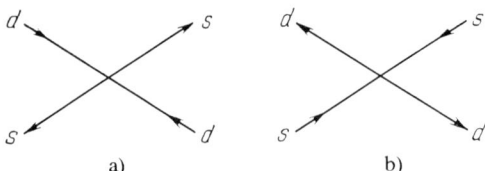

Fig. 12.2.

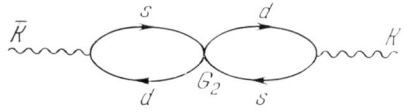

Fig. 12.3.

we conclude that

$$|\text{Im } G_2| \sim 3 \cdot 10^{-3} |\text{Re } G_2|.$$

We have established in Chapter 11 that $|\text{Re } G_2| \sim 10^{-6} - 10^{-7} G$ where G is the universal Fermi coupling constant. Now we have to conclude that $\text{Im } G_2 \sim 10^{-9} G$. An interaction with such a small coupling constant is referred to as superweak. Attempts to observe the superweak interaction in decays with $|\Delta S| = 2$, for instance, in the decay $\Xi^- \to n\pi^-$, are futile because the expected relative width is of the order of 10^{-18}. Despite its superweakness, this interaction gives rise, nevertheless, to effects of the order of 10^{-3} in decays of the K_L^0 meson. This is caused by the resonant enhancement, due to the closeness in masses of the K_L^0 and K_S^0 mesons (see fig. 12.1 and the expression for η_{+-}).

Possible manifestations of the milliweak C-even but P- and CP-odd interaction with $|\Delta S| = 1$ must be searched for, first of all, by experimentally checking the equality $\eta_{+-} = \eta_{00}$ (see Chapter 15, sect. 4).

12.4. On decays of the K_S^0 meson

As a result of the $K_2 \leftrightarrow K_1$ mixing, the wave function of the short-lived kaon becomes

$$K_S^0 = \frac{1}{\sqrt{1 + |\varepsilon|^2}} (K_1^0 + \varepsilon K_2^0) \simeq K_1^0 + \varepsilon K_2^0.$$

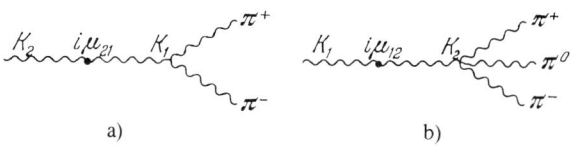

Fig. 12.4.

Note that the sign and the magnitude of ε is the same here as in the case of the K_L^0 meson ($K_L^0 = K_2^0 + \varepsilon K_1^0$). The states K_L^0 and K_S^0 are, therefore, not orthogonal. In order to ascertain that ε is the same in K_L^0 and K_S^0, consider the decays $K_S^0 \to \pi^0\pi^0\pi^0$ and $K_S^0 \to \pi^+\pi^-\pi^0$ (in the second case we mean a decay into a CP-odd state $\pi^+\pi^-\pi^0$ with zero orbital momenta ($l = L = 0$, see chapter 10)). A comparison of figs. 12.4a and b yields

$$\eta_{+-} = \frac{\langle \pi^+\pi^- | K_L \rangle}{\langle \pi^+\pi^- | K_S \rangle} = \frac{i\mu_{21}}{m_2 - m_1 - \tfrac{1}{2}i(\Gamma_2 - \Gamma_1)} = \varepsilon,$$

$$\eta_{+-0} = \frac{\langle \pi^+\pi^-\pi^0 | K_S \rangle}{\langle \pi^+\pi^-\pi^0 | K_L \rangle} = \frac{i\mu_{12}}{m_1 - m_2 - \tfrac{1}{2}i(\Gamma_1 - \Gamma_2)}.$$

Taking into account that $i\mu_{21} = -i\mu_{12}$, we obtain $\eta_{+-0} = \varepsilon$. Likewise,

$$\eta_{000} \equiv \frac{\langle \pi^0\pi^0\pi^0 | K_S \rangle}{\langle \pi^0\pi^0\pi^0 | K_L \rangle} = \varepsilon.$$

Unfortunately, the expected values are far beyond the capabilities of experiment. The experimental upper bounds are very high, $|\eta_{+-0}|^2 < 0.12$, $|\eta_{000}|^2 < 0.28$, for the following reason. The $K_L^0 \to 2\pi$ decay forbidden by CP invariance competes more or less successfully with the decay $K_L \to 3\pi$ suppressed by the phase-space volume. On the other hand, the decay $K_S \to 3\pi$, forbidden by CP and suppressed by the phase-space volume, cannot compete with the intensive decay $K_S \to 2\pi$, so that very high statistics is necessary to find this decay.

A search for $K_S \to 3\pi$ decays is of interest for finding out whether a P-even but C- and CP-odd interaction with $|\Delta S| = 1$ can exist. This interaction could give rise to direct decays $K_1 \to 3\pi$.

12.5. Violation of time-reversal invariance and the neutron dipole moment

Any reasonable physical theory must be CPT invariant. This means that decays of the K_L^0 meson violate not only CP invariance but time-reversal invariance, T, as well. Indeed, an analysis of the CP-odd decays of K_L^0 mesons indicates that the time reversibility in them is violated (violation of CPT and conservation of T would mean a pure imaginary transition parameter μ_{21} and would shift the phase of ε by 90°).

A search for time irreversibility was conducted, in addition to K_L^0 decays, in many other processes. The maximum experimental progress was achieved in a search for the electric dipole moment of the neutron, d_n: since 1964, when the $K_L^0 \to \pi^+ \pi^-$ decay was discovered, the upper bound on d_n was lowered by more than four orders of magnitude. It is customary to write d_n as a product of the elementary electric charge e by a distance l:

$$d_n = el.$$

At the present moment, we know that $l \lesssim 10^{-24}$ cm. This is smaller by approximately ten orders of magnitude than the corresponding length in the neutron magnetic moment.

Why should $d_n \neq 0$ signify that not only P- but T-invariance as well are violated? In order to answer that, let us recall the behavior of a number of physical quantities under space inversion P and time reversal T:

	P	T
coordinates	$r \to -r$	$r \to r$
time	$t \to t$	$t \to -t$
momentum	$p \to -p$	$p \to -p$
energy	$E \to E$	$E \to E$
angular momentum	$J \to J$	$J \to -J$
vector potential	$A \to -A$	$A \to -A$
scalar potential	$\varphi \to \varphi$	$\varphi \to \varphi$
electric field	$E \to -E$	$E \to E$
magnetic field	$H \to H$	$H \to -H$

(The signs are conveniently found by means of the relations $p = m \, dr/dt$, $J = r \times p$, $E = \nabla \varphi$, $H = \nabla \times A$). Both the magnetic μ and electric d dipole moments of a particle must be directed along the particle spin, because the spin is the only vector quantity characterizing the particle. Hence, $\mu \sim J$ and $d \sim J$. We find therefore that the magnetic dipole–magnetic field interaction, $J \cdot H$, is P- and T-invariant, while the electric dipole–electric field interaction, $J \cdot E$, is P- and T-non-invariant.

What is the expected value of d_n? If the superweak interaction $i \, \text{Im} \, G_2(\bar{s}d)(\bar{s}d)$, where $\text{Im} \, G_2 \sim 10^{-9} G$, is the only source of CP violation, then d_n is immeasurably small, namely $l < 10^{-38}$ cm. This estimate is based on the argument that in addition to the superweak interaction, the second order of the weak interaction must also contribute to d_n in order to ensure

$\Delta S = 0$ (and non-conservation of P-parity). Consequently, $l \sim G^2 m^5 \operatorname{Im} G_2$ where m is a mass of the order of one GeV.

However, it seems highly improbable for the superweak interaction (provided it exists) to have only one component, that with $|\Delta S| = 2$. The regularity observed so far is: the weaker the interaction is, the higher the number of conservation laws it breaks. It is therefore logical to expect the superweak interaction to comprise, along with a component with $|\Delta S| = 2$, a P-odd and CP-odd component with $\Delta S = 0$. In this case $l \sim m \operatorname{Im} G_2 \sim 10^{-28}$ cm. If CP invariance is violated because of the phase δ in the matrix of nine charged currents (see Chapter 15), then $l \ll 10^{-33}$ cm. Models in which CP invariance breakdown occurs in the interaction of several Higgs bosons give a value of d_n close to the current experimental limit.

12.6. Gedanken experiments

Physicists were inclined to believe in the absolute character of CP invariance until the $K_L^0 \to \pi^+ \pi^-$ decay was discovered, and at first many were surprised by the CP breakdown. But nowadays people are surprised why the violation of CP is so weak. The principal significance of the discovery of CP non-invariance should nevertheless not be underestimated: we have found out that there are absolute differences between particles and antiparticles, between left and right, and that the microscopic world has its own arrow of time.

Even before CP non-invariance was discovered, a story popular in science fiction books pictured a spaceship of aliens communicating with the Earth and moving towards it. The terrestrials try to determine whether the aliens are built of matter or antimatter. They find that this cannot be done in a CP-invariant world. One cannot distinguish between an electron and a positron, or the left-handed and the right-handed if only information and not material samples are exchanged. This gedanken ship analogy is not a modern invention. Indeed, Galileo illustrated his relativity principle by means of a ship sailing smoothly along the coast. No mechanical experiments aboard the ship can establish whether the ship is at rest or in motion. A 17th century ship demonstrated the invariance of nature's laws with respect to galilean transformations, while the 20th century spaceship demonstrates this with respect to CP transformations. It is of course sufficient to throw a glance at the shore to determine that the ship is moving; and it is sufficient to send a beam of light with left-handed polarization to the spaceship to reach an agreement about what is matter and what is antimatter. Indeed, one can then make use of the left-handed polarization of electrons and right-handed polarization of positrons in β-decay in order to solve the matter-antimatter alternative. However, the light with left-handed polarization is precisely a terrestrial sample mentioned above. The rules of the game forbid looking at the shoreline or sending a sample.

In reality, nature is *CP* non-invariant so that no sample exchange is necessary. For instance, one could use the fact that K_L^0 mesons are the same whether obtained in a laboratory accelerator or in an "anti-laboratory anti-accelerator". The K_L^0 mesons decay more often into positrons than into electrons. This is an absolute difference between particles and antiparticles. Now it becomes elementary to make use of the longitudinal polarization and give an absolute definition of left and right.

The Σ^+ hyperon and its antiparticle, $\bar{\Sigma}^-$, give another interesting example of the absolute difference between a particle and its antiparticle. The total widths are identical by virtue of CPT invariance:

$$\Gamma(\Sigma^+) = \Gamma(\bar{\Sigma}^-).$$

It follows from the *CP* non-invariance, however, that

$$\Gamma(\Sigma^+ \to p\pi^0) \neq \Gamma(\bar{\Sigma}^- \to \bar{p}\pi^0),$$
$$\Gamma(\Sigma^+ \to n\pi^+) \neq \Gamma(\bar{\Sigma}^- \to \bar{n}\pi^-).$$

We leave the proof to the reader as an exercise. We only hint that redistribution among channels is caused by interference of phases due to *CP* non-invariance, and phases due to scattering of decay products on one another (see Chapter 8). The former change sign, and the latter do not under particle-antiparticle conjugation.

CHAPTER 13

Decays of the τ-lepton

The third charged lepton (after e and μ) called triton or τ-lepton (triton means *third* in Greek) was discovered in colliding $e^+ e^-$ beams at Stanford, in 1975. We know now that the τ mass is approximately 1780 MeV, its spin is $\frac{1}{2}$, and that its decays, as those of the muon, are mediated by the V–A interaction.

13.1. ν_τ neutrino

All known decays of the τ-lepton are accompanied by neutrino emission. According to the theory outlined in the present book, this neutrino is different from ν_e and ν_μ, being a specific ν_τ neutrino. Experiments show that $\nu_\tau \neq \nu_\mu$ (otherwise τ-leptons would be produced copiously by muon neutrino beams). So far the possibility of $\nu_\tau = \nu_e$ is not ruled out experimentally (because ν_e beams are much less intense than beams of ν_μ) but seems very unlikely. The experimental upper bound on the ν_τ mass is: $m_{\nu_\tau} < 250$ MeV. As follows from the theory of the hot universe, $m_{\nu_\tau} < 10$–30 eV (see Chapter 27).

So far τ-leptons were produced only in colliding $e^+ e^-$ beams in the reaction

$$e^+ e^- \to \gamma \to \tau^+ \tau^-.$$

It has been found, for instance, that τ-leptons are the products of ψ' meson decay:

$$e^+ e^- \to \gamma \to \psi' \to \tau^+ \tau^-.$$

We shall consider now the main decays of the $\tau^\pm$ leptons (only τ^- will be discussed for the sake of brevity).

13.2. Decays $\tau^- \to \mu^- \bar{\nu}_\mu \nu_\tau$ and $\tau^- \to e^- \bar{\nu}_e \nu_\tau$

These decays are absolutely similar to the $\mu^- \to e^- \nu_e \nu_\mu$ decay. The respective decay rates are obtained from the muon decay rate (see chapter 3) by $m_\mu \to m_\tau$ substitution:

$$\Gamma(\tau \to e\bar{\nu}\nu) = \Gamma(\tau \to \mu\bar{\nu}\nu) = \frac{G^2 m_\tau^5}{192\pi^3} = \left(\frac{m_\tau}{m_\mu}\right)^5 \Gamma(\mu \to e\bar{\nu}\nu)$$

$$\simeq 6.2 \cdot 10^{11} \text{ s}^{-1}.$$

Experimentally, the Michel parameter for these decays is $\rho = 0.73 \pm 0.15$, in good agreement with the V–A interaction.

13.3. Semi-hadronic decays. General remarks

As is well known, the asymptotic freedom of QCD (quantum chromodynamics) enables calculation of the cross section for the process $e^+ e^- \to \gamma \to$ hadrons at $s \to \infty$

$$R = \frac{\sigma(e^+ e^- \to \gamma \to \text{hadrons})}{\sigma(e^+ e^- \to \gamma \to \mu^+ \mu^-)} = 3 \sum_f Q_f^2 \left(1 + \frac{\alpha_s}{\pi} + \cdots\right),$$

where the factor 3 is the number of colors, Q_f is the charge of the fth quark, and the summation covers all flavors that can be produced at a given energy, and α_s is the QCD running coupling constant. Below the charm production threshold, $R \simeq 3(Q_u^2 + Q_d^2 + Q_s^2) \simeq 2$.

We will estimate the total width of the decays of the type $\tau \to \nu_\tau +$ hadrons relying on the analogy with the annihilation $e^+ e^- \to$ hadrons. In the case of τ-lepton decay, we have to take into account the interactions $(\bar{\nu}_\tau \tau)(\bar{d}u)\cos\theta$ and $(\bar{\nu}_\tau \tau)(\bar{s}u)\sin\theta$. A quark estimate is obtained by neglecting all quark masses:

$$\frac{\Gamma(\tau \to \nu_\tau + \text{hadrons})}{\Gamma(\tau \to \nu_\tau \mu \bar{\nu}_\mu)} \simeq \frac{\Gamma(\tau \to \nu_\tau d\bar{u}) + \Gamma(\tau \to \nu_\tau s\bar{u})}{\Gamma(\tau \to \nu_\tau \mu \bar{\nu}_\mu)} \simeq 3.$$

If the deviation from the asymptotic freedom value is taken into account, this ratio changes to nearly 4. The expected lifetime of the τ-lepton is then

$$\tau_\tau \simeq \frac{1}{6\Gamma(\tau \to \nu_\tau \mu \bar{\nu}_\mu)} \simeq \frac{1}{3.1 \cdot 10^{12} \text{ s}^{-1}} \simeq 2.6 \cdot 10^{-13} \text{ s}.$$

So far, only the upper bound was found experimentally:

$$\tau_\tau \lesssim 10^{-12} \text{ s}.$$

Consider now individual channels of the semi-hadronic decays. It seems natural to list them in the order of accuracy with which the corresponding decay rates can be calculated at the present time.

The amplitude of the $\tau \to \pi\nu_\tau$ decay is related directly to that of the $\pi \to \mu\nu$ decay provided the weak interaction lagrangian includes the universal charged current. The amplitude of the $\tau \to \rho\nu$ decay is related directly to that of the $\rho \to e^+ e^-$ decay, owing to the isotopic properties of the ud current. For the same reason, the amplitudes of the $\tau \to \nu + 2n\pi$ decays and the electromagnetic annihilation $e^+ e^- \to 2n\pi$ are related for a given invariant mass of the $2n\pi$ system. As for the τ-lepton decays with emission of an odd number of pions, $\tau \to \nu + (2n + 1)\pi$, they are coupled to the axial ud current and so represent a unique source of pseudoscalar and axial-vector meson states.

13.4. Decay $\tau \to \pi\nu_\tau$

Compare $\pi \to \mu\nu$ and $\tau \to \pi\nu$ decay graphs (see figs. 13.1 and 13.2).

Both decays are due to the same $W\pi^-$ interaction which, in quark terms, reduces to the W-boson coupling to the axial ud current (fig. 13.3). The $\tau \to \pi\nu_\tau$ decay amplitude is given by

$$M = \sqrt{\tfrac{1}{2}}\, Gf_\pi k^\alpha \varphi_\pi \bar{u}_{\nu_\tau} \gamma_\alpha (1 + \gamma_5) u_\tau \cos\theta,$$

where k is the π-meson 4-momentum. It is important to note that both here and in the $\pi \to \mu\nu$ decay amplitude, the factor f_π is the same, that is independent of the type of leptons involved in the decay. This is emphasized in figs. 13.1–13.3 by the presence of intermediate W-bosons. The universal-

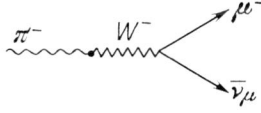

Fig. 13.1.

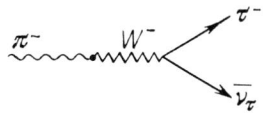

Fig. 13.2.

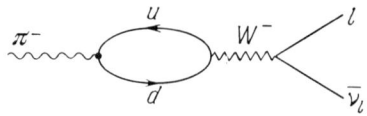

Fig. 13.3.

ity is broken by corrections of the order of α, if virtual photons are taken into account (fig. 13.4), because the τ and μ masses are different. If these corrections are neglected, f_π is a function of k^2 only, with $k^2 = m_\pi^2$ for both decays, since in both cases the π-meson is on the mass shell.

The decay width is

$$\Gamma(\tau \to \pi\nu_\tau) \frac{\overline{|M|^2}}{2 \cdot 2m_\tau} \Phi,$$

where Φ is the phase-space volume:

$$\Phi = \int \frac{\mathrm{d}k_\nu}{2E_\nu(2\pi)^3} \frac{\mathrm{d}k_\pi}{2E_\pi(2\pi)^3} (2\pi)^4 \delta^4(k_\nu + k_\pi - k_\tau) = \frac{E_\nu}{4\pi m_\tau}.$$

The squared modulus of the matrix element averaged over the τ-lepton polarizations is

$$\tfrac{1}{2}\overline{|M|^2} = \tfrac{1}{2} \cdot \tfrac{1}{2} G^2 f_\pi^2 m_\tau^2 \operatorname{Tr}(\hat{k} + m_\tau)(1 - \gamma_5)\hat{k}_\nu(1 + \gamma_5)\cos^2\theta$$

$$= 2G^2 f_\pi^2 m_\tau^2 (k_\tau k_\nu)\cos^2\theta = 2G^2 f_\pi^2 m_\tau^3 E_\nu \cos^2\theta.$$

Finally, we obtain

$$\Gamma(\tau \to \pi\nu_\tau) = \frac{G^2 f_\pi^2 m_\tau E_\nu^2 \cos^2\theta}{4\pi} = \frac{G^2 f_\pi^2 m_\tau^3 \cos^2\theta}{16\pi}\left(1 - \frac{m_\pi^2}{m_\tau^2}\right)^2$$

$$\simeq \frac{12\pi^2 f_\pi^2 \cos^2\theta}{m_\tau^2}\Gamma(\tau \to \mu\bar{\nu}\nu) \simeq 0.6\Gamma(\tau \to \mu\bar{\nu}\nu).$$

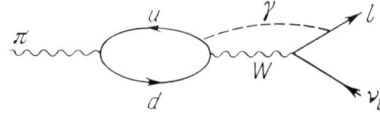

Fig. 13.4.

The $\tau \to \nu_\tau K$ decay rate is obtained from the above expression by substituting $f_\pi \to f_K$, $m_\pi \to m_K$, $\cos\theta \to \sin\theta$.

13.5. Decay $\tau \to \rho\nu_\tau$

The $\tau \to \rho\nu_\tau$ decay amplitude can be written as

$$M = \sqrt{\tfrac{1}{2}}\, G g_\rho \varphi_\alpha \bar{u}_{\nu_\tau} \gamma^\alpha (1+\gamma_5) u_\tau \cos\theta.$$

Here g_ρ is a scalar parameter of the dimension of m^2 (its magnitude will be discussed later), and φ_α is the ρ-meson wave function; φ_α is a four-vector satisfying the transversality condition: $p^\alpha \varphi_\alpha = 0$ where p is the ρ-meson 4-momentum. This condition indicates that the time component of φ_α in the ρ-meson rest frame is zero. Hence, the ρ-meson density matrix is also transverse on the mass shell (see appendix, Chapter 29, sect. 4.3):

$$\varphi_\alpha \varphi_\beta^* = -g_{\alpha\beta} + \frac{p_\alpha p_\beta}{m_\rho^2}.$$

We now find $\overline{|M|^2}$:

$$\overline{|M|^2} = \tfrac{1}{2} G^2 g_\rho^2 \cos^2\theta \left(\frac{p_\alpha p_\beta}{m_\rho^2} - g_{\alpha\beta} \right)$$

$$\times \operatorname{Tr} \hat{p}_\nu \gamma^\alpha (1+\gamma_5)(\hat{p}_\tau + m_\tau)\gamma^\beta(1+\gamma_5)$$

$$= G^2 g_\rho^2 \cos^2\theta \left[\frac{1}{m_\rho^2} \operatorname{Tr} \hat{p}_\nu \hat{p} \hat{p}_\tau \hat{p} + 2 \operatorname{Tr} \hat{p}_\nu \hat{p}_\tau \right]$$

$$= \frac{4 G^2 g_\rho^2 \cos^2\theta}{m_\rho^2} \left[2(p_\nu p)(p_\tau p) + m_\rho^2 (p_\nu p_\tau) \right]$$

$$= \frac{4 G^2 g_\rho^2 \cos^2\theta}{m_\rho^2} (p_\nu p_\tau) \left[2 p_\tau p + m_\rho^2 \right]$$

$$= \frac{4 G^2 g_\rho^2 \cos^2\theta}{m_\rho^2} (p_\nu p_\tau) m_\tau^2 \left(1 + 2 \frac{m_\rho^2}{m_\tau^2} \right).$$

(We have used the equality $p_\nu^2 = (p_\tau - p)^2 = -2 p_\tau p + m_\tau^2 + m_\rho^2 = 0$). Taking into account that the phase-space volume is

$$\Phi = \frac{p_\nu p_\tau}{4\pi m_\tau^2} = \frac{1}{8\pi}\left(1 - \frac{m_\rho^2}{m_\tau^2}\right),$$

we obtain

$$\Gamma(\tau \to \rho\nu_\tau) = \tfrac{1}{2}\overline{|M|^2}\,\frac{\Phi}{2m_\tau} = \frac{G^2 g_\rho^2 \cos^2\theta}{4\pi m_\rho^2 m_\tau}(p_\nu p_\tau)^2\left(1 + 2\frac{m_\rho^2}{m_\tau^2}\right)$$

$$= \frac{1}{16\pi} G^2 \left(\frac{g_\rho}{m_\rho}\right)^2 m_\tau^3 \cos^2\theta \left(1 - \frac{m_\rho^2}{m_\tau^2}\right)^2 \left(1 + 2\frac{m_\rho^2}{m_\tau^2}\right)$$

$$\simeq \tfrac{1}{16\pi} G^2 \left(\frac{g_\rho}{m_\rho}\right)^2 m_\tau^3 \cos^2\theta.$$

We turn now to the parameter g_ρ. In order to find it, compare the quark graphs of a weak decay $\tau \to \rho\nu$ and an electromagnetic decay $\rho \to e^+ e^-$ (fig. 13.5). The sum of graphs (b) and (c) gives the amplitude for $\rho^0 \to e^+ e^-$ decay, normally written in the form

$$M = \frac{m_\rho^2}{\gamma}\varphi_\alpha \frac{4\pi\alpha}{q^2}\bar{u}_e\gamma^\alpha u_e = \frac{4\pi\alpha}{\gamma}\varphi_\alpha \bar{u}_e\gamma^\alpha u_e.$$

Here γ is a dimensionless constant, q is the 4-momentum of a virtual photon, $q^2 = m_\rho^2$. It is now easy to obtain

$$\tfrac{1}{3}\overline{|M|^2} = \left(\frac{4\pi\alpha}{\gamma}\right)^2 \frac{1}{3}\left(-\delta_{\alpha\beta} + \frac{p_\alpha p_\beta}{m_\rho^2}\right)\operatorname{Tr}\hat{p}_1\gamma^\alpha \hat{p}_2\gamma^\beta$$

$$= \left(\frac{4\pi\alpha}{\gamma}\right)^2 \frac{8}{3} p_1 p_2 = \frac{4}{3}\left(\frac{4\pi\alpha}{\gamma}\right)^2 m_\rho^2$$

(the factor $\tfrac{1}{3}$ corresponds to averaging over the ρ-meson polarizations). The expression for the width is

$$\Gamma(\rho^0 \to e^+ e^-) = \frac{1}{2m_\rho}\left(\frac{4\pi\alpha}{\gamma}\right)^2 \frac{4}{3} m_\rho^2 \frac{1}{8\pi} = \tfrac{4}{3}\pi\alpha^2 \frac{m_\rho}{\gamma^2}.$$

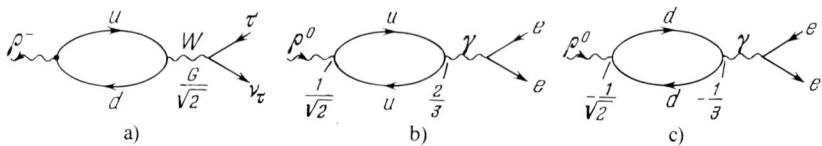

Fig. 13.5.

With standard values of $\Gamma(\rho^0 \to e^+ e^-) \simeq 6.5$ keV and $m_\rho \simeq 770$ MeV, we obtain

$$\frac{\gamma^2}{4\pi} = \frac{\alpha^2}{3} \frac{m_\rho}{\Gamma(\rho \to e^+ e^-)} \simeq 2.1.$$

A detailed theoretical fit of the cross-section data for the $e^+ e^-$ annihilation yields a somewhat larger value:

$$\frac{\gamma^2}{4\pi} = 2.36 \pm 0.18.$$

In order to express g_ρ in terms of γ, let us return again to the graphs of fig. 13.5 and recall that

$$\rho^0 = \sqrt{\tfrac{1}{2}}(u\bar{u} - d\bar{d}), \qquad \rho^+ = u\bar{d}.$$

Taking into account that the charges of the u- and d-quark are $\tfrac{2}{3}$ and $-\tfrac{1}{3}$, respectively, we obtain

$$\frac{m_\rho^2}{\gamma} = \sqrt{\tfrac{1}{2}}\left[\tfrac{2}{3} - (-\tfrac{1}{3})\right]I = \sqrt{\tfrac{1}{2}}\,I$$

for the net contribution of graphs 13.5b and c, and

$$g_\rho = I$$

for that of graph 13.5a, where I is the result of integration over the loop. Hence, $g_\rho/m_\rho = \sqrt{2}\,m_\rho/\gamma$.

Substitution of the expression derived above into that for $\Gamma(\tau \to \rho\nu)$ yields

$$\Gamma(\tau \to \rho\nu) = \frac{G^2 \cos^2\theta}{8\pi\gamma^2} m_\tau^3 m_\rho^2 \left(1 - \frac{m_\rho}{m_\tau}\right)^2 \left(1 + \frac{2m_\rho^2}{m_\tau^2}\right)$$

$$= 6\pi \frac{4\pi}{\gamma^2} \cos^2\theta \left(\frac{m_\rho}{m_\tau}\right)^2 0.91\Gamma(\tau \to e\bar{\nu}_e\nu_\tau)$$

$$= 1.46\,\Gamma(\tau \to e\tilde{\nu}\nu) \quad \text{for } \gamma^2/4\pi = 2.1,$$

$$= 1.30\,\Gamma(\tau \to e\tilde{\nu}\nu) \quad \text{for } \gamma^2/4\pi = 2.36,$$

$$= 1.18\,\Gamma(\tau \to e\tilde{\nu}\nu) \quad \text{for } \gamma^2/4\pi = 2.60.$$

13.6. Decays $\tau \to \nu_\tau + 2n\pi$

The relation between the decays $\tau \to \rho \nu_\tau$ and $\rho \to e^+ e^-$ derived above is a particular case of a more general relationship between the $\tau^- \to \nu_\tau(2n\pi)^-$ decays and the electromagnetic annihilation $e^+ e^- \to (2n\pi)^0$. This relationship appears because the components of the same vector-isovector current are operative in both cases: $\bar{d}\gamma_\alpha u$ in the τ-decay, and $\sqrt{\frac{1}{2}}(\bar{u}\gamma_\alpha u - \bar{d}\gamma_\alpha d)$ in the case of $e^+ e^-$ annihilation. No other currents contribute to it. This is easily demonstrated if we take into account that the $2n\pi$ system is G-even while the isoscalar electromagnetic current $\sqrt{\frac{1}{2}}(\bar{u}\gamma_\alpha u + \bar{d}\gamma_\alpha d)$ and the weak axial current $\bar{d}\gamma_\alpha \gamma_5 u$ are G-odd.

Introduce the notation:

$$v_\alpha = \langle (2n\pi)^0 | \tfrac{2}{3}\bar{u}\gamma_\alpha u - \tfrac{1}{3}\bar{d}\gamma_\alpha d | 0 \rangle,$$

$$v_\alpha^+ = \langle (2n\pi)^+ | \bar{u}\gamma_\alpha d | 0 \rangle.$$

Arguments similar to those which served to show above that $g_\rho = \sqrt{2}m_\rho^2/\gamma$, yield that $|v_\alpha^+| = |\sqrt{2}\, v_\alpha|$. The 4-vector v_α is a function of the invariant mass of the $2n\pi$ system, which we denote by $\sqrt{s}$, and (if $n > 1$) of other kinematic variables characterizing this system. The annihilation amplitude is

$$M(e^+ e^- \to 2n\pi) = \frac{4\pi\alpha}{q^2} \bar{u}_e \gamma^\alpha u_e \cdot v_\alpha.$$

The standard formula for the cross section yields

$$\sigma(e^+ e^- \to 2n\pi) = \int \left(\frac{4\pi\alpha}{q^2}\right)^2 \frac{1}{4 \cdot 2s} \operatorname{Tr} \hat{p}_1 \gamma^\alpha \hat{p}_2 \gamma^\beta v_\alpha v_\beta^* \, d\Phi,$$

where p_1 and p_2 are 4-momenta of the electron and positron, $d\Phi$ is the phase-space volume, $q = p_1 + p_2$, and $q^2 = s$. Because of the transversality of the current ($q^\alpha v_\alpha = 0$), we have

$$\int v_\alpha v_\beta^* \, d\Phi = \pi(q_\alpha q_\beta - g_{\alpha\beta} q^2)\rho(q^2).$$

This expression is a definition of the dimensionless spectral density $\rho(q^2)$.

Calculation of the trace gives

$$\sigma(e^+e^- \to 2n\pi) = \frac{(4\pi\alpha)^2}{8s^3} 4\pi \left[2 p_1^\alpha p_2^\beta - g^{\alpha\beta}(p_1 p_2)\right]$$
$$\times \left(q_\alpha q_\beta - q^2 g_{\alpha\beta}\right)\rho(s)$$
$$= \frac{(4\pi\alpha)^2}{2s^3} \left[2(p_1 q)(p_2 q) + (p_1 p_2)q^2\right] \pi\rho(s)$$
$$= \frac{(4\pi\alpha)^2}{2s} \pi\rho(s).$$

(Here we use the equality $p_1 + p_2 = q$ whence $p_1 p_2 = p_1 q = p_2 q = \frac{1}{2}q^2$.)
If we define the quantity $R(s)$ in a standard manner,

$$\sigma(s) = \frac{4\pi\alpha^2}{3s} R(s),$$

then ρ can be expressed via R:

$$\rho(s) = \frac{R(s)}{6\pi^2}.$$

In the case of a one-particle state (such as the ρ-meson, provided we neglect its width) we have

$$d\Phi = \frac{(2\pi)^4}{(2\pi)^3} \delta^4(q - p_1 - p_2) \delta(q^2 - m_\rho^2) d^4q$$

and

$$\pi\rho(s) = 2\pi\delta(q^2 - m_\rho^2) \frac{m_\rho^2}{\gamma^2},$$

so that

$$\sigma(e^+e^- \to \rho) = 16\pi^3 \frac{\alpha^2}{\gamma^2} \delta(s - m_\rho^2) = 12\pi^2 \frac{\Gamma(\rho \to e^+ e^-)}{m_\rho} \delta(s - m_\rho^2).$$

This last equality is based on the expression derived earlier:

$$\Gamma(\rho \to e^+ e^-) = \frac{4\pi\alpha^2}{3\gamma^2} m_\rho.$$

(Recall that the Breit-Wigner formula for the $e^+ e^- \to \rho \to 2\pi$ annihilation is

$$\sigma(e^+e^- \to \rho \to 2\pi) = \frac{3\pi}{m_\rho^2} \frac{\Gamma(\rho \to ee)\Gamma}{(E - m_\rho)^2 + \frac{1}{4}\Gamma^2}.$$

If $\Gamma \to 0$,

$$\frac{\tfrac{1}{2}\Gamma}{(E-m_\rho)^2+\tfrac{1}{4}\Gamma^2} \Rightarrow \pi\delta(E-m_\rho) = 2\pi m_\rho \delta(s-m_\rho^2).$$

We find therefore that the result obtained above is in agreement with the Breit-Wigner formula as well).

We turn now to τ-lepton decay:

$$M(\tau \to \nu_\tau + 2n\pi) = \sqrt{\tfrac{1}{2}}\, G\cos\theta\, \bar{u}_{\nu_\tau}\gamma^\alpha(1+\gamma_5)u_\tau \cdot v_\alpha^+.$$

The rate of decay into the $2n\pi$ system with mass $\sqrt{s}$ is

$$d^3\Gamma = \frac{G^2\cos^2\theta}{2\cdot 2m_\tau}\,\text{Tr}\,\hat{p}_\nu\gamma^\alpha(1+\gamma_5)(\hat{p}_\tau+m_\tau)\gamma^\beta(1+\gamma_5)\pi\big(q_\alpha q_\beta - g_{\alpha\beta}q^2\big)$$

$$\times \rho(s)\frac{dp_\nu}{2E_\nu(2\pi)^3},$$

where $q = p_\tau - p_\nu$, averaging was carried out over polarizations of the τ-lepton, and $\rho(s)$ is the same as in $e^+e^- \to 2n\pi$ annihilation. After calculation of the trace and integration over the neutrino angle we obtain

$$d\Gamma = \frac{G^2\cos^2\theta}{2\pi m_\tau}\Big[2p_\nu^\alpha p_\tau^\beta - g^{\alpha\beta}(p_\nu p_\tau)\Big]\big(q_\alpha q_\beta - g_{\alpha\beta}q^2\big)\rho(s)E_\nu\, dE_\nu$$

$$= \frac{G^2\cos^2\theta}{2\pi m_\tau}\Big[2(p_\nu q)(p_\tau q) + (p_\nu p_\tau)q^2\Big]\rho(s)E_\nu\, dE_\nu$$

$$= \frac{G^2\cos^2\theta}{2\pi m_\tau}(p_\nu p_\tau)(2p_\tau q + q^2)\rho(s)E_\nu\, dE_\nu$$

$$= \frac{G^2\cos^2\theta}{2\pi}(m_\tau^2+2s)\rho(s)E_\nu^2\, dE_\nu$$

$$= \frac{G^2\cos^2\theta}{2\pi}(3m_\tau - 4E_\nu)m_\tau\rho(s)E_\nu^2\, dE_\nu$$

(here we take into account that $p_\nu q = p_\nu p_\tau$, $2p_\tau q = m_\tau^2 + q^2$, $s = m_\tau^2 - 2E_\nu m_\tau$). Going over to the variable s, we obtain

$$d\Gamma(\tau \to \nu + 2n\pi) = \frac{G^2\cos^2\theta}{2\pi}\frac{(m_\tau^2+2s)(m_\tau^2-s)^2}{8m_\tau^3}\rho(s)\,ds.$$

The formula for $d\Gamma(\tau \to \nu + 2n\pi)$ must be compared with that for $\sigma(e^+e^- \to 2n\pi)$. Both these quantities are expressed, for a given number of pions and a given value of s, via the same $\rho(s)$. In order to find the total decay rate $\Gamma(\tau \to \nu + 2n\pi)$, we have to substitute $\rho(s)$ from $\sigma(e^+ e^- \to 2n\pi)$ into $d\Gamma$ and numerically integrate over ds.

Let us check the formula for $d\Gamma$ using the decay $\tau \to \rho\nu$ as an example. In the case of this decay, $\rho(s) = 2\delta(q^2 - m_\rho^2)m_\rho^2/\gamma^2$ and

$$\Gamma = \frac{G^2 \cos^2\theta}{8\pi\gamma^2} \frac{(m_\tau^2 + 2m_\rho^2)(m_\tau^2 - m_\rho^2)^2 m_\rho^2}{m_\tau^3},$$

which coincides with the result of the preceding section. Let us check our general formula for $d\Gamma$ in the case of free quarks:

$$\int v_\alpha^+ v_\beta^+ * d\Phi \Rightarrow 3 \int \text{Tr}\, \hat{p}_2 \gamma_\alpha \hat{p}_1 \gamma_\beta \frac{d\mathbf{p}_1}{2E_1} \frac{d\mathbf{p}_2}{2E_2} \frac{\delta^4(p_1 + p_2 - q)}{(2\pi)^2}$$

$$= 3 \cdot \frac{1}{6}\pi(q^2 g^{\mu\nu} + 2q^\mu q^\nu)\frac{1}{4\pi^2}(g_{\mu\alpha}g_{\nu\beta} + g_{\mu\beta}g_{\nu\alpha} - g_{\mu\nu}g_{\alpha\beta})$$

$$= 3\frac{1}{6\pi}(q_\alpha q_\beta - q^2 g_{\alpha\beta}) = 2\pi\rho(q^2)(q_\alpha q_\beta - g_{\alpha\beta}q^2),$$

whence $\rho = \frac{3}{2}(1/6\pi^2)$ (the factor 3 appears from summation over quark colors). The total decay rate for $\tau \to \nu + 2n\pi$ then becomes

$$\Gamma \simeq \int_0^{m_\tau/2} 3 \frac{G^2 \cos^2\theta}{24\pi^3} m_\tau(3m_\tau - 4E_\nu)E_\nu^2\, dE_\nu$$

$$= \int_0^1 \frac{G^2 m^5 \cos^2\theta}{192\pi^3} 3(3 - 2\varepsilon)\varepsilon^2\, d\varepsilon = \frac{3}{2}\frac{G^2 m_\tau^5 \cos^2\theta}{192\pi^3}.$$

This result is in agreement with the quark estimate obtained at the beginning of this chapter. Indeed, when quark masses are neglected, the vector current supplies half of the hadronic width of the τ-lepton. Another half comes from the axial current. (These currents do not interfere in the expression for the decay rate since the vector current gives rise to an even number of pions while the axial current gives rise to an odd number.)

It is interesting to compare quark estimates for the $\tau \to \nu + 2n\pi$ decay and for the $e^+ e^- \to 2n\pi$ reaction. In the latter case we must separate an isovector part R_1 out of the total

$$R = R_1 + R_0 = 3(Q_u^2 + Q_d^2) = \tfrac{5}{3},$$

that is to pass from non-interfering quark contributions to non-interfering isotopic ones. Taking into account quark charges, we find that the amplitudes of currents with $T = 1$, $(\bar{u}u - \bar{d}d)$, and of those with $T = 0$, $(\bar{u}u + \bar{d}d)$, are in the proportion of $3:1$, and the corresponding probabilities $9:1$. Consequently,

$$R_1 = \tfrac{9}{10}R = \tfrac{3}{2}.$$

Recalling the relation $R = 6\pi^2\rho$ we find that our new estimate is the same as the earlier one.

13.7. Decays $\tau \to \nu_\tau + (2n+1)\pi$

As we have mentioned above, these decays are due to the axial current $\bar{d}\gamma_\alpha\gamma_5 u$. Consider the amplitude

$$a_\alpha = \langle (2n+1)\pi | \bar{d}\gamma_\alpha\gamma_5 u | 0 \rangle.$$

After calculations similar to those of the preceding section we arrive at

$$d\Gamma(\tau \to \nu_\tau + (2n+1)\pi)$$

$$= \frac{G^2\cos^2\theta}{2\pi m_\tau}\left[2p_\nu^\alpha p_\tau^\beta - g^{\alpha\beta}(p_\nu p_\tau)\right]$$

$$\times \left[(q_{\alpha\beta} - g_{\alpha\beta}q^2)\rho_t(s) + q_\alpha q_\beta \rho_\ell(s)\right] E_\nu \, dE_\nu$$

$$= \frac{G^2\cos^2\theta}{2\pi}\left[(m_\tau^2 + 2s)\rho_t(s) + m_\tau^2\rho_\ell(s)\right] E_\nu^2 \, dE_\nu$$

$$= \frac{G^2\cos^2\theta}{2\pi}\left[(m_\tau^2 + 2s)\rho_t(s) + m_\tau^2\rho_\ell(s)\right]\frac{(m_\tau^2 - s)^2}{8m_\tau^3}\, ds.$$

The following notation is used above:

$$\int a_\alpha a_\beta^* \, d\Phi = 2\pi\left[(q_\alpha q_\beta - g_{\alpha\beta}q^2)\rho_t(s) + q_\alpha q_\beta \rho_\ell(s)\right]$$

(The coefficient 2 is introduced to make the normalization of the axial current spectral density identical to that of the vector current). In the approximation of the rigorously conserved axial current (i.e. $m_u = m_d = 0$), $m_\pi = 0$, $\rho_\ell = 0$. The conservation of the axial current being only partial, we have $\rho_\ell \ne 0$. In the case of $n = 0$ (the $\tau \to \pi + \nu$ decay)

$$a_\alpha = \langle \pi | \bar{d}\gamma_\alpha\gamma_5 u | 0 \rangle = f_\pi q_\alpha,$$

$$\int a_\alpha a_\beta^* \, d\Phi = f_\pi^2 q_\alpha q_\beta 2\pi\delta(q^2 - m_\pi^2),$$

whence

$$\rho_t = 0, \quad 2\rho_\ell = 2f_\pi^2 \delta(s - m_\pi^2).$$

Substitution of these expressions into that for $d\Gamma$ yields

$$\Gamma(\tau \to \nu_\tau + \pi) = \frac{G^2 \cos^2\theta}{16\pi} m_\tau^3 f_\pi^2 \left(1 - \frac{m_\pi^2}{m_\tau^2}\right)^2,$$

which coincides with the result obtained at the beginning of this chapter. The same result could be derived in the limit of pure chiral symmetry when

$$q_\alpha a_\alpha = 0, \quad m_\pi = 0,$$

if we neglect terms of the order of m_π^2. In this case

$$\rho_\ell = 0, \quad 2\rho_t = 2f_\pi^2 \delta(s) + \text{contributions of } 3\pi, 5\pi, \ldots$$

It is easy to demonstrate that we again obtain the familiar result for $\Gamma(\tau \to \pi \nu_\tau)$. As for the states $3\pi, 5\pi, \ldots$ and possible resonances in these systems, their contribution can be shown to satisfy the so-called Weinberg sum rules:

$$\int (\rho - \rho_t) \, ds = 0,$$

$$\int (\rho - \rho_t) s \, ds = 0,$$

which are based on $\rho_t(s) \to \rho(s)$ for $s \to \infty$ in the chiral limit (we recall that $2\pi\rho(q_\alpha q_\beta - g_{\alpha\beta} q^2) = \int v_\alpha^+ v_\beta^+ {}^* d\Phi$).

Further study of $\tau \to 3\pi + \nu_\tau$ decays will help to finally resolve the problem of the existence and properties of the axial A_1 meson ($m_{A_1} \simeq 1100$ MeV, $\Gamma_{A_1} \simeq 300$ MeV; the main decay channel: $A_1 \to \rho\pi$). The second Weinberg sum rule has no contribution from the π-meson (in the $m_\pi = 0$ limit); saturated with ρ and A_1 mesons, it gives the relation

$$\frac{m_\rho^4}{\gamma_\rho^2} = \frac{m_{A_1}^4}{\gamma_{A_1}^2}.$$

Therefore,

$$\frac{\Gamma(\tau \to \rho \nu_\tau)}{\Gamma(\tau \to A_1 \nu_\tau)} = \frac{(m_\tau^2 + 2m_\rho^2)(m_\tau^2 - m_\rho^2)^2 m_{A_1}^2}{(m_\tau^2 + 2m_{A_1}^2)(m_\tau^2 - m_{A_1}^2)^2 m_\rho^2} \simeq 2.7.$$

13.8. Decays $\tau \to \nu_\tau + K + n\pi$

The $\tau \to \nu_\tau + K + n\pi$ decay rates are proportional to $\sin^2 \theta_C$ where θ_C is the Cabibbo angle. Compared to pure pion decays, they are therefore very much suppressed. Decay-rate calculations are similar to those given above, and in some cases relations can be derived on the basis of the SU(3) symmetry of the strong interaction. On the whole, the $\tau \to \nu_\tau + K + n\pi$ decays must come to about 4–5% of all semi-hadronic decays of the τ-lepton.

13.9. Summary of results

To conclude this chapter, we list the theoretical values of partial widths, B.

Decay channel	B (%)	Basis of the estimate
$e\bar{\nu}\nu$	17	QCD + $(V-A)$ for Γ_{tot}
$\mu\bar{\nu}\nu$	17	
$\pi\nu$	10	$\pi \to \mu\nu$ decay
$\rho\nu$	23	CVC + $e^+ e^- \to$ hadrons
$(4\pi)\nu$	10	
$A_1\nu$	9	Weinberg sum rules
$(K + n\pi)\nu$	3	$\tan^2 \theta$
$(3 \text{ or } 5\pi)\nu$	11	remainder
	100	

CHAPTER 14

Decays of charmed particles

Decays of charmed particles are due to the product of currents

$$(\bar{s}c\cos\theta - \bar{d}c\sin\theta)(\bar{\nu}_e e + \bar{\nu}_\mu \mu + \bar{\nu}_\tau \tau + \bar{u}d\cos\theta + \bar{u}s\sin\theta) + \text{h.c.}$$

(color and Lorentz indices are omitted). It can be seen from this expression that decays of charmed non-strange particles must produce predominantly strange particles. This applies both to leptonic and non-leptonic decays.

14.1. Leptonic and non-leptonic decays

Leptonic decays of charmed particles can be described in the following manner:

$$c \to s\ell^+ \nu_\ell \quad \cos^2\theta, \quad \Delta S = \Delta C, \quad D \to \ell^+ \nu \overline{K}, \quad F \to \ell^+ \nu,$$
$$c \to d\ell^+ \nu_\ell \quad \sin^2\theta, \quad \Delta S = 0, \quad D \to \ell^+ \nu, \quad F \to \ell^+ \nu K.$$

Here we give, from left to right: quark decays responsible for leptonic decays of charmed particles, the appropriate Cabibbo factors suppressing decay rates, selection rules, and examples of decays in question. Note that decays into $\tau^+ \nu$ are also kinematically allowed for the F and D^+ mesons, and that

$$\frac{\Gamma(F^+ \to \tau^+ \nu)}{\Gamma(F^+ \to \mu^+ \nu)} = \left(\frac{m_\tau}{m_\mu}\right)^2 \left(\frac{1 - m_\tau^2/m_F^2}{1 - m_\mu^2/m_F^2}\right)^2 \simeq 17,$$

$$\frac{\Gamma(D^+ \to \tau^+ \nu)}{\Gamma(D^+ \to \mu^+ \nu)} = \left(\frac{m_\tau}{m_\mu}\right)^2 \left(\frac{1 - m_\tau^2/m_D^2}{1 - m_\mu^2/m_D^2}\right)^2 \simeq 2.5.$$

The last two decays ($D \to \tau\nu$ and $D \to \mu\nu$) are suppressed by the factor $\sin^2\theta$ (we recall that $m_F \simeq 2040$ MeV, $m_\tau \simeq 1780$ MeV, $m_{D^+} \simeq 1868$ MeV, $m_{D^0} \simeq 1863$ MeV).

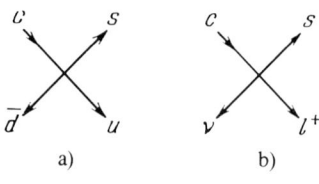

Fig. 14.1.

Creation of charmed particles in neutrino experiments and the subsequent decay of these particles result in the so-called dileptonic events. For example,

$$\nu_\mu + d \to c + \mu^-$$
$$ \hookrightarrow s\mu^+ \nu_\mu \quad \text{or} \quad se^+ \nu_e.$$

Such dileptonic events were observed in high-energy neutrino experiments:

$$\nu + \text{nucleus} \to \mu^- + \mu^+ + \ldots,$$

where the ellipsis stands for the remaining particles created in the reaction; some of these may even escape detection. Dileptons will be discussed in more details in chapter 17.

The three types of non-leptonic decays may be described in the following manner:

$c \to u s \bar{d}$,	$\cos^4 \theta$,	$\Delta S = \Delta C$,	$D \to \bar{K} + n\pi$,	$F \to \eta\pi$, ...
$c \to u d \bar{d}, u s \bar{s}$,	$\cos^2 \theta \sin^2 \theta$,	$\Delta S = 0$,	$D \to n\pi$,	$F \to K\pi$, ...
$c \to u d \bar{s}$,	$\sin^4 \theta$,	$\Delta S = -\Delta C$,	$D \to K + n\pi$,	$F \to KK$, ...

If we neglect decays proportional to $\sin^2 \theta$, and assume approximately $\cos^2 \theta \simeq 1$, then it can be estimated on the basis of the simple quark graphs of fig. 14.1 that

$$\Gamma(c \to su\bar{d}) : \Gamma(c \to s\ell^+ \nu) : \Gamma(c \to s\mu^+ \nu) \simeq 3 : 1 : 1,$$

where the factor 3 stems from summation over colors of the $u\bar{d}$ pair.

14.2. The role of virtual gluons

The branching ratio of non-leptonic decays is slightly enhanced if virtual gluons are taken into account (fig. 14.2).

This enhancement must not be as large as in the case of non-leptonic decays of strange particles. Two factors are responsible here. First, the Compton wavelength of the c-quark is much smaller than that of the

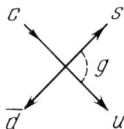

Fig. 14.2.

s-quark: $1/m_c \ll 1/m_s$, and the asymptotic freedom results in smaller α_s at smaller distances. Second, monopole-type transitions (see chapter 7) are practically absent. Their amplitudes are proportional to $\sin\theta\cos\theta$ (see fig. 14.3); moreover, they are considerably suppressed because the masses of the d- and s-quarks are close to one another, much closer than those of the u- and c-quarks (compare figs. 14.3 and 14.4).

In the case of cd → us as well as su → ud transitions, virtual gluons enhance amplitudes in the $\bar{3}_c$ channel and suppress those in the 6_c channel (the subscript c indicates that color is meant). In the case of the su → ud transition, the enhancement coefficient is close to 3 and that of suppression close to 0.6, while in the case of the cd → us transition they are $\simeq 2^{1/2}$ and $\simeq 2^{-1/4}$, respectively. On the whole, the total decay rate of non-leptonic decays of the c-quark increases only slightly, as compared to the estimate based on the bare quark graph. As for the lifetime of, for instance, D-mesons, it can be estimated on the basis of the above arguments as being

$$1/\tau_D \simeq (5-10)\Gamma(c \to se\nu) \simeq (5-10)\left(\frac{\bar{m}_c}{m_\mu}\right)^5 \Gamma(\mu \to e\bar{\nu}\nu),$$

where $\bar{m}_c$ is the effective mass of the decaying c-quark. Recalling now $\tau_\mu \simeq 2\cdot 10^{-6}$ s, and assuming $\bar{m}_c \simeq 1.5$ GeV, we obtain

$$\tau_D \simeq (3-6)\cdot 10^{-13}\text{s}.$$

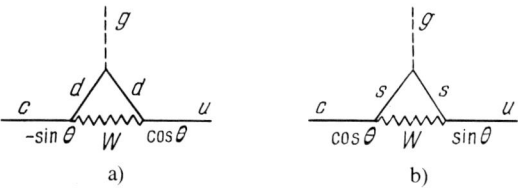

Fig. 14.3.

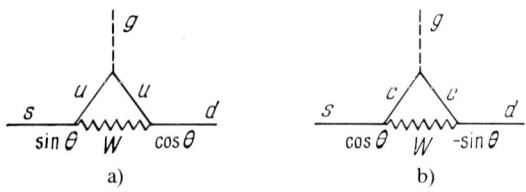

Fig. 14.4.

This estimate is obviously very sensitive to the choice of $\bar{m}_c$, and fairly crude: for instance, the result is identical for the D^+, D^0 and F mesons and charmed baryons*.

14.3. Selection rules for T-, U-, and V-spin

Let us see now what isospin and SU(3) symmetry selection rules govern the main ($\propto \cos^4 \theta$) non-leptonic decays of charmed particles. Obviously, $\Delta T = 1$ in the c → uds decay. One of the corollaries of this is that, for example, the reaction

$$F^+ \to \pi^+ \pi^0$$

is forbidden (isospin of F^+ is zero, while for the $\pi^+\pi^0$ S-wave, $T = 2$).
The interaction $(\bar{u}d)(\bar{s}c)$ is transformed under the group $SU(3)_f$ as

$$8 \times 3 = 3 + \bar{6} + 15 \qquad (t: \times t' = t' + t.. + t::)$$

(see appendix, Chapter 29, sect. 2.7). The state $\bar{6}$ of three quarks $us\bar{d}$ is antisymmetric with respect to u ↔ s exchange. Hence, these quarks form an antisymmetric color antitriplet $\bar{3}_c$, and the corresponding amplitudes are enhanced by virtual gluons. Sometimes this effect is referred to as the sextet enhancement. The sextet enhancement of charmed particle decay is an analog of the octet enhancement in decays of strange particles. (We recall that in the interaction $(\bar{d}u)(\bar{u}s)$, the enhanced octet is 8_f ($8 \times 8 = 1 + 8_f + 8_d + 10 + \overline{10} + 27$), antisymmetrical with respect to the d ↔ ū and u ↔ s exchange. This 8_f octet corresponds to the color antitriplets of the initial and

---

*After this book had been written, experimental data were obtained showing that the D^+ meson lifetime is several times that of the D^0 meson. It is possible that in addition to the c-quark decay, an essential process in the decay of the D^0 and F^+ mesons is the annihilation of quarks of which these mesons consist: $c\bar{u} \to s\bar{d}$ for D^0 and $c\bar{s} \to u\bar{d}$ for F^+ (the annihilation $c\bar{d} \to u\bar{d}$ for the D^+ meson is Cabibbo-suppressed).

final quarks). As a result of the $\bar{u} \leftrightarrow \bar{s}$ antisymmetrization, the sextet transitions satisfy the $\Delta V = 0$ rule (for the definition of V-spin see appendix, Chapter 29, sect. 2.7). This rule would mean, for example, that the decay

$$D^+ \to \bar{K}^0 \pi^+$$

is forbidden. This is easily shown by taking into account that the D^+ meson has zero V-spin, π^+ and $\bar{K}^0$ are components of a V-doublet, and that by virtue of the generalized Bose principle the S-wave $\pi \bar{K}^0$ system has $V = 1$.

It must be emphasized that the $\Delta V = 0$ rule must be very strongly violated, much more so than the $\Delta T = \frac{1}{2}$ rule for non-leptonic decays of strange particles, since the sextet enhancement is weaker than the octet one (we have already mentioned this fact). This conclusion is consistent with the experimental data. Thus, the decay $D^+ \to \bar{K}^0 \pi^+$ which is forbidden owing to V conservation, has $B \simeq 1.5 \pm 0.6\%$, while the V-allowed decay $D^0 \to K^- \pi^+$ has $B \simeq 2.2 \pm 0.6\%$. No straightforward comparison of these figures is possible since these widths are relative and not absolute quantities. However, it is unlikely that the total widths of D^+ and D^0 differ very much; therefore, the absolute widths of the V-allowed and V-forbidden decays probably differ not more than by an order of magnitude.

The following relations between amplitudes of two-particle decays can be derived on the basis of only the SU(3) invariance of the strong interaction, without taking into account the sextet enhancement:

$$M(D^0 \to \bar{K}^0 \pi^0) = -\sqrt{3} M(D^0 \to \bar{K}^0 \eta^0),$$

$$M(D^0 \to K^- \pi^+) + \sqrt{2} M(D^0 \to \bar{K}^0 \pi^0) = M(D^+ \to \bar{K}^0 \pi^+),$$

$$M(F^+ \to \bar{K}^0 K^+) - \sqrt{\tfrac{3}{2}} M(F^+ \to \pi^+ \eta^0) = M(D^+ \to \bar{K}^0 \pi^+).$$

So far these relations were not tested experimentally.

Let us consider now the non-leptonic interaction with $\Delta S = 0$,

$$\cos\theta \sin\theta \left[-\bar{u} O_\alpha d \cdot \bar{d} O^\alpha c + \bar{u} O_\alpha s \cdot \bar{s} O^\alpha c \right], \qquad O_\alpha = \gamma_\alpha(1 + \gamma_5).$$

This interaction is a component of the U-vector with $U_3 = 0$; therefore for this interaction

$$\Delta U = 1, \qquad \Delta U_3 = 0.$$

This yields, in particular, that the reactions $D^0 \to K^0 \bar{K}^0$, $\pi_U \pi_U$, $\eta_U \eta_U$, where

$$\pi_U = \tfrac{1}{2}(\pi^0 - \sqrt{3}\eta^0), \qquad \eta_U = \tfrac{1}{2}(\eta^0 + \sqrt{3}\pi^0),$$

are forbidden. It can readily be shown that the sextet component in the weak non-leptonic lagrangian satisfies the rule $\Delta T = \frac{1}{2}$ for decays with $\Delta S = 0$. In its turn, this leads to the relations

$$\Gamma(F^+ \to K^0 \pi^+) \simeq 2\Gamma(F^+ \to K^+ \pi^0),$$
$$\Gamma(D^0 \to \pi^+ \pi^-) \simeq 2\Gamma(D^0 \to \pi^0 \pi^0),$$

which must be strongly violated in experiment.

CHAPTER 15

Weak decays of b- and t-quarks

Weak decays of the b- and t-quarks occur because they participate in the weak charged current. In the general form, this current is given by

$$j_\mu = \sum_i \bar{\alpha}_L^i \gamma_\mu V^{ik} \kappa_L^k,$$

where $\alpha^1 = u$, $\alpha^2 = c$, $\alpha^3 = t$; $\kappa^1 = d$, $\kappa^2 = s$, $\kappa^3 = b$; α (for the Greek $\alpha\nu\omega$ meaning top) symbolizes "top" quarks, and κ (for the Greek $\kappa\alpha\tau\omega$ meaning bottom) symbolizes "bottom" quarks;

$$\bar{\alpha}_L = \bar{\alpha}\tfrac{1}{2}(1 - \gamma_5), \qquad \kappa_L = \tfrac{1}{2}(1 + \gamma_5)\kappa;$$

V is a unitary 3×3 matrix. If V were a unit matrix, $V^{ik} = \delta^{ik}$, the structure of the charged current would be $\bar{u}d + \bar{c}s + \bar{t}b$. The s- and b-quarks would then be stable, and with them the strange particles and b-hadrons. We know, however, that in reality these particles are unstable. Therefore we need to consider the matrix V in the most general form. In the case of six quarks this matrix has four parameters and is called the Kobayashi-Maskawa matrix.

15.1. Unitary $n \times n$ matrix

For the sake of generality, let us consider the current in which n α-quarks are transformed into n κ-quarks. Let us find the number of parameters which in this case determine the form of the unitary current matrix V. In the general case, the matrix V comprises n^2 complex numbers or $2n^2$ real parameters. The unitarity condition

$$V^+ V = 1$$

reduces the number of parameters to n^2. Indeed, we have n conditions of the type

$$V_{1k} V_{k1}^+ = V_{1k} V_{1k}^* = 1$$

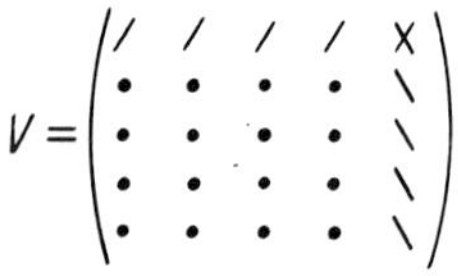

$$V = \begin{pmatrix} / & / & / & / & \times \\ \cdot & \cdot & \cdot & \cdot & \backslash \\ \cdot & \cdot & \cdot & \cdot & \backslash \\ \cdot & \cdot & \cdot & \cdot & \backslash \\ \cdot & \cdot & \cdot & \cdot & \backslash \end{pmatrix}$$

Fig. 15.1.

on the diagonal, as well as $\frac{1}{2}n(n-1)$ conditions of the type $\operatorname{Re} V_{1k}V_{2k}^* = 0$ and $\frac{1}{2}n(n-1)$ conditions of the type $\operatorname{Im} V_{1k}V_{2k}^* = 0$, the last two sets of equations resulting from the condition

$$V_{1k}V_{k2}^+ = V_{1k}V_{2k}^* = V_{2k}V_{1k}^* = V_{2k}V_{k1}^+ = 0.$$

Of the remaining n^2 parameters, $2n - 1$ are non-physical phases which can be "eliminated" by redefining, or "rotating", the (unobservable) phases of n α-quarks and $(n-1)$ κ-quarks. The total number of such non-physical phases is $2n - 1$, and not $2n$, since one of the matrix elements of V cannot be "κ-rotated" as it is already "α-rotated" (in fig. 15.1 this matrix element is at the intersection of the upper row and the right-hand column). The total number of physical parameters is equal therefore to

$$n^2 - (2n - 1) = (n - 1)^2.$$

Let us find now how many of these $(n-1)^2$ parameters are angles of orthogonal rotations, and how many are phase factors. The number of independent rotations in an n-dimensional space is $n_\theta = \frac{1}{2}n(n-1)$, and therefore the number of physical phase parameters is

$$n_\delta = (n-1)^2 - \tfrac{1}{2}n(n-1) = \tfrac{1}{2}(n-1)(n-2).$$

Let us consider the simplest examples:

$n = 2$, four unitary matrices $(1, \tau)$, $\quad n_\theta = 1, n_\delta = 0$,

$n = 3$, nine unitary matrices $(1, \lambda)$, $\quad n_\theta = 3, n_\delta = 1$,

$n = 4$, sixteen unitary matrices, $\quad n_\theta = 6, n_\delta = 3$.

15.2. Angles θ_1, θ_2, θ_3 and phase δ

Let us construct the V-matrix explicitly for the case of six quarks (u, c, t, d, s, b). The angles $\theta_1, \theta_2, \theta_3$ can be chosen in several ways. We shall use the Euler angles and the notation usually adopted in the literature.

Consider a triad of coordinates (z, y, x) and a triad of quarks (d, s, b) put in correspondence with it. Now we make three rotations: by an angle θ_3 around the z-axis (fig. 15.2), then by an angle θ_1 around a new x-axis, and then by an angle θ_2 around a new z-axis. The sequence of these rotations is described by a product of three matrices:

$$\begin{pmatrix} 1 & 0 & 0 \\ 0 & c_2 & s_2 \\ 0 & -s_2 & c_2 \end{pmatrix} \begin{pmatrix} c_1 & s_1 & 0 \\ -s_1 & c_1 & 0 \\ 0 & 0 & 1 \end{pmatrix} \begin{pmatrix} 1 & 0 & 0 \\ 0 & c_3 & s_3 \\ 0 & -s_3 & c_3 \end{pmatrix},$$

where $c_i = \cos \theta_i$, $s_i = \sin \theta_i$. Now we only need to insert the phase factor $e^{i\delta}$. Clearly, it cannot be put either at the beginning or at the end, since in these positions it will be non-physical, being identifiable with the unobservable phase of one of the quarks. The result will be non-trivial if we write, for example,

$$(\bar{u}, \bar{c}, \bar{t}\,) \begin{pmatrix} 1 & 0 & 0 \\ 0 & c_2 & s_2 \\ 0 & -s_2 & c_2 \end{pmatrix} \begin{pmatrix} c_1 & s_1 & 0 \\ -s_1 & c_1 & 0 \\ 0 & 0 & e^{i\delta} \end{pmatrix} \begin{pmatrix} 1 & 0 & 0 \\ 0 & c_3 & s_3 \\ 0 & -s_3 & c_3 \end{pmatrix} \begin{pmatrix} d \\ s \\ b \end{pmatrix}.$$

Multiplication of these matrices yields

$$(\bar{u}, \bar{c}, \bar{t}\,) \begin{pmatrix} c_1 & s_1 c_3 & s_1 s_3 \\ -s_1 c_2 & c_1 c_2 c_3 - e^{i\delta} s_2 s_3 & c_1 c_2 s_3 + e^{i\delta} s_2 c_3 \\ s_1 s_2 & -c_1 s_2 c_3 - e^{i\delta} c_2 s_3 & -c_1 s_2 s_3 + e^{i\delta} c_2 c_3 \end{pmatrix} \begin{pmatrix} d \\ s \\ b \end{pmatrix}.$$

The matrix can be simplified if we assume that all $s_i \ll 1$:

$$V = \begin{pmatrix} 1 & s_1 & s_1 s_3 \\ -s_1 & 1 & s_3 + s_2 e^{i\delta} \\ s_1 s_2 & -s_2 - s_3 e^{i\delta} & e^{i\delta} \end{pmatrix}.$$

So far no b-hadrons with single b-quarks have been found (this statement reflects the state of the art at the beginning of 1979), and not even

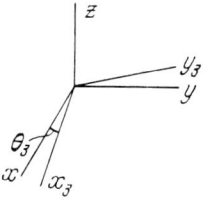

Fig. 15.2.

quarkonium, $t\bar{t}$, has been observed; nevertheless, it can be said even at this juncture that the b- and t-quarks are unstable, that is $s_3 \neq 0$, $s_2 \neq 0$. We shall give substantiation to this statement in the next section.*

15.3. Metastability of the b-quark

If we had $s_2 = s_3 = 0$, the b-quark would be absolutely stable, and with it the lightest of the B-hadrons, such as the mesons $B^- = b\bar{u}$ and $B^+ = \bar{b}u$. The anticipated masses of these mesons are of the order of 5 GeV. A search for stable particles with such masses in pp collisions has been unsuccessful. Moreover, the upper bound on the cross section of production of $B^\pm$ with lifetime above 10^{-8} s was found to be approximately 10^{-37} cm², which is by an order of magnitude smaller than the cross section of the reaction $pp \to \Upsilon + \ldots$. The anticipated cross section of $B^+ B^-$ pair production must not be so small; hence, these experiments indicate that $\tau_B \lesssim 10^{-8}$ s. This means in its turn, that s_3 and/or $s_2 \gtrsim 10^{-4}$.

Possible values of s_3 can also be derived from the comparison of values of the Cabibbo angle found from $\cos \theta$ (in nuclear β-decay) and from $\sin \theta$ (in non-leptonic decays of K-mesons and hyperons). It is easy to show that in the transition from four to six quarks

$$\cos \theta \Rightarrow c_1,$$
$$\sin \theta \Rightarrow s_1 c_3.$$

The Fermi coupling constant G_μ for the muon decay can be obtained from the experimentally measured muon lifetime τ_μ if we use the theoretical expression for the decay rate:

$$\tau_\mu^{-1} = \frac{G_\mu^2 m_\mu^5}{192 \pi^3} f\left(\frac{m_e^2}{m^2}\right)\left[1 - \frac{\alpha}{2\pi}\left(\pi^2 - \tfrac{25}{4}\right)\right],$$

where

$$f(x) = 1 - 8x + 8x^3 - x^4 + 12x^2 \ln(1/x).$$

This yields

$$G_\mu = (1.16632 \pm 0.00004) \cdot 10^{-5} \text{ GeV}^{-2}$$
$$= (1.43582 \pm 0.00004) \cdot 10^{-49} \text{ erg} \cdot \text{cm}^3.$$

---

*According to the data obtained at the $e^+ e^-$ colliding beam facility PETRA, the lower bound on the t-quark mass is $\simeq 18$ GeV. As for the discovery of mesons comprising single b-quarks see footnote on p. 5.

On the other hand, the data on superallowed β-transitions $0^+ \to 0^+$ in nuclei yield

$$G_\beta = (1.398 \pm 0.003) \cdot 10^{-49} \text{ erg} \cdot \text{cm}^3,$$

whence $c_1 = G_\beta/G_\mu = 0.974 \pm 0.003$, $s_1 = 0.227 \pm 0.013$. (the uncertainty in G_β reflects the experimental errors and, to approximately the same degree, theoretical ambiguity in covering radiative, Coulomb, and weak corrections, the last of which takes into account the second order of perturbation theory in weak interactions). If we compare $s_1 = 0.227 \pm 0.013$ with the value $\sin\theta = 0.219 \pm 0.011$ yielded by analyzing leptonic decays of K-mesons and hyperons, we can derive from the equality $c_3 = \sin\theta/s_1$, that c_3 is in the interval $0.87-1.00$, the mean value being $c_3 = 0.965$. Hence,

$$s_3 \simeq 0.25 \pm 0.25.$$

15.4. Contribution of t-quarks to $K^0 \leftrightarrow \bar{K}^0$ transitions

Let us analyze the contribution of the u-, c-, and t-quarks to transitions with $|\Delta S| = 2$ (fig. 15.3). The integrand for the loop comprises a weighted sum of three propagators:

$$(1 - \gamma_5)\left[V_{11}^+ V_{12} \frac{1}{\hat{k} - m_u} + V_{12}^+ V_{22} \frac{1}{\hat{k} - m_c} + V_{13}^+ V_{32} \frac{1}{\hat{k} - m_t}\right](1 + \gamma_5)$$

$$= 2\hat{k}\left[V_{11}^+ V_{12} \frac{1}{k^2 - m_u^2} + V_{12}^+ V_{22} \frac{1}{k^2 - m_c^2} + V_{13}^+ V_{32} \frac{1}{k^2 - m_t^2}\right](1 + \gamma_5)$$

$$= 2\hat{k}\left[V_{12}^+ V_{22}\left(\frac{1}{k^2 - m_c^2} - \frac{1}{k^2 - m_u^2}\right)\right.$$

$$\left. + V_{13}^+ V_{32}\left(\frac{1}{k^2 - m_t^2} - \frac{1}{k^2 - m_u^2}\right)\right](1 + \gamma_5)$$

$$\simeq 2\hat{k}\left[V_{21}^* V_{22} \frac{m_c^2}{k^2(k^2 - m_c^2)} + V_{31}^* V_{32} \frac{m_t^2}{k^2(k^2 - m_t^2)}\right](1 + \gamma_5)$$

$$\simeq -2\hat{k}s_1\left[\frac{m_c^2}{k^2(k^2 - m_c^2)} + s_2(s_2 + s_3 e^{i\delta})\frac{m_t^2}{k^2(k^2 - m_t^2)}\right](1 + \gamma_5).$$

The above transformations were based on the unitarity of the V-matrix ($V_{11}^+ V_{12} + V_{12}^+ V_{22} + V_{13}^+ V_{32} = 0$) and on the equality $(1 - \gamma_5)(1 + \gamma_5) = 0$; we have neglected the u-quark mass and used the simplified form of the V-matrix valid in the case of all $s_i \ll 1$ (see above).

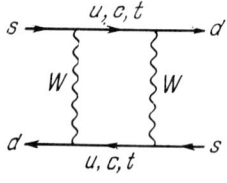

Fig. 15.3.

Discussing the Glashow-Iliopoulos-Maiani scheme in chapter 11, we gave an expression for Δm_{LS} taking into account contributions of only u- and c-quarks. Now it is not difficult to demonstrate, by calculating integrals corresponding to the loop in fig. 15.3, that inclusion of the t-quark into the picture results in multiplication of the above expression by a correction factor $\mathrm{Re}\, r$, where

$$r = 1 + x^2 \xi + 2x \ln \xi,$$

in which $\xi = m_t^2/m_c^2$ and $x = s_2(s_2 + s_3 e^{i\delta})$ (in deriving the expression for r we made use of the equalities

$$m_c^2 \int_0^\infty \frac{dk^2}{(k^2 + m_c^2)^2} = 1, \quad m_t^2 \int_0^\infty \frac{dk^2}{(k^2 + m_t^2)^2} = 1,$$

$$m_t^2 \int_0^\infty \frac{dk^2}{(k^2 + m_t^2)(k^2 + m_c^2)} = \frac{m_t^2}{m_t^2 - m_c^2} \ln \frac{m_t^2}{m_c^2} \simeq \ln \frac{m_t^2}{m_c^2},$$

and assumed that $m_c^2/m_t^2 \ll 1$).

If the "dressing" of the graph of fig. 15.3 with virtual gluons is taken into account, then the first two terms in r are multiplied by $\simeq 0.6$, and the third by $\simeq 0.4$. Keeping this in mind, let us recall the contribution of the bare graph of fig. 11.6 with u- and c-quarks only. This contribution (see p. 90) is equal to $\simeq 2 \cdot 10^7 (m_c/m_\mu)^2 \, \mathrm{s}^{-1}$, which is to be compared with the experimental value $\Delta m_{LS} \simeq 5 \cdot 10^9 \, \mathrm{s}^{-1}$. By using the known value $m_c \simeq 1.3$ GeV, we find that the contribution of t-quarks is essential: $\mathrm{Re}(r - 1) \gtrsim 1$. This means, in turn, that angle θ_2 cannot be very small.

The value of $\mathrm{Im}\, r$ determines the violation of CP invariance in the $K^0 - \overline{K}^0$ system. In particular, we find for the parameter $i\mu_{12}$ characterizing the transitions $K_2 \leftrightarrow K_1$ (see chapter 12):

$$\frac{\mu_{12}}{\Delta m_{LS}} = \frac{\mathrm{Im}\, r}{\mathrm{Re}\, r}.$$

Experimentally, this ratio is about $2 \cdot 10^{-3}$. Consequently,

$$s_2 s_3 \sin \delta \gtrsim 10^{-3}$$

for not too high values of m_t/m_c. From this we can conclude that the lifetime of b-hadrons does not exceed 10^{-12} s.

It must be emphasized that if the angle θ_3 is sufficiently small, the experimentally observed smallness of CP-odd effects does not necessarily require that phase δ be small. For example, if $s_3 \simeq 10^{-3}$ and $s_2 \simeq s_1$, then $\delta \simeq 1$. (This variant will be eliminated once it is established experimentally that the relative widths of the $b \to u\ell\bar{\nu}$, $b \to ud\bar{u}$, or $b \to us\bar{c}$ decays are not vanishingly small.) If, however, $s_3 \simeq s_2 \simeq s_1$, then $\delta \simeq 10^{-2}$.

Note that all CP-noninvariant amplitudes are proportional to $s_1 s_2 s_3 \delta$, both in the decays of charmed particles and in the decays of the b- and t-hadrons. (They must vanish if at least one of the angles θ_i is zero.)*

15.5. Cascade decays of b- and t-hadrons

It has been mentioned above that the b-quark must decay mostly owing to the interaction

$$(s_3 + s_2 e^{i\delta})(\bar{c}b)(\bar{d}u + \bar{s}c + \bar{e}\nu_e + \bar{\mu}\nu_\mu + \bar{\tau}\nu_\tau).$$

The expected lifetime of B-mesons obtained by scaling the muon lifetime, must then be of the order of 10^{-14} s. This scaling takes into account the factor $(m_b/m_\mu)^5 \simeq (4.750/105)^5 \simeq 2 \cdot 10^8$ and a factor of about 5 corresponding to five channels, two of which are color-tripled (but four out of nine are especially strongly suppressed by the phase space); it is also assumed that $|s_3 + s_2 e^{i\delta}| \simeq 0.5$. B-mesons must then produce cascade decays, in accordance with the quark decay cascade. For example,

$$b \to c\mu^- \bar{\nu} \qquad B^- \to D^0 \mu^- \bar{\nu}.$$
$$\hookrightarrow s\mu^+ \nu, \qquad \hookrightarrow K^- \mu^+ \nu.$$

In particular, such cascade decays must produce the so-called trimuon events in neutrino experiments.

As a rule, decays of t-hadrons must produce b-hadrons (see the expression for the V matrix) and therefore the decay cascade must include one additional step.

---

*It is interesting to note that the phase δ enters the amplitude of the CP-violating K-meson decays not only through the box diagram of fig. 15.3, which gives superweak K_2^0-K_1^0 mixing, but also through a milliweak transition between s- and d-quarks with an emission of a gluon. This milliweak transition with $|\Delta S| = 1$ and $\Delta T = 1/2$ is described by the graphs of fig. 7.5 (and 7.6) (p. 53) with virtual u-, c- and t-quarks. The interplay of the two mechanisms leads to an interesting prediction: $|\eta_{+-} - \eta_{00}| \simeq 2 \cdot 10^{-5}$. This quantity may be even larger if CP is violated by Higgs bosons (for references see Chapter 28).

CHAPTER 16

Neutrino–electron interactions

In this and in the subsequent chapter we discuss neutrino reactions. These reactions played an important role both in establishing the properties of weak interaction and in the determination of the parton structure of nucleons. This chapter deals with reactions occurring in neutrino-electron collisions. It is preferable to begin the discussion with these processes since they are much simpler from a theoretical point of view than neutrino-nucleon interactions. Unfortunately, they are much less studied experimentally, owing to a sort of "complementarity principle".

16.1. Kinematics of $\nu + e \to \nu + \ell$ reactions

Before calculating cross sections, we consider the kinematics of reactions. We start with the reaction

$$\nu(k_1) + e(p_1) \to \nu(k_2) + \ell(p_2),$$

where ν is the electron or muon neutrino or antineutrino, and ℓ (e or μ) is a charged lepton. Quantities in parentheses are 4-momenta of the particles involved: $k_1^2 = k_2^2 = 0$, $p_1^2 = m^2$, $p_2^2 = \mu^2$, where m is the electron mass and μ is the mass of the final lepton. In the laboratory reference frame

$$k_i = \omega_i, \mathbf{k}_i, \qquad p_i = E_i, \mathbf{p}_i, \qquad i = 1, 2.$$

In the center of mass

$$k_i = \omega_i^0, \mathbf{k}_i^0, \qquad p_i = E_i^0, \mathbf{p}_i^0, \qquad i = 1, 2,$$

$$\mathbf{k}_1^0 + \mathbf{p}_1^0 = 0, \qquad \mathbf{k}_2^0 + \mathbf{p}_2^0 = 0 \qquad \text{(see fig. 16.1)}.$$

Introduce the standard invariants s and t. As usual, s is the total squared

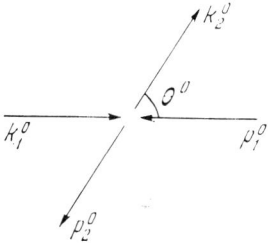

Fig. 16.1.

energy in the center-of-mass frame:

$$s = (k_1 + p_1)^2 = (\omega_1^0 + E_1^0)^2 = 2m\omega_1 + m^2$$
$$= (k_2 + p_2)^2 = (\omega_2^0 + E_2^0)^2,$$
$$\omega_1^0 = \frac{(k_1, k_1 + p_1)}{\sqrt{s}} = \frac{k_1 p_1}{\sqrt{s}} = \frac{s - m^2}{2\sqrt{s}}, \qquad E_1^0 = \frac{s + m^2}{2\sqrt{s}},$$
$$\omega_2^0 = \frac{(k_2, k_2 + p_2)}{\sqrt{s}} = \frac{(k_2 p_2)}{\sqrt{s}} = \frac{s - \mu^2}{2\sqrt{s}}, \qquad E_2^0 = \frac{s + \mu^2}{2\sqrt{s}}.$$

The threshold for the $\nu e \to \nu \mu$ reaction is

$$s = \mu^2, \qquad \omega_1 = \frac{\mu^2 - m^2}{2m} \approx 11 \text{ GeV}.$$

In the case of elastic scattering, $\mu = m$:

$$\omega_2^0 = \omega_1^0 = \omega^0 = \frac{s - m^2}{2\sqrt{s}}, \qquad E_1^0 = E_2^0 = E^0 = \frac{s + m^2}{2\sqrt{s}}.$$

The 4-momentum transfer squared is defined in the standard manner:

$$t = (k_1 - k_2)^2$$
$$= -2\omega_1^0 \omega_2^0 (1 - \cos\theta^0) = -2\omega_1 \omega_2 (1 - \cos\theta) = (p_1 - p_2)^2$$
$$= 2m^2 - 2E_1^0 E_2^0 (1 - v_1^0 v_2^0 \cos\theta^0)$$
$$= -2mE_2 + m^2 + \mu^2 = (\mu - m)^2 - 2mT,$$

where $T = E_2 - \mu$ is the kinetic energy of the recoil lepton in the laboratory reference frame,

$$v_1^0 = \frac{|\mathbf{p}_1^0|}{E_1^0} = \frac{s - m^2}{s + m^2}, \qquad v_2^0 = \frac{|\mathbf{p}_2^0|}{E_2^0} = \frac{s - \mu^2}{s + \mu^2}.$$

In the case of elastic scattering
$$v_1^0 = v_2^0 = v^0 = \frac{s - m^2}{s + m^2}.$$

Determine now the physical region of T. The minimum value of T corresponds to $\theta^0 = 0$; consequently, $t = 0$, and

$$T_{min} = \frac{(\mu - m)^2}{2m}.$$

In the case of elastic scattering $T_{min} = 0$. The maximum value of T corresponds to $\theta^0 = \pi$,

$$T_{max} = \frac{(\mu - m)^2 + 4\omega_1^0 \omega_2^0}{2m}.$$

In the case of elastic scattering

$$T_{max} = \frac{2(\omega^0)^2}{m} = \frac{(s - m^2)^2}{2sm} = \frac{2m\omega_1^2}{2m\omega_1 + m^2} = \frac{\omega_1}{1 + m/2\omega_1}.$$

Neglecting the m/μ ratio, we obtain for the $\nu e \to \nu \mu$ reaction:

$$T_{max} \simeq s/2m.$$

In addition to the relations

$$2p_1k_1 = s - m^2, \qquad 2p_2k_2 = s - \mu^2,$$

two more relations are useful:

$$2p_1k_2 = s + t - \mu^2, \qquad 2p_2k_1 = s + t - m^2.$$

The experimental observation of the processes in question, and their analysis against the background of other processes, are very difficult because cross sections are small and only one particle, namely the recoil lepton, is recorded. The energy of the impinging neutrino, ω_1, can be found by measuring the energy E_2 of the recoil lepton and its emission angle with respect to the neutrino beam, θ:

$$2E_2\omega_1(1 - v_2 \cos \theta) = \mu^2 - m^2 + 2m(\omega_1 + m - E_2).$$

The result for elastic νe scattering at high energies ($\omega_1 \gg m$, $E_2 \gg m$) is

$$\omega_1 \simeq \frac{2mE_2}{2m - \theta^2 E_2}$$

(note that at high energies the kinematically allowed angles θ are small: $\theta \lesssim \sqrt{m/\omega_1}$). Determination of ω_1 by the above formula facilitates identification of an event as νe scattering, provided characteristic neutrino energies in the beam are known in advance (at least roughly).

16.2. Cross section of the $\nu_\mu e^- \to \nu_e \mu^-$ reaction

The theory of the $\nu_\mu e^- \to \mu^- \nu_e$ and $\bar{\nu}_e e^- \to \mu^- \bar{\nu}_\mu$ reactions is especially simple because in contrast to elastic νe scattering, only charged currents are involved (see fig. 16.2). We begin with the reaction

$$\nu_\mu(k_1) + e(p_1) \to \nu_e(k_2) + \mu^-(p_2).$$

Quantities in parentheses are 4-momenta of the particles. The process amplitude is

$$M = \sqrt{\tfrac{1}{2}} G \left[\bar{u}_{\nu_e} \gamma_\alpha (1+\gamma_5) u_e \right] \left[\bar{u}_\mu \gamma^\alpha (1+\gamma_5) u_{\nu_\mu} \right].$$

The calculation of the squared amplitude summed over spin variables is standard (see chapter 3); the result is

$$\overline{|M|^2} = 128 G^2 (p_1 k_1)(p_2 k_2).$$

As dictated by the crossing symmetry, this expression is absolutely identical to that for muon decay.

The cross section is given by the formula

$$\sigma = \frac{1}{2 \cdot 4(k_1 p_1)} \int \overline{|M|^2} \frac{d k_2}{(2\pi)^3 2\omega_2} \frac{d p_2}{(2\pi)^3 2 E_2} (2\pi)^4 \delta^4(k_2 + p_2 - k_1 - p_1).$$

Averaging of $|M|^2$ in the integrand is carried out over polarizations of the initial electron and not over those of the neutrino since the latter is longitudinally polarized. Calculating now the center-of-mass phase-space

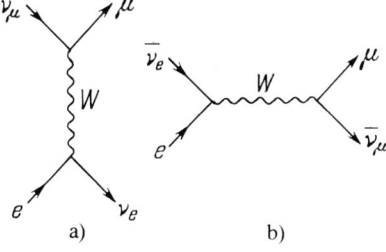

Fig. 16.2.

volume, we obtain

$$d\sigma = -\overline{|M|^2}\frac{(p_2k_2)}{(p_1k_1)s}\frac{d\cos\theta^0}{64\pi}$$

$$= -\frac{2G^2}{\pi}\frac{(p_1k_1)(p_2k_2)(p_2k_2)}{(p_1k_1)s}d\cos\theta^0 = -\frac{2G^2}{\pi}(\omega_2^0)^2 d\cos\theta^0.$$

We find that the cross section in the center-of-mass reference frame is isotropic. Making use of the expression $dt = 2\omega_1^0\omega_2^0 d\cos\theta^0$, we obtain

$$d\sigma = -\frac{\overline{|M|^2}}{(p_1k_1)^2}\frac{dt}{128\pi} = -\frac{G^2(p_1k_1)(p_2k_2)}{\pi(p_1k_1)^2}dt = -\frac{G^2}{\pi}\frac{s-\mu^2}{s-m^2}dt$$

$$= \frac{G^2}{\pi}\frac{s-\mu^2}{s-m^2}2m\,dT.$$

The total cross section is found by integration over $d\cos\theta^0$ (or dt, or dT):

$$\sigma = \frac{4G^2}{\pi}(\omega_2^0)^2 = \frac{G^2}{\pi}\frac{(s-\mu^2)^2}{s}.$$

The cross section vanishes at the threshold. In the case $s \gg \mu^2$ we have

$$\sigma \approx \frac{G^2 s}{\pi}.$$

16.3. Cross section of the $\bar{\nu}_e e^- \to \bar{\nu}_\mu \mu^-$ reaction

Calculation of $\overline{|M|^2}$ for the reaction

$$\bar{\nu}_e(k_1) + e^-(p_1) \to \bar{\nu}_\mu(k_2) + \mu^-(p_2)$$

need not be repeated. It is sufficient to make a substitution $k_1 \leftrightarrow k_2$ in the expression for $\overline{|M|^2}$ for the reaction $\nu_\mu e^- \to \nu_e \mu^-$. This substitution is based on the two reactions being coupled by the crossing operation: emission of ν_e is substituted by absorption of $\bar{\nu}_e$, and absorption of ν_μ by emission of $\bar{\nu}_\mu$. Consequently,

$$\overline{|M|^2} = 128G^2(p_1k_2)(p_2k_1),$$

and the differential cross section becomes

$$d\sigma = -\frac{2G^2}{\pi} \frac{(p_1k_2)(p_2k_1)(p_2k_2)}{(p_1k_1)s} d\cos\theta^0$$

$$= -\frac{2G^2}{\pi} \frac{(\omega_2^0)^2 E_1^0 E_2^0}{s}(1 + v_1^0 \cos\theta^0)(1 + v_2^0 \cos\theta^0) d\cos\theta^0.$$

Note that

$$\frac{d\sigma(\nu_\mu e \to \mu \nu_e)}{d\cos\theta^0}\bigg|_{\theta^0=0} = \frac{d\sigma(\nu_e e \to \mu^- \bar{\nu}_\mu)}{d\cos\theta^0}\bigg|_{\theta^0=0} = -\frac{2G^2}{\pi}(\omega_2^0)^2.$$

In the invariant notation

$$d\sigma = -\frac{G^2}{\pi} \frac{(p_1k_2)(p_2k_1)}{(p_1k_1)^2} dt = -\frac{G^2}{\pi} \frac{(s+t-\mu^2)(s+t-m^2)}{(s-m^2)^2} dt,$$

$$\frac{d\sigma}{dt}(\nu_\mu e \to \nu_e \mu)\bigg|_{t=0} = \frac{d\sigma}{dt}(\bar{\nu}_e e \to \bar{\nu}_\mu \mu)\bigg|_{t=0}.$$

Integration over θ^0 from 0 to π yields

$$\sigma = \frac{4G^2}{\pi}(\omega_2^0)^2 \frac{E_1^0 E_2^0}{s}\left(1 + \tfrac{1}{3}v_1^0 v_2^0\right) = \frac{G^2}{\pi} \frac{4(\omega_2^0)^2}{s}\left(E_1^0 E_2^0 + \tfrac{1}{3}\omega_1^0 \omega_2^0\right).$$

The reaction $\nu_\mu e \to \nu_e \mu$ has been observed experimentally only recently, owing to its high threshold value and low cross section. This observation has not yielded any principally new data on weak interactions since the amplitude of this reaction is trivially related to the thoroughly analyzed amplitude of the $\mu \to e\bar{\nu}_e \nu_\mu$ decay. Nevertheless, several reasons justify a discussion of the reactions $\nu_\mu e \to \nu_e \mu$ and $\bar{\nu}_\mu e \to \bar{\nu}_e \mu$ in this book: first, cross sections of these processes are usually not given in books devoted to weak interactions; second, similar formulas are used in the framework of the parton model to calculate production cross sections of heavy quarks (c, b, t) in neutrino experiments; third, these formulas can be used in the $\mu = m$ limit to analyse $\nu e \to \nu e$ elastic scattering.

16.4. Elastic νe and $\bar{\nu} e$ scattering induced by the charged current

Neutrino elastic scattering on electrons is caused by neutral currents in the case of ν_μ and $\bar{\nu}_\mu$, and by a combination of neutral and charged current in the case of ν_e and $\bar{\nu}_e$. We begin with the idealized case of the elastic

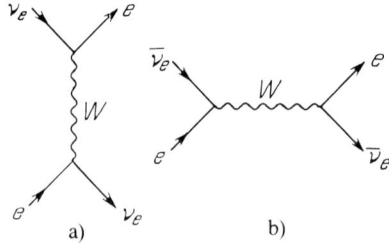

Fig. 16.3.

scattering cross sections of the processes $\nu_e e \to \nu_e e$ and $\bar{\nu}_e e \to \bar{\nu}_e e$, ignoring for a moment neutral currents. We shall obtain cross sections in the form in which they were written in textbooks before the discovery of neutral currents. They are of more than just historical interest, however. Cross sections for neutrino interactions with light quarks are of a very similar form. Moreover, it is very convenient to use this simple case as a basis for discussing the behavior of neutrino processes at high energies ($s \gg m^2$).

Let us start with the graphs of fig. 16.3. These graphs reduce to the four-fermion graphs of fig. 16.4 at sufficiently low energies, i.e. for $s \ll m_W^2$ (the case of $s \sim m_W^2$ being unfeasible for laboratory neutrino experiments). Assuming $\mu = m$, we easily derive from the formulas obtained above:

$$\frac{d\sigma_{\nu e}}{dt} = -\frac{G^2}{\pi}, \quad \frac{d\sigma_{\bar{\nu} e}}{dt} = -\frac{G^2}{\pi} \frac{(s+t-m^2)^2}{(s-m^2)^2}, \quad 0 > t > -\frac{(s-m^2)^2}{s},$$

$$\sigma_{\nu e} = \frac{G^2}{\pi} \frac{(s-m^2)^2}{s} = \frac{2G^2}{\pi} \frac{m\omega_1}{1+m/2\omega_1},$$

$$\sigma_{\bar{\nu} e} = \frac{G^2}{3} \frac{(s-m^2)^3 - [(s-m)^2 m^2/s]^3}{(s-m^2)^2} = \frac{G^2}{3\pi}(s-m^2)\left[1 - \left(\frac{m^2}{s}\right)^3\right]$$

$$= \frac{2G^2 m\omega_1}{3\pi}\left[1 - \left(\frac{m/2\omega_1}{1+m/2\omega_1}\right)^3\right].$$

Fig. 16.4.

§4] Elastic νe and $\bar{\nu} e$ scattering induced by the charged current

The formulas are simplified for $s \gg m^2$:

$$\frac{d\sigma_{\nu e}}{dt} \approx -\frac{G^2}{\pi}, \quad \frac{d\sigma_{\bar{\nu} e}}{dt} \approx -\frac{G^2}{\pi}\left(1 + \frac{t}{s}\right)^2, \quad 0 \geq t \geq -s,$$

$$\sigma_{\nu e} \approx \frac{G^2 s}{\pi} \approx \frac{G^2 2 m \omega_1}{\pi} \approx 1.68 \cdot 10^{-41}(E_\nu/\text{GeV})\text{cm}^2,$$

$$\sigma_{\bar{\nu} e} \approx \frac{G^2 s}{3\pi} \approx \frac{G^2 2 m \omega_1}{3\pi} \approx 0.56 \cdot 10^{-41}(E_\nu/\text{GeV})\text{cm}^2.$$

This behavior of cross sections is easily understood in qualitative terms. The linear dependence of the cross section on s stems from the dimension of G: the only other dimensional parameter available to us at $s \to \infty$ is the total energy in the center-of-mass frame. The isotropy of νe scattering in this frame is equally clear. Indeed, the total angular momentum J of both the ingoing and outgoing ν and e is zero so that, for the point interaction, all directions of scattering are equally good (see fig. 16.5 where long arrows symbolize momenta, and short ones spins of the particles). Mutual orientations of spins and momenta shown in fig. 16.5 are characteristic for the charged current. Indeed, it includes only the left-handed components of the neutrino and of the electron. The configuration of spins and momenta in $\bar{\nu} e$ scattering is shown in fig. 16.6; it is clear that in this case the total angular momentum J of the colliding electron and neutrino equals unity and is directed along the momentum of the ingoing neutrino: $J_z = 1$. Owing to conservation of J_z, only one of the three possible final states of $\bar{\nu} e$ ($J_z = \pm 1, 0$) is realized. This corresponds to the coefficient $\frac{1}{3}$ in the expression for $\sigma_{\bar{\nu} e}$. (Note that the $\nu_e e$ and $\bar{\nu} e$ cross sections are equal for $\omega_1 \ll m$:

$$\sigma_{\nu e} = \sigma_{\bar{\nu} e} = \frac{4 G^2 \omega_1^2}{\pi}.$$

In this limiting case the cross section is determined by the available energy only, and does not include the electron mass).

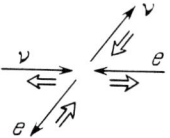

Fig. 16.5.

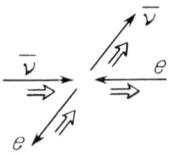

Fig. 16.6.

The cross sections for νe and $\bar{\nu} e$ scattering for $s \gg m^2$ being different, a question arises concerning a violation of the Pomeranchuk theorem in this case (the theorem states the equality of cross sections in the interaction of a particle and its antiparticle with a given target at asymptotically high energies). The answer to this question is that in a certain sense the energies for which νe scattering was discussed so far, were superlow: $s \ll m_W^2 \ll G^{-1}$. The quantity Gs is a natural dimensionless parameter for the four-fermion interaction. As long as $Gs \ll 1$, the amplitude is purely real, and only one partial wave is present (the interaction is point-like). The Pomeranchuk theorem is applicable in the range $Gs \gg 1$ where multiparticle inelastic processes are important and many partial waves are involved. Expressions for cross sections derived in this chapter are not valid in this range.

16.5. Cross sections for the νe and $\bar{\nu} e$ scattering: general formulas

With neutral currents taken into account (see chapter 22) the νe scattering amplitude takes the form

$$M = \sqrt{\tfrac{1}{2}}\, G \bar{u}_{e_2}\bigl[g_L \gamma_\alpha (1 + \gamma_5) + g_R \gamma_\alpha (1 - \gamma_5) \bigr] u_{e_1} \cdot \bar{u}_{\nu_2} \gamma_\alpha (1 + \gamma_5) u_{\nu_1}.$$

Here $\nu = \nu_e$ or ν_μ (scattering of $\bar{\nu}_e$ and $\bar{\nu}_\mu$ is discussed at the end of this section). As will be shown in chapter 22, the constants g_L and g_R in the case of the standard model of the electroweak interaction could be written in the form

$$\left.\begin{array}{l} g_L = \tfrac{1}{2} + \sin^2 \theta_W \\ g_R = \sin^2 \theta_W \end{array}\right\} \text{ for } \nu_e e \text{ scattering,}$$

$$\left.\begin{array}{l} g_L = -\tfrac{1}{2} + \sin^2 \theta_W \\ g_R = \sin^2 \theta_W \end{array}\right\} \text{ for the } \nu_\mu e \text{ scattering.}$$

(We recall that $g_V = g_L + g_R$, $g_A = g_L - g_R$.) The total amplitude of $\nu_\mu e$ scattering is due to the coupling of the neutral currents $\bar{\nu}_\mu \nu_\mu$ and $\bar{e} e$; the $\nu_e e$

scattering amplitude is the sum of two terms, one of which is a product of neutral currents $\bar{\nu}_e \nu_e$ and $\bar{e}e$, and the other term is a product of charged currents $\bar{\nu}_e e$ and $\bar{e}\nu_e$. In writing the total amplitude, the Fierz transformation was used for the last of these terms:

$$\bar{\nu}_e \gamma_\alpha (1 + \gamma_5) e \cdot \bar{e} \gamma^\alpha (1 + \gamma_5) \nu_e = \bar{e} \gamma_\alpha (1 + \gamma_5) e \bar{\nu}_e \gamma^\alpha (1 + \gamma_5) \nu_e.$$

(The charged current corresponds therefore to the coefficients $g_L = 1$, $g_R = 0$). For $\overline{|M|^2}$ we obtain

$$\overline{|M|^2} = 128 G^2 \left[g_L^2 (p_1 k_1)^2 + g_R^2 (p_1 k_2)^2 - g_L g_R m^2 (k_1 k_2) \right].$$

Here the terms proportional to g_L^2 and g_R^2 are the same as in scattering due to the charged currents $\nu_e e \to \nu_e e$ and $\bar{\nu}_e e \to \bar{\nu}_e e$, respectively (see above). It is convenient to use the Fierz transformation when calculating interference terms proportional to $g_L g_R$. For example, the first interference term is conveniently written in the form

$$g_L g_R \bar{u}_{e_2} \gamma_\alpha (1 + \gamma_5) u_{e_1} \bar{u}_{\nu_2} \gamma^\alpha (1 + \gamma_5) u_{\nu_1} \bar{u}_{\nu_1} (1 - \gamma_5) u_{e_2} \bar{u}_{e_1} (1 + \gamma_5) u_{\nu_2}.$$

Both interference terms give identical contributions. Taking into account that $p_1 k_1 = m\omega_1$, $p_1 k_2 = m\omega_2 = m(\omega_1 - T)$, $k_1 k_2 = p_1 p_2 - m^2 = T \cdot m$, the final expression for the cross section for νe scattering becomes:

$$\frac{d\sigma}{dT} = \frac{2G^2 m}{\pi} \left[g_L^2 + g_R^2 \left(1 - \frac{T}{\omega_1} \right)^2 - g_L g_R \frac{mT}{\omega_1^2} \right].$$

(Recall that m is the electron mass, T is the electron kinetic energy after scattering in the laboratory frame, and ω_1 is the energy of the ingoing neutrino.) The interference term tends to zero in the $\omega_1 \gg m$ limit. This is a natural result since the interference of left-handed and right-handed helicity states must vanish for ultra-relativistic particles.

The following symmetry arguments could be used to obtain cross sections for $\bar{\nu}_e e$ and $\bar{\nu}_\mu e$ scattering. The $\bar{\nu}e^- \to \bar{\nu}e^-$ scattering amplitude is equal to that for $\nu e^+ \to \nu e^+$. But the left-handed and right-handed components must change places when going from the electrons to the positron. Hence,

$$\frac{d\sigma_{\bar{\nu}e}}{dT} = \frac{2G^2 m}{\pi} \left[g_R^2 + g_L^2 \left(1 - \frac{T}{\omega_1} \right)^2 - g_L g_R \frac{mT}{\omega_1^2} \right].$$

16.6. Other effects of the νe interaction

The neutrino-electron interaction must result in a number of processes, for instance, in $e^+ e^- \to \nu \bar{\nu}$ annihilation (fig. 16.7). It is practically impossible

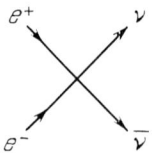

Fig. 16.7.

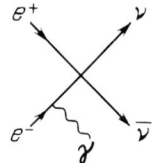

Fig. 16.8.

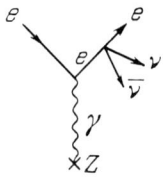

Fig. 16.9.

to observe this process in laboratory conditions. Hopefully, bremsstrahlung photons accompanying this annihilation (fig. 16.8) may be detected in the future in colliding high-energy beams of e^+ and e^-. The process of bremsstrahlung neutrino emission (fig. 16.9) must play an important role in the dynamics of hot stars. This process may be significant, despite its small cross section, because the small interaction cross section of the neutrino allows it to leave the star much more easily than a photon.

To conclude this chapter, let us turn to the neutrino-muon interaction.

16.7. Creation of muon pairs by neutrinos in the Coulomb field of the nucleus

The diagonal interaction $G_{\mu\mu}(\bar{\nu}\mu)(\bar{\mu}\nu)$ must result in the elastic scattering $\nu\mu \to \nu\mu$. There being no muonic targets, this process cannot be observed directly in experiments. Information about this interaction can be extracted by studying the process

$$\nu + Z \to \nu + \mu^+ + \mu^- + Z,$$

taking place in the Coulomb field of a nucleus with charge Z (fig. 16.10). The cross section for this process can be estimated rather easily at asymptotically high energies if calculations are carried out in two steps. The first step consists in calculating the muon pair production cross section in the neutrino collision with a real photon, $\nu\gamma \to \nu\mu^+\mu^-$. It is found to be

$$\sigma_\phi \approx \frac{2\alpha G_{\mu\mu}^2 s}{9\pi^2}\left(\ln\frac{s}{4\mu^2}\right).$$

Here $G_{\mu\mu}$ is the interaction constant: $G_{\mu\mu}(\bar{\nu}_\mu\gamma_\alpha(1+\gamma_5)\mu)(\bar{\mu}\gamma^\alpha(1+\gamma_5)\nu_\mu)$, $s = (k+q)^2$ where k and q are the neutrino and photon 4-momenta, respectively, and μ is the muon mass. This expression is valid for $s \gg 4\mu^2$.

At the second stage the cross section obtained must be multiplied by the probability of finding a virtual photon γ in the Coulomb field of the nucleus with such 4-momenta q that the total energy of ν and γ in their center-of-mass frame is $\sqrt{s}$. It can be shown that this probability is

$$d^2n(q^2, s) \approx \frac{Z^2\alpha}{\pi}\frac{dq^2}{q^2}\frac{ds}{s}.$$

Integration over the photon mass (over dq^2) yields

$$dn(s) = \frac{Z^2\alpha}{\pi}ds\ln\frac{Q_{max}^2}{Q_{min}^2}.$$

Here Q_{min} is determined by the threshold of the $\nu\gamma_v \to \nu\mu^+\mu^-$ reaction (the subscript v indicates that the photon is virtual). If we take into account that the energy transferred by the photon is much smaller than its momentum ($q_0 \sim q^2/2M$ where M is the nucleus mass) then the threshold condition

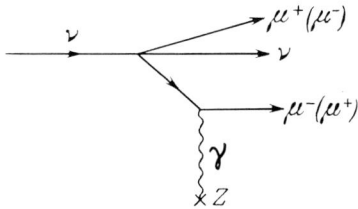

Fig. 16.10.

takes the form

$$4\mu^2 = (k + q_{min})^2 \approx 2kq_{min} \approx -2k \cdot q_{min} \leq 2\omega Q_{min},$$

where ω is the neutrino energy in the laboratory reference frame, and $Q = |q|$.

The quantity Q_{max} characterizes the maximum momentum that can be transferred to the nucleus without causing it to disintegrate (this requirement means that the process must occur in the Coulomb field of the nucleus as a whole). By order of magnitude, $1/Q_{max}$ is equal to the nucleus radius:

$$Q_{max} = m_\pi A^{-1/3},$$

where m_π is the π-meson mass, and A is the total number of nucleons in the nucleus. Finally,

$$\sigma(\nu Z \to \nu \mu^+ \mu^- Z) = \int_{4\mu}^{2\omega Q_{max}} \sigma_\phi n(s) \, ds$$

$$\approx \frac{2Z^2 \alpha^2 G_{\mu\mu}^2 (2\omega) m_\pi A^{-1/3}}{9\pi^3} \ln \frac{2\omega m_\pi A^{-1/3}}{4\mu^2}.$$

This expression holds only when the logarithm is substantially greater than unity, that is when the energy is high enough above threshold. We see that the modulus of the photon effective mass is of the order of $Q_{max} = m_\pi A^{-1/3}$. Numerical calculations of the cross section $\sigma(\nu Z \to \nu \mu^+ \mu^- Z)$, not for the asymptotic but for really feasible neutrino energies, are available in the literature. According to these calculations the production cross section for a muon pair in a neutrino beam with energy 50 GeV colliding with an iron nucleus ($Z = 26$, $A = 56$) is approximately $3 \cdot 10^{-40}$ cm² $(G_{\mu\mu}/G)^2$. (Our asymptotic formula gives the cross section smaller by a factor of about 1.5. One of the factors causing this difference is the fact that the value of the radius assumed in the numerical calculation was not exactly equal to $m_\pi A^{-1/3}$.)

So far the process in question has not been observed. The data available yield only an upper bound for $G_{\mu\mu}$: $G_{\mu\mu} \lesssim 7G$. Observation of the process $\nu_\mu Z \to \mu^+ \mu^- \nu Z$, measurement of $G_{\mu\mu}$ and of the ratio of the vector to axial constants in the expression $G_{\mu\mu} \bar{\nu} \gamma_\alpha (1 + \gamma_5) \nu \bar{\mu} [\gamma_\alpha + (g_A/g_V) \gamma_\alpha \gamma_5] \mu$ (to simplify the expressions we have assumed $g_A/g_V = 1$) are of great interest from the standpoint of testing μ–e universality.

CHAPTER 17

Neutrino–nucleon interactions

This chapter is mostly devoted to neutrino-nucleon interactions which are due to charged currents (to specify the analysis, we single out the reactions produced by the muon neutrino).

17.1. Kinematics

The process of the neutrino-nucleon interaction is usually described in terms of kinematic variables shown in fig. 17.1 where $k = (E, \mathbf{k})$ and $k' = (E', \mathbf{k}')$ are the 4-momenta of the initial and final lepton, p is the nucleon 4-momentum, p' is the 4-momentum of the final hadron state h, and $q = k - k' = p' - p$ is the 4-momentum carried by the W-boson. Usually three invariant variables, ν, Q^2 and W, are introduced.

$$\nu = qp/m = E - E',$$

where m is the nucleon mass. Consequently, ν is the total energy carried away by the final hadrons (in the literature the symbol ν sometimes stands for qp).

$$Q^2 \equiv -q^2 = -(k - k')^2 = 2EE'(1 - v'\cos\theta),$$

where θ is the angle between the muon and neutrino momenta in the laboratory reference frame, v' is the muon velocity ($v' \simeq 1$ at high energies).

$$W^2 = (p')^2 = (p + q)^2 = m^2 + 2m\nu - Q^2.$$

Thus, W is the mass of the final hadron state. We observe that the variable W^2 is an explicit function of ν and Q^2. Distinct kinematic regions are shown in fig. 17.2 in the $(Q^2, \nu m)$ plane. The line $W = m$ corresponds to quasi-elastic scattering reactions (h = N):

$$\nu_\mu + n \to \mu^- + p,$$
$$\bar{\nu}_\mu + p \to \mu^+ + n.$$

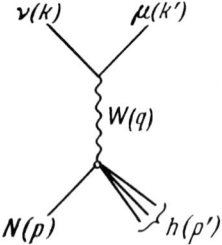

Fig. 17.1.

To the left of this line, the region is kinematically forbidden. The line $W = m + m_\pi$ traces the threshold of inelastic reactions of the type

$$\nu + N \to \mu + N + \pi.$$

Dashed lines correspond to production of a number of hadronic resonances ($h = N^*, \Delta$), for instance

$$\nu_\mu + p \to \mu^- + \Delta^{++}.$$

The hatched area gives the lower left-hand part of the deep-inelastic region. In this region both ν and Q^2 are large and multiple production of hadrons takes place.

17.2. Quasi-elastic scattering

The quasi-elastic scattering cross section at low energies (for instance, that of the reactor-produced neutrino) is written without difficulty by analogy with the νe-scattering cross section (see Chapter 16):

$$\sigma(\bar{\nu}_e p \to n e^+) = \frac{G^2(g_V^2 + 3g_A^2)E^2}{\pi}$$

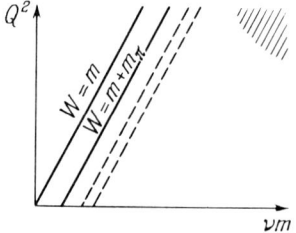

Fig. 17.2.

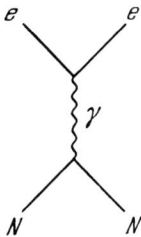

Fig. 17.3.

(we have assumed above that $m_p \gg E \gg m_e$). The constants g_V and g_A are those of the neutron β-decay: $g_V = 1$, $g_A = 1.25$. If nucleons were structureless point-like particles, the quadratic growth of the cross section would change to be linear at $E \gg m_p$. However, when E is of the order of 1 GeV, the nucleon form factor becomes effective and cuts off high values of Q^2. Experiment indicates that both the vector and axial form factors are dipole in form, namely

$$g_V^2 \frac{1}{\left(1 + Q^2/m_V^2\right)^2}, \quad g_A^2 \frac{1}{\left(1 + Q^2/m_A^2\right)^2},$$

with $m_V \simeq 0.84$ GeV and $m_A \simeq 0.9$ GeV. As a result, beginning with energies of the order of 1 GeV, the neutrino-nucleon quasi-elastic interaction cross section asymptotically tends to the level of approximately 10^{-38} cm^2. (Let us remark that approximately the same axial form factor with the same value of m_A was found in the case of $\nu_\mu p \to \mu^- \Delta^{++}$).

It is important to emphasize that by virtue of the isotopic properties of the ud current the vector form factor in quasi-elastic neutrino reactions must coincide with the electromagnetic isovector form factor of the nucleon, measured in experiments on scattering of electrons by nucleons (fig. 17.3). Experiment confirms this prediction of the theory. Experiments on deep-inelastic scattering have demonstrated a still more profound relationship between electromagnetic and weak processes. This is discussed in the next section.

17.3. Partons

The experiments on deep-inelastic electroproduction carried out of Stanford in 1967 have led to the discovery that the cross section at high Q^2 and ν does not decrease with increasing Q^2, and is a function of the dimensionless

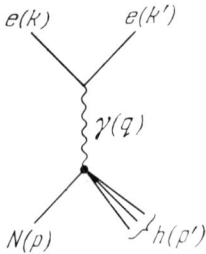

Fig. 17.4.

variables $x = Q^2/2m\nu$ and $y = 2m\nu/s$, where $s = (k + p)^2$ (kinematic variables are defined as in the case of neutrino reactions (compare figs. 17.1 and 17.4)). In a sense, this discovery resembled that of the atomic nucleus in the Rutherford-Marsden experiments when it was found that the cross section of α-particle scattering by an atom at high momentum transfer is not small. In the case of α-particle scattering, high momentum is transferred to a "point-like" atomic nucleus; in the case of electroproduction, high momentum is transferred to point-like particles forming the nucleon. Feynman coined for these parts of the nucleon the name "partons".

The partons in the nucleon are quarks (and gluons). Mostly, nucleons consist of three quarks, called valence quarks. However, according to the uncertainty relation, the three valence quarks could be joined for short duration by quark-antiquark pairs. Such pairs form the so-called quark-antiquark sea. In addition, gluons are also present in the nucleon. The momentum of a fast nucleon is therefore carried by quarks, antiquarks, and gluons. Deep-inelastic scattering occurs when a lepton collides with a quark or an antiquark. By virtue of the asymptotic freedom of quantum chromodynamics, at high Q^2 and ν a light quark can be considered in the zero approximation as a free and practically massless particle.

17.4. Kinematics of lepton-parton collisions

Let us consider a collision of a fast nucleon and a lepton. As compared to the quark longitudinal momentum, its transverse momentum at high energies can be neglected. Denote the quark 4-momentum by xp where p is the nucleon 4-momentum, and $0 \le x \le 1$. First find $\hat{s}$, that is the squared center-of-mass total energy of the lepton and quark:

$$\hat{s} = (xp + k)^2 \simeq x \cdot 2pk \simeq x \cdot s.$$

As a result of the collision, the quark momentum changes to

$$k + px - k' = q + px.$$

The parton mass being zero, we obtain

$$(q + px)^2 = q^2 + x \cdot 2pq = 0,$$

hence,

$$x = -\frac{q^2}{2pq} = \frac{Q^2}{2\nu m}.$$

We find that in terms of the parton model, the quantity x introduced in the preceding section has a straightforward interpretation. The quantity y also has a simple meaning. Indeed, $y = 2m\nu/s = 2pq/2pk = \nu/E$, that is y gives the fraction of the initial neutrino energy carried away by the hadrons. Denoting the lepton scattering angle in the lepton-quark center-of-mass frame by $\hat{\theta}$, we obtain

$$y = \frac{2m\nu}{s} = \frac{Q^2}{x \cdot s} = -\frac{(k-k')^2}{x \cdot s} = -\frac{\hat{t}}{\hat{s}}$$

$$= \frac{\hat{s}(1-\cos\hat{\theta})}{2\hat{s}} = \frac{1-\cos\hat{\theta}}{2} = \sin^2\tfrac{1}{2}\hat{\theta}.$$

What is the mechanism by which the elastic scattering of a lepton by a quark results in the inelastic scattering of the lepton by the nucleon? Let us resort to an analogy: imagine an atom instead of a nucleon, and one of the electrons of this atom instead of the parton. A knocked-out electron is a result of the inelastic ionization of the atom. The emitted electron is observable; but what is the fate of a knocked-out parton (quark)? The lepton-quark collision occurs at a small distance (Q^2 is high) where quarks are nearly free; as the quarks move away, however, they interact more and more strongly, exchanging soft gluons and producing $q\bar{q}$ pairs in vacuum. As a result, a fast colored quark generates a white hadron jet. This very complicated process, closely related to the mechanism of color confinement, is referred to as the hadronic fragmentation of the parton.

17.5. Lepton–parton collision cross sections

The partons being point-like particles, the lepton-parton interaction is described by the same formulas as the lepton-lepton interaction. We recall that the νe and $\bar{\nu} e$ charged current scattering cross sections for the total center-of-mass energy $\sqrt{s}$ are equal, respectively, to

$$\frac{d\sigma}{dy} = \frac{G^2 s}{\pi} \quad \text{for } \nu e \to \nu e,$$

and

$$\frac{d\sigma}{dy} = \frac{G^2 s}{\pi}(1-y)^2 \quad \text{for } \bar{\nu} e \to \bar{\nu} e,$$

where $y = E_2/\omega$. These differential cross sections are plotted in fig. 17.5. We easily notice that cross sections for collisions of ν and $\bar{\nu}$ with quarks and antiquarks differ only in the trivial factors $\cos^2 \theta_C$ and $\sin^2 \theta_C$ (θ_C is the Cabibbo angle) and in the substitution $s \to sx$:

Processes		$\dfrac{d\sigma}{dy} \Big/ \dfrac{G^2 sx}{\pi}$	$\sigma \Big/ \dfrac{G^2 sx}{\pi}$
$\nu_\mu d \to u\mu^-$,	$\bar{\nu}_\mu \bar{d} \to \bar{u}\mu^+$	$\cos^2 \theta_C$	$\cos^2 \theta_C$
$\nu_\mu s \to u\mu^-$,	$\bar{\nu}_\mu \bar{s} \to \bar{u}\mu^+$	$\sin^2 \theta_C$	$\sin^2 \theta_C$
$\nu_\mu \bar{u} \to \bar{d}\mu^-$,	$\bar{\nu}_\mu u \to d\mu^+$	$(1-y)^2 \cos^2 \theta_C$	$\tfrac{1}{3}\cos^2 \theta_C$
$\nu_\mu \bar{u} \to \bar{s}\mu^-$,	$\bar{\nu}_\mu u \to s\mu^+$	$(1-y)^2 \sin^2 \theta_C$	$\tfrac{1}{3}\sin^2 \theta_C$
$\nu_\mu u \to d\mu^+$,	$\bar{\nu}_\mu \bar{u} \to \bar{d}\mu^-$	0	0
$\nu_\mu u \to s\mu^+$,	$\bar{\nu}_\mu \bar{u} \to \bar{s}\mu^-$	0	0
$\nu u \to d\mu^-$,	$\bar{\nu}_\mu \bar{u} \to \bar{d}\mu^+$	0	0
$\nu u \to s\mu^-$,	$\bar{\nu}_\mu \bar{u} \to \bar{s}\mu^+$	0	0

Here $\sigma = \int_0^1 (d\sigma/dy) dy$. Also note that the factors $\tfrac{1}{3}$ appear because $\int_0^1 (1-y)^2 dy = \tfrac{1}{3}$. Selection rules in the last two rows stem from electric charge conservation, and those in the two preceding rows from leptonic charge conservation.

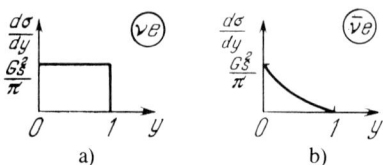

Fig. 17.5.

§6] Parton distributions

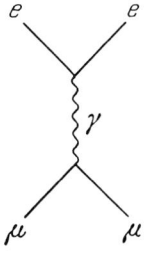

Fig. 17.6.

Now let us consider the electromagnetic cross sections. The electron-muon scattering differential cross section (see fig. 17.6) has the form

$$\frac{d\sigma^\gamma}{dy} = \frac{4\pi\alpha^2}{2s}\frac{s^2 + u^2}{t^2} = \frac{2\pi\alpha^2}{s}\frac{1 + (1-y)^2}{y^2}.$$

The cross section for electron scattering by a u- (or $\bar{u}$-) quark differs from the above expression by a factor of $\frac{4}{9}$, and that by d-, s-, $\bar{d}$-, and $\bar{s}$-quarks by a factor of $\frac{1}{9}$. (Let us trace the relationship between the cross sections of the $e^-\mu^- \to e^-\mu^-$ and $\nu e \to \nu e$ processes. Four helicity configurations, LL, RR, LR, and RL, contribute to the former. Their contributions do not interfere at $v \to c$ and are proportional to s^2 for LL and RR, and to u^2 for LR and RL. The factor t^2 in the denominator appears because of the photon propagator, and s because of the relativistic normalization of the cross section. The coefficient $2\pi\alpha^2$ is easily obtained if we compare the contributions of LL to the cross sections of νe and $e\mu$ scattering. Namely, $(G/\sqrt{2})^2$ is replaced by $(4\pi\alpha)^2/t^2 2^5$, since $\gamma_\mu = \frac{1}{2}\gamma_\mu(1 + \gamma_5) + \frac{1}{2}\gamma_\mu(1 - \gamma_5)$, and because averaging takes place over spin states of the initial electron but not over those of the neutrino. Hence, the contribution of the LL configuration is $4\pi\alpha^2 s/4t^2$, and the total contribution of LL + RR is twice as large).

17.6. Parton distributions

In order to give a formula for the cross section for the deep-inelastic lepton-nucleon interaction, we need the distributions of quarks and antiquarks in nucleons. The number of u-quarks in the range from x to $x + dx$ is usually denoted by $u(x)dx/x$ (and sometimes by $u(x)dx$). In our notation, $xu(x)dx/x = u(x)dx$ is the fraction of the total proton momentum carried by u-quarks with values of x within the dx interval. Similarly,

$d(x)$ and $s(x)$ describe the corresponding quantities for d- and s-quarks, and $\bar{u}(x)$, $\bar{d}(x)$, and $\bar{s}(x)$ for the respective antiquarks (all this for the proton!).

The isotopic symmetry (symmetry with respect to the $T_3 \to -T_3$ operation) states that in the neutron, $u(x)$ refers to d-quarks, $d(x)$ to u-quarks, $s(x)$ to s-quarks, and $\bar{u}(x)$, $\bar{d}(x)$, $\bar{s}(x)$ to $\bar{d}$, $\bar{u}$, and $\bar{s}$-quarks, respectively.

A number of sets of six quark and antiquark distribution functions, $q_i(x)$, can be found in the literature. These sets are basically similar and differ only in minor details. All were obtained by processing the deep-inelastic experimental data by means of parton formulas given in the next section. One of such sets is schematically shown in fig. 17.7.

Denote the total relative momentum carried by all u-quarks in the proton by $U \equiv \int u(x)\,\mathrm{d}x$, and introduce similar symbols for all other quarks and antiquarks. The distributions shown in fig. 17.7 correspond to the following values of the total relative momenta:

$$U \simeq 0.28, \qquad D \simeq 0.15, \qquad \bar{U} \simeq \bar{D} \simeq 0.03, \qquad S \simeq \bar{S} \simeq 0.01.$$

The sum of relative momenta of all quarks and antiquarks are, respectively:

$$Q = U + D + S \equiv \int_0^1 [u(x) + d(x) + s(x)]\,\mathrm{d}x$$

$$\equiv \int_0^1 q(x)\,\mathrm{d}x \simeq 0.44,$$

$$\bar{Q} = \bar{U} + \bar{D} + \bar{S} \equiv \int_0^1 [\bar{u}(x) + \bar{d}(x) + \bar{s}(x)]\,\mathrm{d}x$$

$$\equiv \int_0^1 \bar{q}(x)\,\mathrm{d}x \simeq 0.07,$$

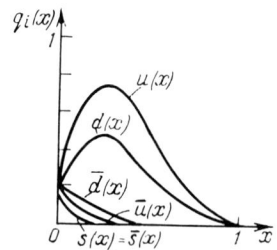

Fig. 17.7.

so that approximately one half of the proton momentum is carried by quarks and antiquarks, and the other half by gluons. Usually the quark-antiquark ratio is characterized by $\alpha = \bar{Q}/(Q + \bar{Q})$ or by $B = (Q - \bar{Q})/(Q + \bar{Q}) = 1 - 2\alpha$. High-energy neutrino experiments yield

$$\alpha \simeq 0.15, \qquad B \simeq 0.70.$$

The same experiments demonstrate that

$$q(x) - \bar{q}(x) \sim \sqrt{x}\,(1 - x)^n, \qquad \text{where } n \cong 3.5,$$

$$\bar{q}(x) + \bar{s}(x) \sim (1 - x)^m, \qquad \text{where } m \cong 6.5.$$

Note that quark distributions must satisfy the following obvious conditions:

$$\int_0^1 [u(x) - \bar{u}(x)] \frac{dx}{x} = 2,$$

$$\int_0^1 [d(x) - \bar{d}(x)] \frac{dx}{x} = 1,$$

$$\int_0^1 [s(x) - \bar{s}(x)] \frac{dx}{x} = 0.$$

The right-hand sides of these equations yield the number of u, d, and s valence quarks in the proton.

17.7. Cross sections of deep-inelastic processes

Let us write down the cross sections $d^2\sigma/dx\,dy$ of some deep-inelastic processes occurring in collisions of leptons with the proton (p), neutron (n), and "mean nucleon" (N) (the last of these refers to a nucleus with an equal number of protons and neutrons):

$$\text{ep} \to \text{eh}: \quad \frac{2\pi\alpha^2}{s} \frac{1 + (1 - y)^2}{x^2 y^2} \Big[\tfrac{4}{9}(u(x) + \bar{u}(x)) + \tfrac{1}{9}(d(x) + \bar{d}(x)) + \tfrac{1}{9}(s(x) + \bar{s}(x)) \Big],$$

$$\text{en} \to \text{eh}: \quad \frac{2\pi\alpha^2}{s} \frac{1 + (1 - y)^2}{x^2 y^2} \Big[\tfrac{4}{9}(d(x) + \bar{d}(x)) + \tfrac{1}{9}(u(x) + \bar{u}(x)) + \tfrac{1}{9}(s(x) + \bar{s}(x)) \Big],$$

$$\nu p \to \mu^- h: \quad \frac{G^2 s}{\pi}\left[d(x) + s(x) + \bar{u}(x)(1-y)^2\right],$$

$$\nu p \to \mu^+ h \quad \frac{G^2 s}{\pi}\left[u(x)(1-y)^2 + \bar{d}(x) + \bar{s}(x)\right],$$

$$\nu n \to \mu^- h: \quad \frac{G^2 s}{\pi}\left[u(x) + s(x) + \bar{d}(x)(1-y)^2\right],$$

$$\bar{\nu} n \to \mu^+ h: \quad \frac{G^2 s}{\pi}\left[d(x)(1-y)^2 + \bar{u}(x) + \bar{s}(s)\right],$$

$$\nu N \to \mu^- h: \quad \frac{G^2 s}{2\pi}\left[u(x) + d(x) + 2s(x)\right.$$
$$\left. + \left(\bar{u}(x) + \bar{d}(x)\right)(1-y)^2\right],$$

$$\bar{\nu} N \to \mu^+ h: \quad \frac{G^2 s}{2\pi}\left[(u(x) + d(x))(1-y)^2\right.$$
$$\left. + \bar{u}(x) + \bar{d}(x) + 2\bar{s}(x)\right].$$

These expressions hold at energies substantially above the production threshold of charmed particles. It is for this reason that they do not contain $\cos^2\theta_C$ or $\sin^2\theta_C$. By integrating over dx and dy we obtain the total cross sections of the $\bar{\nu}$ and ν interaction with the "mean nucleon":

$$\sigma(\bar{\nu} N \to \mu^+ h) \cong \frac{G^2 s}{2\pi}\left[\tfrac{1}{3}(Q+S) + (\bar{Q}-\bar{S})\right] \cong \frac{G^2 s}{\pi} 0.10,$$

$$\sigma(\nu N \to \mu^- h) \cong \frac{G^2 s}{2\pi}\left[(Q+S) + \tfrac{1}{3}(\bar{Q}-\bar{S})\right] \cong \frac{G^2 s}{\pi} 0.22.$$

The ratio of these cross sections is close to 0.5. Note that it would be nearly $\tfrac{1}{3}$ if nucleons contained no antiquarks.

17.8. Production of strange and charmed particles

This section is devoted to the production of a hadronic "spray" containing one strange or one charmed particle. Denote by h_s ($h_{\bar{s}}$) a final hadronic state containing a single strange particle with strangeness $S = -1$ ($S = +1$) and an arbitrary number of ordinary hadrons (π-mesons and nucleons). Let us consider the cross sections $d^2\sigma/dx\,dy$ of the following processes:

$$\nu p \to \mu^- h_{\bar{s}}: \quad \frac{G^2 s}{\pi} \bar{u}(x)(1-y)^2 \sin^2\theta_C,$$

$$\nu n \to \mu^- h_{\bar{s}}: \quad \frac{G^2 s}{\pi} \bar{d}(x)(1-y)^2 \sin^2\theta_C,$$

$$\bar{\nu} p \to \mu^+ h_s: \quad \frac{G^2 s}{\pi} u(x)\sin^2\theta_C,$$

$$\bar{\nu} n \to \mu^+ h_s: \quad \frac{G^2 s}{\pi} d(x)\sin^2\theta_C.$$

We notice that in $\bar{\nu}$-beams, strange particles are produced on u- and d-quarks, and in ν-beams on $\bar{u}$- and $\bar{d}$-quarks. The momentum carried by quarks is much larger than that of antiquarks, so that production of strange particles must be much more abundant in antineutrino beams. This conclusion is born out by experiment.

We shall discuss now the production of charmed particles. Denote the final hadronic state, containing a single charmed particle with $C = +1$ ($C = -1$), by h_c ($h_{\bar{c}}$). We have for $d^2\sigma/dx\,dy$:

$$\nu p \to \mu^- h_c: \quad \frac{G^2 s}{\pi}\left[d(x)\sin^2\theta_C + s(x)\cos^2\theta_C\right],$$

$$\nu n \to \mu^- h_c: \quad \frac{G^2 s}{\pi}\left[u(x)\sin^2\theta_C + s(x)\cos^2\theta_C\right],$$

$$\bar{\nu} p \to \mu^+ h_{\bar{c}}: \quad \frac{G^2 s}{\pi}\left[\bar{d}(x)\sin^2\theta_C + \bar{s}(x)\cos^2\theta_C\right](1-y)^2,$$

$$\bar{\nu} n \to \mu^+ h_{\bar{c}}: \quad \frac{G^2 s}{\pi}\left[\bar{u}(x)\sin^2\theta_C + \bar{s}(x)\cos^2\theta_C\right](1-y)^2.$$

The formulas are valid only at energies so high that the c-quark mass can be neglected, that is when $x \gg m_c^2/s$. At smaller x, the threshold supression of the cross section takes place. The corresponding formulas are easily derived by using the expression for the cross section of the $\nu_\mu e \to \mu^- \nu_e$ and $\bar{\nu}_e e \to \mu^+ \nu_\mu$ processes (see chapter 16).

We see that a neutrino beam is a better generator of charmed particles than a beam of antineutrinos. Production of charmed particles by a neutrino beam on the "mean nucleon" must come to about 10% of the total cross section, that is to $\simeq \sin^2\theta_C + 2S/(U+D)$. It has been mentioned in chapter 14 that the production and subsequent decay of the charmed quark results in so-called dileptonic events. In most cases these events involve a strange particle since the c-quark decay mostly produces the s-quark.

17.9. Phenomenology of deep-inelastic processes

Consider now the general form of the cross section of the deep-inelastic neutrino-nucleon interaction without resorting to the parton model. We shall use only the argument that the weak interaction is due to the product of two left-handed currents, leptonic and hadronic. In calculating the squared modulus of the amplitude, we deal with the product of the leptonic and hadronic tensors. The leptonic tensor is written in the form

$$L^{\mu\nu} = \tfrac{1}{8}\mathrm{Tr}\gamma^{\mu}(1 \pm \gamma_5)\hat{k}\gamma^{\nu}(1 \pm \gamma_5)\hat{k}'$$
$$= k^{\mu}k'^{\nu} + k^{\nu}k'^{\mu} - g^{\mu\nu}kk' \pm i\varepsilon^{\mu\nu\alpha\beta}k_{\alpha}k'_{\beta}.$$

Here the upper sign $(+)$ refers to the neutrino and the lower sign $(-)$ to the antineutrino (see the momentum notation in fig. 17.1). If we neglect the muon mass, then those terms of the hadronic tensor $H_{\mu\nu}$ which are proportional to $q_{\mu} = (k - k')_{\mu}$ and q_{ν} give zero after multiplication by $L^{\mu\nu}$. Only three terms give non-zero contributions, and the cross section can be written as

$$\frac{d^2\sigma}{m\,dQ^2\,d\nu} = \frac{G^2}{\pi(2mE)^2} L^{\mu\nu} H_{\mu\nu}$$

$$= \frac{G^2}{\pi(2mE)^2} L^{\mu\nu}\left[-W_1 g_{\mu\nu} + W_2 \frac{p_\mu p_\nu}{m^2} - \frac{i}{2m^2} W_3 \varepsilon_{\mu\nu\rho\sigma} p^\rho q^\sigma \right],$$

where p is the nucleon 4-momentum, m is its mass, E is the neutrino energy, $2mE = s - m^2$, and W_1, W_2, W_3 are dimensionless functions of $Q^2 = -q^2$ and $\nu = qp/m$. Multiplication of the tensors yields

$$\frac{d^2\sigma}{m\,dQ^2\,d\nu} = \frac{G^2}{(2mE)^2}\left[W_1 Q^2 + W_2 \frac{2(pk)(pk') - m^2(kk')}{m^2} \right.$$
$$\left. \mp W_3 \frac{(pk)(qk') - (pk')(qk)}{m^2} \right]$$

$$= \frac{G^2}{2\pi m^2}\frac{E'}{E}\left[2W_1 \sin^2\tfrac{1}{2}\theta + W_2 \cos^2\tfrac{1}{2}\theta \right.$$
$$\left. \mp W_3 \frac{E + E'}{m} \sin^2\tfrac{1}{2}\theta \right]$$

$$\simeq \frac{G^2}{\pi s^2}\left[W_1 Q^2 + W_2 \frac{s(s - 2m\nu)}{2m^2} \mp W_3 \frac{Q^2(s - \nu m)}{2m^2} \right].$$

(We have made use of the fact that $Q^2 = 4EE' \sin^2 \tfrac{1}{2}\theta$, $s \cong 2mE$, $qk' = -qk = +kk' = -\tfrac{1}{2}q^2 = +\tfrac{1}{2}Q^2$.) For high Q^2 and ν, that is in the range of scaling invariance (the so-called Bjorken scaling), the quantities W_1, $W_2\nu/m$, $W_3\nu/m$ are functions of a single dimensionless variable $x = Q^2/2m\nu$:

$$W_1 = F_1(x),$$

$$\frac{\nu}{m} W_2 = F_2(x),$$

$$\frac{\nu}{m} W_3 = F_3(x).$$

We then easily obtain that in this case

$$\frac{d^2 \sigma^{\nu,\bar\nu}}{dx\,dy} = \frac{G^2 s}{2\pi} \left[F_1(x)xy^2 + F_2(x)(1-y) \mp F_3(x)xy(1-\tfrac{1}{2}y) \right].$$

This expression coincides with the parton cross sections for the "mean nucleon" if

$$2xF_1(x) = F_2(x) = q(x) + \bar{q}(x),$$
$$-xF_3(x) = q(x) - \bar{q}(x).$$

(We have neglected the contribution of the s-quarks.) The first of these formulas is referred to as the Callan-Gross relation. It is a corollary of the quark spin being equal to $\tfrac{1}{2}$.

17.10. The parton model and quantum chromodynamics

According to quantum chromodynamics, scaling is not absolute even for $Q^2 \to \infty$, $\nu \to \infty$, and $F_i(x)$ must be a logarithmic function of Q^2. The necessity of this violation of scaling is easily understood if we take into account that quarks are not completely free and may emit bremsstrahlung gluons. Indeed, the strong-interaction constant α_s falls off only logarithmically as Q^2 increases. This violation of scaling results in the total momentum of quarks being a slowly decreasing function, and that of gluons and antiquarks a slowly increasing function of Q^2.

CHAPTER 18

Renormalizability

The preceding chapters were devoted to weak processes at low energies ($Gs \ll 1$), and the calculations were limited to first order perturbation theory in the Fermi constant G. Now we start considering weak processes at high energies ($Gs \gtrsim 1$). The structure of the weak interaction was unimportant at low energies, but at high energies (at short distances) the situation is different and the intermediate vector bosons $W^\pm$ and Z^0 mentioned at the beginning of the book now play a crucial role. We shall present the standard model of the unified electromagnetic and weak interaction (Chapter 21) based on intermediate vector bosons (Chapter 23) and on scalar (so-called Higgs) bosons (Chapter 24). This standard model gives a very good description of all data on neutral currents (Chapter 22).

Before considering the standard model, however, we shall discuss separately its main components. The present chapter deals with the requirement of renormalizability, which states that all divergences in the theory must be removable by renormalizing several physical quantities (charges and masses). This in turn requires that the cross sections of physical processes (calculated in the framework of perturbation theory) decrease sufficiently rapidly as the momentum transfer increases. This decrease is provided by the intermediate bosons $W^\pm$ and Z^0. To achieve this, the bosons must represent gauge fields of a non-abelian group (Chapter 19) and acquire masses through spontaneous symmetry breaking (Chapter 20).

18.1. Why do we need renormalizability?

One could say, with a measure of snobbishness, that the requirement of renormalizability of a theory is imposed by the "handycraft" approach. The purpose of renormalizability is to give meaning to perturbation theory calculations. But why should nature worry about weak processes being adequately represented by perturbation theory? After all, perturbation theory is not more than a "poor man's" method of calculation.

At least two arguments exist which support both perturbation theory and renormalizability. The first is purely empirical: were perturbation theory invalid, one would see no basis for explaining why the vector constant in neutron β-decay is equal to the muon decay constant times $\cos\theta_C$. This equality would be violated by a large contribution due to virtual particles.

This line of reasoning can be countered, of course, by declaring this equality as absolutely accidental; but does this equality not form the foundation of the very idea of Cabibbo universality? The small width of the $K \to \mu^+ \mu^-$ decay, and the small difference, Δm_{LS}, of the K_L and K_S meson masses are two additional empirical manifestations of the smallness of the higher orders of perturbation theory. However, both the $K_L \to 2\mu$ decay rate and Δm_{LS} are almost independent of the physics at small distances. This became clear after the discovery of the fourth quark and after the confirmation of the Glashow-Illiopoulos-Maiani scheme. The fact is, according to this scheme, that the smallness of these effects is due to the compensation of contributions of the u- and c-quarks.

Another argument in favor of perturbation theory is of an aesthetic nature: in spite of its "non-aristocratic", "earthbound" origin (and possibly owing to it), the renormalizable theory of weak interactions is very much alive, beautiful, and enables us to obtain a number of predictions.

Renormalizability of quantum electrodynamics is known to give high accuracy in calculations of electron-photon interactions, by "burying" all ultraviolet divergences in the renormalized mass and charge of electron. This cannot be done in the case of the four-fermion interaction, since the number of ultraviolet-divergent amplitudes is infinite. This results from the m^{-2} dimension of the constant G. Therefore, a factor G is added in the next order of perturbation theory, together with the energy squared of virtual particles, E^2, which makes the product dimensionless, and has to be cut off "by hand" at a threshold Λ^2.

18.2. Unitarity limit

The divergence of integrals over the energy of virtual particles is closely related to the use of cross sections yielded by perturbation theory in the case of four-fermion interactions (for example, νe scattering and $e^+ e^- \to \nu \bar{\nu}$ annihilation). The fact that these processes proceed in a channel with fixed angular momentum results in violation of the unitarity condition at a certain value of energy which is called the unitarity limit.

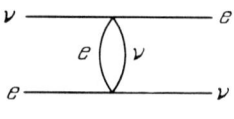

Fig. 18.1.

For example, the unitarity condition for the νe scattering amplitude with $J = 0$, f_0, is: $|\operatorname{Re} f_0|^2 \leq \operatorname{Im} f_0 - |\operatorname{Im} f_0|^2$. Its right-hand side is at a maximum if $\operatorname{Im} f_0 = \frac{1}{2}$, which gives $|\operatorname{Re} f_0| \leq \frac{1}{2}$. In the lowest order of perturbation theory $f_0 = Gs/2\sqrt{2}\,\pi$ (only charged currents are taken into account here for the sake of simplicity) and therefore $Gs < \sqrt{2}\pi$. The energy $\sqrt{s} = \sqrt{\sqrt{2}\,\pi/G} \simeq 600$ GeV is the unitarity limit energy for "standard" νe scattering.

The unitarity limit energy for $\bar{\nu} e$ scattering is of approximately the same order of magnitude. At energies above the unitarity limit, higher angular momenta must become important, owing to the second (fig. 18.1) and higher orders in G of perturbation theory. In the general case, the contribution of higher orders to the low-energy amplitude will not be small. Perturbation theory will be effective if the weak cross sections stop increasing long before the unitarity limit is reached.

18.3. Intermediate bosons

The above situation is realized if the weak interaction is mediated by the intermediate W-boson (fig. 18.2). In this case the differential cross section takes the form

$$\frac{d\sigma}{d|t|} = \frac{G^2}{\pi} \frac{m_W^4}{\left(m_W^2 + |t|\right)^2},$$

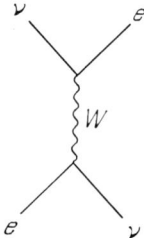

Fig. 18.2.

§3] *Intermediate bosons* 159

where the denominator corresponds to the propagator of the W-boson, and the numerator can easily be written if we demand that the "four-fermion expression"

$$\frac{d\sigma}{d|t|} = \frac{G^2}{\pi},$$

which gives $\sigma = G^2 s/\pi$ when integrated over $|t|$ from 0 to s, be valid at $t \to 0$. As for the total cross section corresponding to the graph of fig. 18.2, it is

$$\sigma = \int_0^s \frac{d\sigma}{d|t|} d|t| = \frac{G^2 m_W^2}{\pi}.$$

We see that introduction of the intermediate boson with mass m_W stops the growth of the weak interaction at energies of the order of m_W.

The coupling constant of the W-boson with the weak current being dimensionless, like electric charge, it would seem logical for the theory operating with W-bosons to be renormalizable, like quantum electrodynamics.

This is not the case, however, owing to several serious differences between the W-boson and the photon: the W-boson is massive, it is charged, and (the point of paramount importance) it is coupled to a non-conserved current. The axial current of fermions is not conserved because fermion masses are distinct from zero; and the vector current is not conserved because of the non-zero mass difference between current-producing particles. Additionally, beyond the lowest order of perturbation theory, the weak fermion current would not be conserved even if the fermion masses were zero. To demonstrate that this is so, it is sufficient to compare a weak vertex $\bar{\nu}eW^-$ with an electromagnetic vertex $\bar{e}e\gamma$ (fig. 18.3).

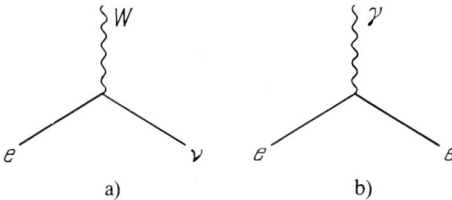

Fig. 18.3.

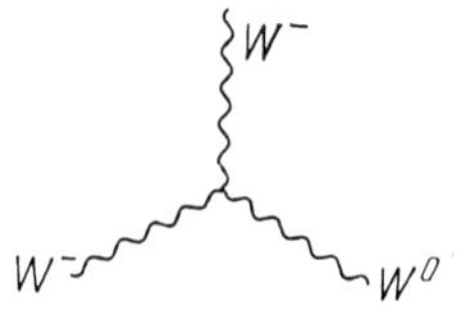

Fig. 18.4.

After emission of a photon, an electron conserves its electric charge and can emit new photons with the same efficiency. As for the weak charge, it is not transferred from the electron to the neutrino, and the neutrino cannot emit another W^- meson since this is forbidden by conservation of the electric (vertex $\bar{e}\nu W^-$) or leptonic charge ($\bar{e}^+ \nu W^-$).

This means that emission of a W^- boson in the graph of fig. 18.3a is accompanied by the sudden disappearance of the weak charge responsible for the emission of the W^- boson. In order to avoid this, we have to assume that the weak charge is transferred to the W^- boson (see fig. 18.3a) which now can emit a W^- boson via the process shown in fig. 18.4. But this requires the existence, together with charged $W^\pm$ bosons, of neutral W^0 bosons as well. (In one special class of theories the role of the W^0 boson is played by the photon; such theories are not discussed in this book). Before we resume the discussion of neutral bosons, let us analyse in more detail what is wrong with a non-conserved current.

18.4. Wave function of the vector boson

Let us consider the wave function of a vector particle represented by a four-dimensional vector A_α. This vector must satisfy the condition $k_\alpha A^\alpha = 0$ where k is the momentum of the vector boson. In the boson rest frame this condition has a very simple meaning. In this frame $k_0 A_0 = mA_0 = 0$ and hence $A_0 = 0$, that is the wave function of the vector particle is represented by a three-dimensional vector $\mathbf{A}$ and thus has three independent spatial components. From the normalization condition, namely $A^2 = 1$, we obtain $\mathbf{A}^2 = -1$. Let us now find what form the vector $\mathbf{A}$ will take if the boson moves, say, along the z-axis. Its 4-momentum then is

$$k = (k_x, k_y, k_z, E) = (0, 0, |\mathbf{k}|, E),$$

where

$$E = \sqrt{m^2 + k^2}.$$

Obviously, a 4-vector A_α satisfying the conditions $A^2 = -1$ and $Ak = 0$, can be represented as a sum of a longitudinal and a transverse term

$$A_\alpha = L_\alpha \cos\theta + T_\alpha \sin\theta,$$

where θ is the angle between the vector A in the particle rest frame and the z-axis,

$$L_\alpha = \frac{1}{m}(0, 0, E; |\mathbf{k}|),$$

$$T_\alpha = (t_x, t_y, 0, 0),$$

$$t_x^2 + t_y^2 = 1.$$

The longitudinal vector L_α, with both of its components increasing with increasing energy, behaves dangerously at high energies. If this growth of longitudinal components is not suppressed, it will result in the growth of matrix elements and, finally, in non-renormalizability of the theory. To guard against this, we need conservation of the current responsible for the boson emission. Conservation of the current, $\partial_\alpha j^\alpha(x) = 0$, means that the boson emission vertex, $L_\alpha \Gamma^\alpha$, is transverse: $k_\alpha \Gamma^\alpha = 0$. Let us write L_α in the following form:

$$L_\alpha = \frac{k_\alpha}{m} + \delta_\alpha,$$

where

$$\delta_\alpha = L_\alpha - \frac{k_\alpha}{m} = \frac{E - |\mathbf{k}|}{m}(0, 0, 1, -1) = \frac{m}{E + |\mathbf{k}|}(0, 0, 1, -1).$$

If the boson emission vertex is transverse, then for $m \to 0$ the transverse component is eliminated:

$$\Gamma^\alpha A_\alpha = \Gamma^\alpha T_\alpha \sin\theta + \Gamma^\alpha \delta_\alpha \cos\theta,$$

and the amplitude has no terms increasing with energy as E/m.

18.5. Digression on the photon mass

It will be instructive in this connection to discuss the problem of the photon mass. If the photon possessed a mass m_γ, then by virtue of the conservation of the electromagnetic current the amplitudes of all physical processes would contain no terms of the ω/m_γ type (where ω is the photon frequency) which tend to infinity for $m_\gamma \to 0$, and would only have correction terms of the m_γ/ω type tending to zero for $m_\gamma \to 0$. With the current conserved, there is therefore a continuous transition to the zero mass limit. The non-zero photon mass would be revealed in a number of phenomena. For example, with $m_\gamma \neq 0$, the field of a magnetic dipole m would have an additional Yukawa-type exponential factor:

$$A = \frac{\boldsymbol{m} \cdot \boldsymbol{r}}{r^3}(1 + m_\gamma r)e^{-m_\gamma r}.$$

Schrödinger has concluded that, with the field of the earth extending over 10^4 km, the Compton wave length of the photon

$$\lambda_\gamma = \frac{1}{m_\gamma} \gtrsim 10^4 \text{ km}.$$

Launchings of satellites have raised this bound to 30 000 km, and space probe measurements of Jupiter's magnetic field increase this estimate by another order of magnitude.

The non-zero photon mass would also change the Coulomb potential:

$$V(r) = \frac{Qe^{-m_\gamma r}}{r}.$$

As a result, the field inside a charged sphere would not be zero. This field was searched for by Plimpton and Lowton who measured the potential difference in two concentric spheres with radii $R_1 = 75$ cm and $R_2 = 60$ cm. At $V_1 = 3000$ V they obtained $V_1 - V_2 \lesssim 10^{-6}$ V. It is easy to show that

$$\frac{V_1 - V_2}{V_1} \simeq \tfrac{1}{6} m_\gamma^2 (R_1^2 - R_2^2).$$

The lower bound found from this estimate is lower than those given above, namely $\lambda_\gamma \gtrsim 10$ km.

The best restriction on the photon mass ($\lambda_\gamma \gtrsim 10^{22}$ cm) is derived from observations of galactic magnetic fields.

But let us return to intermediate vector bosons.

18.6. Vector boson propagator

The longitudinal part of the vector boson wave function is an increasing function of energy, and corresponds to a term in the propagator of the virtual vector boson which also increases with increasing energy. The propagator of a vector particle is obtained from the Proca equation,

$$\partial^\mu F_{\mu\nu} + m^2 A_\nu = j_\nu,$$

where
$$F_{\mu\nu} = \partial_\mu A_\nu - \partial_\nu A_\mu.$$
By taking the derivative of the left-hand and right-hand sides, we obtain
$$\partial^\nu \partial^\mu F_{\mu\nu} + m^2 \partial^\nu A_\nu = \partial^\nu j_\nu.$$
As $F_{\mu\nu}$ is antisymmetric, this equation yields
$$\partial^\nu A_\nu = \frac{\partial^\nu j_\nu}{m^2}.$$
Let us rewrite the initial equation, $\partial^\mu \partial_\mu A_\nu - \partial_\nu \partial^\mu A_\mu + m^2 A_\nu = j_\nu$, in the form
$$\left[(\partial_\mu)^2 + m^2\right] A_\nu = j_\nu + \frac{\partial_\nu \partial_\mu j^\mu}{m^2} = \left(g_{\mu\nu} + \frac{\partial_\nu \partial_\mu}{m^2}\right) j^\mu,$$
which yields the propagator in the form
$$D_{\mu\nu} = -\frac{g_{\mu\nu} - k_\mu k_\nu/m^2}{k^2 - m^2}.$$
Pay attention to the common minus sign in the vector particle propagator. It is this sign that leads to repulsion of like charges.

Unless the contribution of the term $k_\mu k_\nu/m^2$ vanishes owing to the transversality of the vertices, it immediately results in non-renormalizability of a theory containing massive vector bosons. Therefore we have to introduce, along with charged bosons, neutral vector bosons, and envisage conservation of the current coupled to these bosons. A theory which to a certain degree possesses the required properties was proposed by Yang and Mills as long ago as 1954, without any relation to the weak interaction. This theory is treated in the next chapter.

CHAPTER 19

Gauge invariance

This chapter is mostly devoted to a discussion of the Yang-Mills theory which describes an isotopic triplet of massless vector fields coupled to a conserved current. Before presenting the theory, we shall remark on the classification of symmetries and shall also discuss several theories more simple than that of Yang and Mills. All symmetries can be separated into two large groups: global symmetries and local symmetries.

19.1. Global abelian symmetry U(1)

The simplest example of global symmetry is charge conservation, expressed as the invariance of the lagrangian with respect to transformations of the type

$$\psi \to \psi' = e^{i\alpha Q}\psi,$$

$$\bar{\psi} \to \bar{\psi}' = \bar{\psi} e^{-i\alpha Q},$$

where Q is the charge of the particle described by the field ψ, and α is an arbitrary number independent of space-time coordinates of the particle. Not only the electric charge Q, but other charges as well (baryonic charge B, leptonic charge L, etc.) may play the part of the above charge. The group of these phase transformations is called U(1).

Various transformations of the group U(1) commute. Such groups are called abelian. If the parameter α is independent of x_μ, the group is called global. We have thus defined the global abelian symmetry U(1).

19.2. Global non-abelian symmetry SU(2)

Another example of global symmetry is the usual isotopic invariance. Under the isotopic transformation, SU(2),

$$\psi \to \psi' = e^{i\boldsymbol{\alpha}\cdot\boldsymbol{T}}\psi,$$
$$\bar{\psi} \to \bar{\psi}' = \bar{\psi}e^{-i\boldsymbol{\alpha}\cdot\boldsymbol{T}},$$

where $\boldsymbol{T}$ are matrices, and $\boldsymbol{\alpha}$ are, as before, parameters independent of coordinates. The values of $\boldsymbol{\alpha}$ are identical be it in Moscow, New York, or on the moon. This is the reason for referring to the symmetry as global. In the case of the simplest non-trivial isotopic multiplet (i.e. doublet)

$$\psi = \begin{pmatrix} \psi^1 \\ \psi^2 \end{pmatrix}, \qquad \boldsymbol{T} = \tfrac{1}{2}\boldsymbol{\tau},$$

where $\boldsymbol{\tau}$ stands for three Pauli matrices

$$\tau_1 = \begin{pmatrix} 0 & 1 \\ 1 & 0 \end{pmatrix}, \quad \tau_2 = \begin{pmatrix} 0 & -i \\ i & 0 \end{pmatrix}, \quad \tau_3 = \begin{pmatrix} 1 & 0 \\ 0 & -1 \end{pmatrix}.$$

The symmetry SU(2) is called non-abelian since the matrices τ are non-commuting.

19.3. Local abelian symmetry

Let us consider now the lagrangian of quantum electrodynamics which describes electrons, photons, and their interaction:

$$\mathcal{L} = \bar{\psi}(i\hat{\partial} + e\hat{A} - m)\psi - \tfrac{1}{4}F_{\mu\nu}F^{\mu\nu}.$$

This lagrangian is readily shown to be invariant with respect to the transformation

$$\psi \to \psi' = e^{i\alpha(x)}\psi \equiv S\psi, \qquad \bar{\psi} \to \bar{\psi}' = \bar{\psi}e^{-i\alpha(x)} \equiv \bar{\psi}S^+,$$

where $\alpha(x)$ is a parameter depending on the world point x. Under this transformation

$$A_\mu \to A'_\mu = A_\mu + \frac{1}{e}\frac{\partial\alpha(x)}{\partial x^\mu} \equiv A_\mu - \frac{i}{e}\left(\frac{\partial}{\partial x^\mu}S\right)S^+.$$

The field strength,

$$F_{\mu\nu} = \frac{\partial}{\partial x^\mu}A_\nu - \frac{\partial}{\partial x^\nu}A_\mu,$$

is invariant with respect to this transformation. It should be emphasized that in the case of non-zero photon mass the local U(1) symmetry would be broken since the term $m_\gamma^2 A_\mu A^\mu$ does not transform into itself after $\partial_\mu \alpha(x)/e$ has been added to A_μ.

We find that local invariance requires that the conserved charge be a source of a massless vector field. In this respect the photon-generating electric charge is radically different from the baryonic, leptonic, and muonic charges which, as far as we know, generate no specific massless fields, that is, there exist no baryonic, leptonic or muonic "photons". Hence, these charges correspond to global and not to local U(1) symmetries.

The observable transverse components of the photon field, $A_x(k)$ and $A_y(k)$, are not transformed under gauge transformations (we assume here that the photon momentum is directed along the z-axis); only non-physical longitudinal components of the photon 4-vector, $A_z(k)$ and $A_t(k)$, are transformed. Unobservability of these components is a result of conservation of the vector current.

From a purely formal point of view, local invariance can be achieved with a non-physical longitudinal field only, without introducing observable vector fields and without interpreting the longitudinal field as a non-physical component of the vector field. We shall use, however, the terminology generally accepted in the literature, and refer to a theory as a local one only if it involves a massless vector gauge field.

19.4 Digression on baryonic and leptonic photons

We have made the remark above that baryonic photons do not exist. This statement calls for clarification.

It is not difficult to show that if baryonic photons exist, their interaction with baryons must be extremely weak: $\alpha_B \leq 10^{-47}$ (this should be compared to $\alpha = \frac{1}{137}$ for ordinary photons). This restriction stems from the equality of the inertial and gravitational mass verified to an accuracy of 10^{-12} in experiments pioneered by Eötvös. Indeed, baryonic photons would produce a sort of "Coulomb" field around the earth, repulsing baryons away from it. For a given body, the force of this repulsion would be proportional to the number of baryons in this body and not to its mass, and would be different in, for example, a lead and copper sphere with identical masses. The force applied to the ith body with mass M_i containing A_i

nucleons is

$$F = \kappa \frac{M_i M}{r^2} + \alpha_B \frac{A_i A}{r^2} = \kappa \frac{M_i M}{r^2}\left\{1 + \frac{\alpha_B}{\kappa}\left(\frac{A}{M}\right)\left(\frac{A_i}{M_i}\right)\right\},$$

where M is the earth mass, A is the number of nucleons in the earth ($M/A \simeq m_p$), and $\kappa = 6 \cdot 10^{-39} m_p^{-2}$ is the gravitational constant where m_p is the proton mass. The nucleon mass in the lead nucleus is greater than that in the copper nucleus by approximately 1 MeV (($M_{Pb}/A_{Pb} - M_{Cu}/A_{Cu}) = 10^{-3} m_p$).

Experiment gives

$$\frac{\alpha_B}{\kappa m_p}\left[\frac{A_{Cu}}{M_{Cu}} - \frac{A_{Pb}}{M_{Pb}}\right] < 10^{-12},$$

whence

$$\alpha_B < 10^{-9}\kappa m_p^2 \le 10^{-47}.$$

A similar argument is possible for the constant α_L of the interaction between hypothetical leptonic "photons" and electrons. In this case

$$\frac{\alpha_L}{\kappa m_p}\left[\frac{Z_{Cu}}{M_{Cu}} - \frac{Z_{Pb}}{M_{Pb}}\right] < 10^{-12},$$

where Z is the charge of the nucleus; hence,

$$\alpha_L \le 10^{-49}.$$

As for the constant α_μ of interaction between hypothetical muonic "photons" and the muonic charge, the corresponding upper bound is many orders of magnitude higher than for α_B and α_L, owing to the muon instability.

It is natural to conclude on the basis of the above estimates that massless vector particles coupled to the leptonic and baryonic charges do not exist. These charges are possibly coupled to some massive vector particles.

19.5. Local SU(2) symmetry

Let us now return to Yang-Mills theory which is a local realization of isotopic invariance. The theory is invariant with respect to local isotopic rotations:

$$\psi \to \psi' = S\psi,$$
$$\bar\psi \to \bar\psi' = \bar\psi S^+.$$

Here $S = e^{i\alpha(x) \cdot T}$, where T are three matrices of isotopic rotations, and $\alpha(x)$ denotes three parameters of these rotations which in the general case are different in different world points. Realization of such symmetry requires the existence of a triplet of massless vector fields A_μ coupled to fields ψ.

The lagrangian has the form

$$\mathcal{L} = \bar\psi\,(i\hat D - m)\psi - \tfrac{1}{4} \boldsymbol{G}_{\mu\nu} \cdot \boldsymbol{G}^{\mu\nu},$$

where $D_\alpha = \partial_\alpha - igA_\alpha$ is the covariant, or generalized, derivative (or simply the long derivative, in the lingo of some students). In contrast to the abelian case, A_α is a matrix,

$$(A_\alpha)_i^k = \sum_1^3 A_\alpha (T)_i^k.$$

Let us consider a field transforming under a specific representation of the group SU(2). This representation determines the specific form of the matrices T and therefore, of A_α. If the field ψ is an isodoublet, then

$$(T_a)_\alpha^\beta = (\tfrac{1}{2}\tau_a)_\alpha^\beta, \qquad \alpha, \beta = 1, 2,$$

and if the field ψ is an isotriplet, then

$$(T_a)_i^k = \varepsilon_{iak},$$

where $a, i, k = 1, 2, 3$.

The isovector of the field strength has the form

$$G_{\mu\nu} = \boldsymbol{G}_{\mu\nu} \boldsymbol{T} = \partial_\mu A_\nu - \partial_\nu A_\mu - ig \big[A_\mu A_\nu - A_\nu A_\mu \big].$$

(Usually the matrices chosen for matrices T are $T = \tfrac{1}{2}\boldsymbol{\tau}$.) This expression which we shall clarify later, shows that in contrast to the abelian case, the field strength of a non-abelian gauge field is a non-linear function of the field. (The commutator $A_\mu A_\nu - A_\nu A_\mu$ is zero in the abelian case.) As a result, the lagrangian of a free Yang-Mills field contains, along with $(A)^2$ terms, terms $(A)^3$ and $(A)^4$. In contrast to the usual photons, Yang-Mills photons have isovector "charges"; these photons are self-emitting, and thus behave as "radiating light".

Gravitons, that is quanta of the gravitation field, have a similar property. The source of gravitons is known to be the energy-momentum tensor. Therefore even massless particles, such as photons, interact with the gravitational field, the stronger the higher their energy is. The photon-graviton interaction vertex is shown in fig. 19.1a. Since the graviton has energy and momentum, it might also emit gravitons (fig. 19.1b). The usual abelian

§5] Local SU(2) symmetry 169

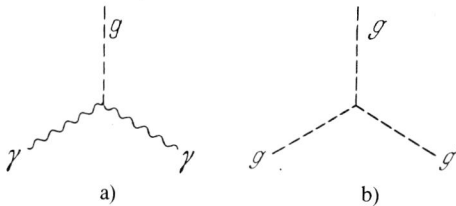

Fig. 19.1.

photon has no such triple vertex, $\gamma\gamma\gamma$, while for non-abelian photon the vertex $A^+ A^- A^0$ is non-vanishing. This vertex represents scattering of A^+ and A^- "photons" on the A^0 field, $A^+ A^- \to A^0$ annihilation, and other processes related to the above-mentioned process through crossing symmetry. Recall that we needed a vertex of just this type in order to provide conservation of the weak current in higher orders of perturbation theory (see chapter 18).

How are A_μ and $F_{\mu\nu}$ transformed? In order to find how the A_μ field is transformed, consider, for instance, the expression $\bar{\psi}\hat{D}\psi = \bar{\psi}(\hat{\partial} - ig\hat{A})\psi$. Let $\psi' = S\psi$, $\bar{\psi}' = \bar{\psi}S^+$. What should be the form of A' for the expression $\bar{\psi}\hat{D}\psi$ to be gauge-invariant, that is for the gauge transformation to give

$$\bar{\psi}\hat{D}\psi \to \bar{\psi}'\hat{D}'\psi' = \bar{\psi}\hat{D}\psi?$$

Obviously the necessary and sufficient condition for this is

$$A_\mu \to A'_\mu = SAS^+ - \frac{i}{g}(\partial_\mu S)S^+.$$

(Recall that in the abelian case

$$A_\mu \to A'_\mu = A_\mu - \frac{i}{g}(\partial_\mu S)S^+.)$$

Thus the non-abelian field is first isotopically rotated and only after that "elongated" by the term $-(i/g)(\partial_\mu S)S^+$. When the gauge matrix $S = \exp(i\alpha(x)T)$ is nearly a unit matrix, that is when the angle of rotation $\alpha(x)$ is small, the explicitly isovector form of field transformation gives a clearer picture than the matrix form used above. In this case $S = 1 + i\alpha(x)T$:

$$A_\mu \to A'_\mu = A_\mu - \alpha \times A_\mu + \frac{1}{g}\frac{\partial \alpha}{\partial x^\mu}.$$

(We have used above that

$$SA_\mu S^+ = (1 + i\boldsymbol{\alpha}\cdot\boldsymbol{T})(A_\mu\cdot\boldsymbol{T})(1 - i\boldsymbol{\alpha}\cdot\boldsymbol{T}) = A_\mu \boldsymbol{T} + i[\boldsymbol{T\alpha}, \boldsymbol{T}A_\mu]$$
$$= (A_\mu)_l T_l + i\alpha_i(A_\mu)_k [T_i T_k]$$
$$= [(A_\mu)_l - \alpha_i(A_\mu)_k \varepsilon_{ikl}]T_l = (A_\mu - \boldsymbol{\alpha}\times A_\mu)\boldsymbol{T},$$

and also that for small $\boldsymbol{\alpha}$: $(\partial S/\partial x^\mu)S^+ = i\boldsymbol{T}\partial\boldsymbol{\alpha}(x)/\partial x^\mu$.)

As for the field strength $G_{\mu\nu}$, it immediately follows from the definition

$$G_{\mu\nu} = \partial_\mu A_\nu - \partial_\nu A_\mu - ig[A_\mu A_\nu - A_\nu A_\mu],$$

that

$$G_{\mu\nu} \to G'_{\mu\nu} = S G_{\mu\nu} S^+ .$$

We see that $G_{\mu\nu}$ is transformed as a usual isovector. In particular, for small $\boldsymbol{\alpha}$ we have: $G_{\mu\nu} \to G'_{\mu\nu} = G_{\mu\nu} - \boldsymbol{\alpha}\times G_{\mu\nu}$. The field strength $G_{\mu\nu}$ is constructed in such a clever manner that it is transformed in the same way by the global and by the local isotopic rotations.

19.6. Panegyric to Yang-Mills theory

We are quite used to symmetries imposing certain restrictions on masses of particles and on the corresponding coupling constants. For example, isotopic invariance of strong interactions requires that the proton and neutron masses be equal, and that their interaction with π-mesons be described by the same coupling constant. A spectacular property of non-abelian gauge symmetry is that in addition to restricting masses and coupling constants of particles, this symmetry determines the dynamics of interactions of gauge fields (non-linear terms of the $(A)^3$ and $(A)^4$ types). Non-abelian gauge fields carry "isotopic charge" which determines their self-interaction as well as their interaction with other fields. The form of these interactions is given unambiguously by gauge symmetry.

In this sense Yang-Mills theory very much resembles general relativity theory in which the dynamics of the gravitational interaction is determined to a great extent by the requirement of invariance with respect to the most general transformations of coordiantes. Hence, the similarity of non-linear interactions of non-abelian photons and gravitons mentioned above, is only one of the specific manifestations of the profound similarity of these two theories. Yang-Mills theory is a sufficiently simple model in which to attempt to understand some consequences of quantization in such an essentially non-linear theory as the general theory of relativity. More and more attention is being paid nowadays to the development of the quantum theory of gravitation.

In the sixties Yang-Mills theory was subjected to a thorough theoretical scrutiny. The rules for constructing Feynman graphs were formulated, and renormalizability was proved. In Yang-Mills theory the renormalizability is a corollary of the constant g being dimensionless, of the conservation of isotopic currents, and of the masslessness of non-abelian photons.

Today Yang-Mills fields, which at the beginning were considered as a mere theoretical curiosity, have become the central object of theoretical studies. In fact, all our hopes of arriving at the theory of elementary particles are pivoted on non-abelian gauge fields. This is valid for the unified theory of weak and electromagnetic interactions, for the gluonic theory of strong interactions, and finally, for the possible future synthesis of these theories. (Recall that the theory of colored gluons coupled to colored quarks possesses the local SU(3) symmetry.)

At this juncture a thoughtful reader should have interrupted the panegyric to put the natural question: how are masses taken into account?

19.7. How to take masses into account?

The first question to appear is how to introduce into the theory the masses of intermediate bosons. Indeed, we know from experiment that these particles must have finite (and quite large!) masses, while Yang-Mills gauge fields are massless. It would seem innocuous at first glance to insert a term $m^2 A^2$ into the lagrangian "by hand". We have established in the discussion of the photon mass that this causes no harm in the case of abelian gauge fields. There is a "soft" transition from $m = 0$ to $m \neq 0$ for the photon, and quantum electrodynamics remains renormalizable.

One easily finds that this is not the case for non-abelian gauge fields: introduction of mass "by hand" destroys renormalizability. Let us consider the amplitude for the emission of n photons:

$$A^{\alpha_1} A^{\alpha_2} \ldots A^{\alpha_n} M_{\alpha_1 \alpha_2 \ldots \alpha_n},$$

where A^i is the wave function of the ith photon.

In the abelian case the matrix element $M_{\alpha_1 \alpha_2 \ldots \alpha_n}$ is transverse with respect to any combination of indices. For instance,

$$k^{\alpha_1} M_{\alpha_1 \alpha_2 \ldots \alpha_n} = 0,$$

$$k^{\alpha_1} k^{\alpha_2} M_{\alpha_1 \alpha_2 \ldots \alpha_n} = 0,$$

and so on.

Moreover, k_i^2 may be zero or non-zero, that is photons may be real as well as virtual. In contrast to this, in the non-abelian case, transversality for any of the "photons" is realized only if all the remaining "photons" are real (are on the mass shell), so that in the amplitude their emission is taken into account explicitly by their wave functions:

$$k^{\alpha_1} M_{\alpha_1 \alpha_2 \alpha_3 \ldots \alpha_n} A^{\alpha_2} A^{\alpha_3} \ldots A^{\alpha_n} = 0.$$

In all other cases transversality is absent; in particular,

$$k^{\alpha_1} k^{\alpha_2} M_{\alpha_1 \alpha_2 \alpha_3 \ldots \alpha_n} A^{\alpha_3} \ldots A^{\alpha_n} \neq 0.$$

This is caused by the isotopic charge of non-abelian photons. By definition, a virtual non-abelian photon transfers this charge from one part of a Feynman graph to another. In essence, by considering a graph in which one of the photons is virtual, we consider only a segment of a physical graph in which the isotopic current is not conserved and therefore transversality is absent. (This is not so for usual (abelian) photons because they are electrically neutral.)

Recall now the expression for the longitudinal part of the wave function of a massive vector particle:

$$L_\alpha = \frac{k_\alpha}{m} + \delta_\alpha, \qquad \delta_\alpha = \frac{m}{E + |k|}(0, 0, 1, -1),$$

where m, E, k are the mass, energy, and momentum of the particle, and k_α is its 4-momentum. It then follows that the amplitude for emission of n longitudinal abelian photons tends to zero as $(m/E)^n$ for $E \to \infty$. In contrast to this, the amplitude for emission of n longitudinal non-abelian photons is proportional to $(m/E)^{-n+2}$, and for $n > 2$ grows to infinity for $E \to \infty$. Indeed, by multiplying the matrix element by "photon" wave functions A^i, we can retain the δ_α term only in one of them, and therefore have to retain the term k_α/m in the rest of the wave functions. Consequently, we shall have at high energies:

$$\frac{m}{E}\left(\frac{E}{m}\right)^{n-1} = \left(\frac{E}{m}\right)^{n-2}.$$

This means that in the non-abelian case we again deal with the non-renormalizable theory.

Although the gauge origin of intermediate bosons made it possible to improve the high-energy behavior of amplitudes by slightly slowing down their growth, the power of energy is reduced only by two. This growth will be eliminated completely if the introduction of a non-zero mass for intermediate bosons is accompanied by the appearance of additional fields whose contribution cancels the divergence in question. Such a "soft" introduction of the intermediate boson mass is a feature of the spontaneous breaking of gauge symmetry to be discussed in the next chapter.

CHAPTER 20

Spontaneous symmetry breaking

The purpose of this chapter is to introduce W-boson masses by means of the Higgs mechanism of spontaneous breaking of local SU(2) symmetry. This method of mass introduction does not upset the renormalizability of the theory. Spontaneous symmetry breaking results from the degeneracy of the vacuum.

We begin with several simpler examples of spontaneous breaking of various symmetries: discrete, global and local U(1), and global SU(2).

20.1. Spontaneous breaking of discrete symmetry

Let us look first at the simplest case: the ordinary, scalar real field with the lagrangian

$$\mathcal{L} = \frac{1}{2}\left(\frac{\partial \varphi}{\partial x_\mu}\right)^2 - \tfrac{1}{2}m^2\varphi^2 - \tfrac{1}{4}\lambda^2\varphi^4$$

and the hamiltonian

$$H = \frac{1}{2}\left(\frac{\partial \varphi}{\partial t}\right)^2 + \frac{1}{2}\left(\frac{\partial \varphi}{\partial x}\right)^2 + \tfrac{1}{2}m^2\varphi^2 + \tfrac{1}{4}\lambda^2\varphi^4,$$

where m is the mass of particles described by the field φ, λ being a dimensionless constant characterizing the interaction between these particles. Consider a field which is constant in time and space. For this field

$$H(\varphi) = V(\varphi) \equiv \tfrac{1}{2}m^2\varphi^2 + \tfrac{1}{4}\lambda^2\varphi^4.$$

The function $V(\varphi)$ is plotted in fig. 20.1. It has a minimum at $\varphi = 0$ which means that there is no field in vacuum (in the minimum-energy state). The expression for the lagrangian and fig. 20.1 show that the lagrangian is symmetric with respect to a discrete transformation $\varphi \to -\varphi$.

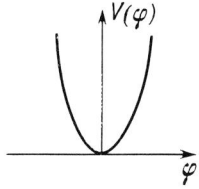

Fig. 20.1.

Now let us consider the same lagrangian but with a different sign in front of m^2. At first glance, we obtain a particle for which $E^2 = p^2 - m^2$ and which therefore moves at a speed exceeding that of light:

$$v = \frac{|p|}{E} = \frac{\sqrt{E^2 + m^2}}{E} > 1.$$

In the literature such particles are called tachyons. In fact no tachyon is implied since the state with $\varphi = 0$ is not the vacuum any more. In order to demonstrate this, let us turn to $V(\varphi)$. Fig. 20.2 gives the potential energy

$$V(\varphi) = -\tfrac{1}{2}m^2\varphi^2 + \tfrac{1}{4}\lambda^2\varphi^4.$$

We see that at $\varphi = 0$ this function has a maximum, not a minimum. Small perturbations of the field φ in the vicinity of $\varphi = 0$ cannot therefore be considered as particles. Here the system is unstable, tending to slip into one of the stable minima, $\varphi = m/\lambda$ or $\varphi = -m/\lambda$. Instead of a single vacuum at $\varphi = 0$, as in the usual case of positive sign in front of $\tfrac{1}{2}m^2$, the system has two degenerate vacua at $\varphi = \pm m/\lambda$. We call them degenerate because they have equal energies.

By adding a constant to $V(\varphi)$ (this does not change the field equations!) we rewrite it in a form in which the degenerate vacua are found at the points where $V(\varphi) = 0$. The new expression for $V(\varphi)$ takes the form (fig. 20.3)

$$V(\varphi) = \tfrac{1}{4}\lambda^2(\varphi^2 - \eta^2)^2,$$

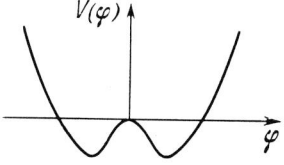

Fig. 20.2.

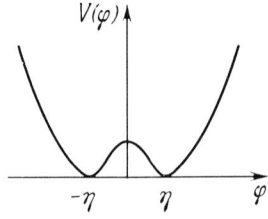

Fig. 20.3.

where $\eta = m/\lambda$. It seems more logical to consider $V(\varphi)$ as shown in fig. 20.3 and not as in fig. 20.2, since in the latter case the energy density of the vacuum is $-m^4/4\lambda^2$. This is a fantastically high negative energy (10^{41} GeV/cm³ for $\lambda \simeq 1$ and $m \simeq 1$ GeV) as compared with the mean energy density in the universe, 10^{-6} GeV/cm³ (one proton per cubic meter, i.e. 10^{-30} g/cm³). It can be concluded from the observed data on the expansion of the universe that the vacuum energy density (the so-called cosmological term in Einstein's equations) is unlikely to be much larger than the observed density of matter (hardly by a factor more than 10). Therefore in what follows we shall use the form $V(\varphi) = \frac{1}{4}\lambda^2(\varphi^2 - \eta^2)^2$.

The choice of a specific vacuum is decided by microscopic perturbations in the first moments of the life of the universe. But after the system has spontaneously "slipped" into one of the vacua, it cannot change to another one. The amplitude of a below-barrier transition from the state $\varphi = +\eta$ to the state $\varphi = -\eta$ is zero. It has the form e^{iS} where S is the action. In the case in question the action is imaginary (the transition is below-barrier, that is classically forbidden) and infinitely large (since the action must be found as an integral over the whole space of the universe), so that we obtain $e^{-\infty}$.

Could it be possible to realize the transition to another vacuum in only a part of the universe? For example, could we create the new vacuum in a volume of about 1 m³, in laboratory conditions? Unfortunately, it is very difficult to generate even a very small bubble of the new vacuum, and even a small bubble is unstable. It must collapse on a nuclear time scale, emitting mesons. This would happen because the interface between the two vacua must be a wall with very high surface density σ of the order of $\lambda\eta^3$ and with thickness δ of the order of $1/\lambda\eta$. For $\lambda \simeq 1$ and $\eta = 1$ GeV, we obtain $\delta \simeq 10^{-14}$ cm, and $\sigma \simeq 1$ kg/cm². These estimates are easily derived if we minimize the wall energy density which is equal to the sum of two terms:

$$\sigma \simeq \left\{ \frac{1}{2}\left(\frac{\partial \varphi}{\partial x}\right)^2 + V(\varphi) \right\} \delta.$$

By substituting into it the estimates

$$\left(\frac{\partial \varphi}{\partial x}\right)^2 \simeq \left(\frac{\eta}{\delta}\right)^2 \quad \text{and} \quad V(\varphi) \simeq \lambda^2 \eta^4,$$

we find that the minimum of δ is reached at $\delta \simeq 1/\lambda\eta$, so that $\sigma \simeq \lambda\eta^3$.

In the case $V(\varphi) = \tfrac{1}{4}\lambda^2(\varphi^2 - \eta^2)^2$ we therefore encounter a lagrangian with mirror symmetry (with respect to the $\varphi \to -\varphi$ transformation) and a vacuum (let it be $\varphi = \eta$) which has no such symmetry. This is a typical example of the so-called spontaneous symmetry breaking.

Abdus Salam once gave an illustration of spontaneous breaking of a discrete symmetry. Imagine a banquet with guests seated at a large round table. A person could equally well take a serviette from his right or from his left. But this symmetry is spontaneously broken once one of the guests decides to pick up the serviette, say, on his left; other guests are left no choice. Obviously, the symmetric state is unstable, especially with hungry guests.

If we write the field φ in the form $\varphi = \eta + \chi$, then χ will describe excitations of the field (particles) with respect to the vacuum $\varphi = \eta$. In the new variables the lagrangian does not possess the mirror symmetry:

$$\mathcal{L} = \frac{1}{2}\left(\frac{\partial \varphi}{\partial x_\mu}\right)^2 - \tfrac{1}{4}\lambda^2(\varphi^2 - \eta^2)^2$$

$$= \frac{1}{2}\left(\frac{\partial \chi}{\partial x_\mu}\right)^2 - \lambda^2\eta^2\chi^2 - \lambda^2\eta\chi^3 - \tfrac{1}{4}\lambda^2\chi^4.$$

Note that the field χ is not tachyonic: its mass μ is equal to $\sqrt{2}\,\lambda\eta$, and the mass term has the usual sign (minus in the lagrangian and plus in the hamiltonian).

20.2. Spontaneous breaking of global U(1) symmetry

Spontaneous breaking of a continuous (not discrete) global symmetry produces massless particles, the so-called Goldstone bosons. Let us analyze this effect by considering the lagrangian

$$\mathcal{L} = \left|\frac{\partial \varphi}{\partial x_\mu}\right|^2 - V(|\varphi|),$$

where φ is a complex scalar field

$$V(|\varphi|) = \tfrac{1}{2}\lambda^2\left(|\varphi|^2 - \tfrac{1}{2}\eta^2\right)^2.$$

(The coefficient $\tfrac{1}{2}$ of η^2 is introduced to reduce the number of factors $\sqrt{2}$ in further formulas). This lagrangian is invariant with respect to the global U(1) gauge transformation

$$\varphi \to \varphi' = \varphi e^{i\alpha}.$$

The potential $V(|\varphi|)$ is plotted in fig. 20.4. In this situation we deal with an infinite number of degenerate vacua. The shape of the $V(\varphi)$ potential surface in fig. 20.4 resembles that of a bottom of a beer bottle. All the vacua satisfy the condition $\varphi = \eta/\sqrt{2}$, and lie in the trough. We choose the gauge phase value $\alpha = 0$ common for the whole world, and present φ in the form:

$$\varphi(x) = (\eta + \chi(x) + i\psi(x))/\sqrt{2}.$$

Here $\chi(x)$ and $\psi(x)$ are two real fields describing the excitations of the system with respect to the stable vacuum $\varphi = \eta/\sqrt{2}$. The choice of a stable vacuum breaks the U(1) invariance by fixing the phase of the function φ (fig. 20.5). In the new variables our lagrangian has the form

$$\mathcal{L} = \frac{1}{2}\left(\frac{\partial \chi}{\partial x_\mu}\right)^2 + \frac{1}{2}\left(\frac{\partial \psi}{\partial x_\mu}\right)^2 - \tfrac{1}{8}\lambda^2(\chi^2 + 2\eta\chi + \psi^2)^2.$$

This lagrangian contains a massive scalar field χ (its mass is $\mu = \lambda\eta$) and a massless scalar field ψ (massless because of the absence of a term proportional to ψ^2). As we see from fig. 20.5, the field χ describes small radial oscillations of the system with respect to a point $\varphi = \eta/\sqrt{2}$, and the field ψ describes tangential oscillations (along the vacuum trough). The field ψ is

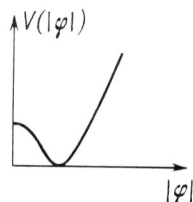

Fig. 20.4.

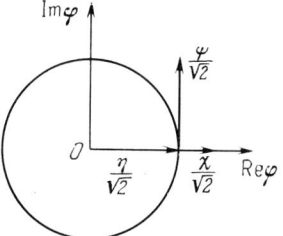

Fig. 20.5.

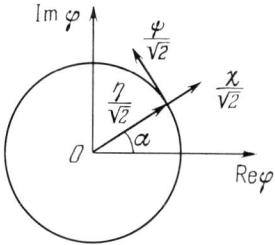

Fig. 20.6.

called the Goldstone field, and particles corresponding to this field are Goldstone bosons. Different values of the gauge phase α define different vacuae and different physical fields (see fig. 20.6).

20.3. Spontaneous breaking of global SU(2) symmetry

The lagrangian in this case is

$$\mathcal{L} = \left(\frac{\partial \bar\varphi}{\partial x_\mu}\right)\left(\frac{\partial \varphi}{\partial x_\mu}\right) - V(\bar\varphi\varphi).$$

The field φ is now an isotopic doublet similar to the K-meson doublet:

$$\varphi = \begin{pmatrix} \varphi_+ \\ \varphi_0 \end{pmatrix},$$

and $\bar\varphi\varphi$ is an isotopic scalar: $\bar\varphi\varphi = \varphi_+^* \varphi_+ + \varphi_0^* \varphi_0$. In fact we deal with four real scalar fields since $\varphi_0 \neq \tilde\varphi_0$. Spontaneous breaking of the isotopic symmetry produces one massive field and three Goldstone massless fields, all of them scalar and real.

We are now very close to what will be necessary to construct a renormalizable model of the weak interaction, and the spontaneous breaking of local symmetry.

20.4. Spontaneous breaking of abelian gauge symmetry

We shall write the lagrangian with local abelian symmetry and degenerate vacuum in the form:

$$\mathcal{L} = |D_\mu \varphi|^2 - \tfrac{1}{2}\lambda^2 \left(|\varphi|^2 - \tfrac{1}{2}\eta^2\right)^2 - \tfrac{1}{4}F_{\mu\nu}F^{\mu\nu},$$

where φ is a complex scalar field, $D_\mu = \partial_\mu - ieA_\mu$, $F_{\mu\nu} = \partial_\mu A_\nu - \partial_\nu A_\mu$, and A_μ is the 4-vector potential of the photon field. We shall write $\varphi(x) = \phi(x)\exp(i\theta(x))$ and choose $\theta(x)$ in such a manner that $\phi(x)$ is real at all world points. Now we make use of the gauge transformation to eliminate the phase $\theta(x)$, and introduce a field $\chi(x)$ which describes excitations in the vicinity of the stable vacuum $\eta/\sqrt{2}$: $\phi(x) = (\eta + \chi(x))/\sqrt{2}$.

After spontaneous symmetry breaking the lagrangian changes to

$$\mathcal{L} = \frac{1}{2}\left(\frac{\partial \chi}{\partial x_\mu}\right)^2 - \tfrac{1}{8}\lambda^2 \chi^2 (2\eta + \chi)^2 + e^2 A_\mu^2 (\eta + \chi)^2 - \tfrac{1}{4}F_{\mu\nu}F^{\mu\nu},$$

and describes a real scalar field χ with mass $\mu = \lambda\eta$ and a massive vector field A_μ with mass $m = e\eta$. All non-linear interactions of these fields $(\chi^3, \chi^4, A^2\chi, A^2\chi^2)$ are renormalizable since the dimensions of the corresponding coupling constants are m^1 and m^0, respectively.

We observe that spontaneous symmetry breaking resulted in a redistribution of fields: one of the two real fields forming the complex scalar field was transformed into the third (longitudinal) component of the vector particle which in its turn was transformed from a massless two-component Maxwell photon into a massive three-component Proca boson. Note that this did not alter the total number of scalar and vector components: $2 + 2 = 3 + 1$.

The above example illustrates the general fact that the spontaneous breaking of local symmetry does not produce Goldstone bosons, but that the vector gauge field becomes massive. In field theory this effect is called the Higgs effect and the scalar particles described by the field χ are called Higgs bosons.

A non-relativistic analog of the Higgs effect was known for a long time. It is the magnetic field penetration into a superconductor. As a result of the spontaneous breaking of gauge symmetry, the magnetic field gains a mass whose inverse value characterizes the depth of field penetration into the superconductor. A phenomenological description of this phenomenon is given by the Ginzburg-Landau equation.

This is a good moment to emphasize that in recent years field theory has come into very close contact with the non-relativistic theory of phase transitions in many-body systems, and borrowed a number of fruitful ideas. The very idea of spontaneous symmetry breaking (a seemingly negligible cause triggering an avalanche-like phase transition), has its origin in statistical physics (recall, e.g., spontaneous magnetization of ferromagnets).

Another significant trend is the growing interpenetration of the theory of elementary particles and cosmology. On one hand, the knowledge of particle properties leads to a more reliable relativistic statistical picture of the first moments of the big bang (see Chapter 27). On the other hand, it becomes clearer that these first moments have determined not only the cosmological pattern of the world as a whole but, possibly, the properties of the microcosm as well, by fixing the physical vacuum in which elementary particles have to exist.

20.5. On the conservation of electric charge

The fact that the charged scalar field φ of the above example gained a non-zero vacuum-expectation value implies non-conservation of the electric charge. This will be shown more clearly if we imagine that the field φ interacts with leptons, that is with electrons and neutrinos:

$$\sqrt{2} f(\bar{e}\nu\varphi + \bar{\nu}e\varphi^+).$$

As a result of spontaneous symmetry breaking this term generates a "non-diagonal mass" $f\eta$ transforming an electron into a neutrino,

$$f\eta(\bar{e}\nu + \bar{\nu}e),$$

and an interaction violating electric charge conservation:

$$f\chi(\bar{e}\nu + \bar{\nu}e)$$

(we recall that the field χ is neutral). It has already been mentioned above that non-conservation of electric charge creates a non-zero photon mass: $m_\gamma = e\eta$. As for the scalar field, its mass $\mu = \lambda\eta$ cannot be much larger than the photon mass. An important point is that if we want the lagrangian to be physically meaningful, the upper bound on λ must be of the order of unity (otherwise the simplest graph representing $\chi + \chi \to \chi + \chi$ scattering contradicts unitarity). This restriction on λ means that $\mu \lesssim \eta$ and, as a result, $\mu \simeq m_\gamma$ (recall that $m_\gamma = \sqrt{4\pi\alpha}\,\eta \approx \eta$). We have therefore constructed a theory in which a massive vector field (photon) is coupled to a non-conserved current (electric current). It can be shown that this theory is renormalizable and that it manifests reasonable behaviour at asymptotically high energies, which distinguishes it from theories in which the vector field mass is introduced into the lagrangian "by hand" from the very beginning.

The mass term added directly to the massless lagrangian is called the hard mass term. The mass appearing because of spontaneous symmetry breaking is called the soft mass. The theory with the hard mass is non-renormalizable, while that with the soft mass is renormalizable. The renormalizability in the soft case is due to the possibility of neglecting the η^2 term in the hamiltonian in the high-energy limit. This can be expressed in a different manner, by saying that the dangerous terms including longitudinal components of the massive vector field are compensated by the contribution of the Higgs bosons.

Experimentally, conservation of electric charge has been verified with an accuracy much lower than that of baryonic charge conservation: the lower limit of the electron lifetime with respect to $e \to \nu\gamma$ type decays is of the order of 10^{22} years, while the lower limit of the proton lifetime with respect to $p \to \mu\pi$ type decays is of the order of 10^{30} years. It is then natural to ask what restriction on the constant f of the interaction $\sqrt{2}f(\bar{e}\nu\varphi + \bar{\nu}e\varphi^+)$ follows from the fact that $\tau_e > 10^{22}$ years. The answer is quite unexpected. Namely, it can be shown that the theory of spontaneous breaking of local $U(1)$ symmetry discussed in this and earlier sections, cannot be used at all as a realistic model of electric charge non-conservation; it cannot be enforced on nature. It is essential that as follows from the existing upper bound on the photon mass, $m_\gamma \lesssim 10^{-22}$ cm^{-1}, the quantity $\eta = m_\gamma/e$ is also of the order of 10^{-22} cm^{-1}, and the mass of the neutral Higgs boson μ is nearly the same or even smaller. Under terrestrial conditions, even with practically static fields, we have to deal with frequencies much higher than these values of m_γ, μ and η. All these quantities can therefore be neglected under such conditions, so that the Higgs boson and the longitudinal component of the photon, γ_l, are emitted coherently as a practically massless charged scalar particle φ (see fig. 20.7). However, the excellent

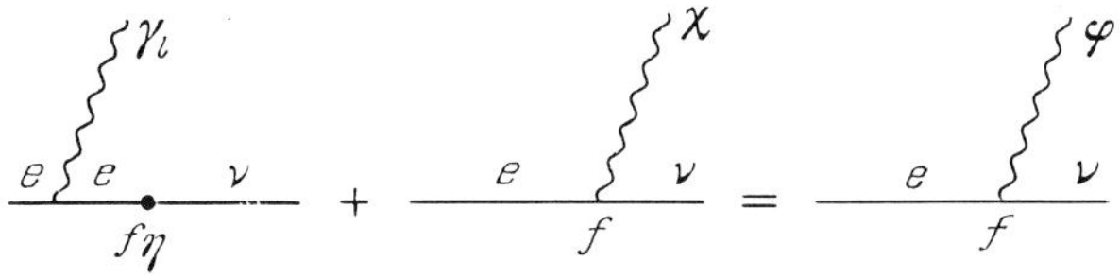

Fig. 20.7.

agreement of quantum electrodynamics with experiment absolutely excludes the possibility of the existence of such a practically massless charged particle. Indeed, creation of pairs of such particles by photons would be observed as the most spectacular of all electromagnetic effects.

Hence, non-conservation of the electric charge is not achievable in a soft manner: the practical masslessness of the photon is an insurmountable obstacle. In contrast to photons, intermediate bosons are very heavy particles, so that the soft mechanism of mass generation for these bosons is quite possible.

As the last step to constructing a renormalizable theory of weak interactions, we shall discuss the spontaneous breaking of non-abelian gauge symmetry, in which massless non-abelian photons become massive and turn into both neutral and charged intermediate bosons.

20.6. Spontaneous breaking of local SU(2) symmetry

Consider the lagrangian

$$\mathcal{L} = |D_\mu \varphi|^2 - \tfrac{1}{2}\lambda^2\bigl(|\varphi|^2 - \tfrac{1}{2}\eta^2\bigr)^2 - \tfrac{1}{4}G_{\mu\nu}G^{\mu\nu}.$$

Here $\varphi = \begin{vmatrix} \varphi_+ \\ \varphi_0 \end{vmatrix}$ is an isotopic spinor, $|\varphi|^2 = \varphi_+^*\varphi_+ + \varphi_0^*\varphi_0$,

$$D_\mu \varphi = \bigl(\partial_\mu - \tfrac{1}{2}ig\boldsymbol{\tau}\cdot\boldsymbol{A}_\mu\bigr)\varphi,$$

$$G_{\mu\nu} = \boldsymbol{G}_{\mu\nu}\cdot\boldsymbol{T} = \partial_\mu \boldsymbol{A}_\nu - \partial_\nu \boldsymbol{A}_\mu - ig\bigl[A_\mu A_\nu - A_\nu A_\mu\bigr]$$
$$= \bigl(\partial_\mu \boldsymbol{A}_\nu - \partial_\nu \boldsymbol{A}_\mu + g[\boldsymbol{A}_\mu \times \boldsymbol{A}_\nu]\bigr)\boldsymbol{T}.$$

This lagrangian is locally SU(2) invariant, and describes the infinitely degenerate vacuum at $\varphi = \eta/\sqrt{2}$. Let us use the local invariance and choose φ in the form

$$\varphi = \frac{1}{\sqrt{2}}\begin{pmatrix} 0 \\ \eta + \chi \end{pmatrix}.$$

With this choice of gauge, three of four real scalar fields are sent to the vector sector to be kept there forever; then we choose the vacuum at $|\varphi| = \eta/\sqrt{2}$, thereby spontaneously breaking the symmetry. Only one Higgs field χ is left, and the three vector fields became massive. The lagrangian now takes the form

$$\mathcal{L} = \frac{1}{2}\left(\frac{\partial \chi}{\partial x_\mu}\right)^2 + \tfrac{1}{8}g^2\left|\boldsymbol{\tau}\cdot\boldsymbol{A}\begin{pmatrix}0\\ \eta+\chi\end{pmatrix}\right|^2 - \tfrac{1}{8}\lambda^2\chi^2(2\eta+\chi)^2 - \tfrac{1}{4}G_{\mu\nu}G^{\mu\nu}.$$

The second term can be rewritten explicitly, taking into account that

$$\boldsymbol{\tau}\cdot\boldsymbol{A} = \tau_+ A_- + \tau_- A_+ + \tau_3 A_3,$$

where

$$\tau_+ = (\tau_1 + i\tau_2)/\sqrt{2} = \sqrt{2}\begin{pmatrix}0 & 1\\ 0 & 0\end{pmatrix},$$

$$\tau_- = (\tau_1 - i\tau_2)/\sqrt{2} = \sqrt{2}\begin{pmatrix}0 & 0\\ 1 & 0\end{pmatrix},$$

$$A_+ = (A_1 + iA_2)/\sqrt{2}, \qquad A_- = (A_1 - iA_2)/\sqrt{2}.$$

Then

$$\tfrac{1}{8}g^2\left|\boldsymbol{\tau}\cdot\boldsymbol{A}\begin{pmatrix}0\\ \eta+\chi\end{pmatrix}\right|^2 = \tfrac{1}{8}g^2\left|A_3\begin{pmatrix}0\\ -\eta-\chi\end{pmatrix} + A_-\sqrt{2}\begin{pmatrix}\eta+\chi\\ 0\end{pmatrix}\right|^2$$

$$= \tfrac{1}{8}g^2(A_3^2 + 2A_- A_+)(\eta+\chi)^2.$$

The masses of the three vector particles (A_-, A_+, and A_3) are thus shown to be identical and equal to $\tfrac{1}{2}g\eta$.

This concludes the preliminaries to the presentation of the standard unified model of electromagnetic and weak interactions.

CHAPTER 21

Standard model of the electroweak interaction

To begin the description of the standard unified model of electroweak interactions, we shall first write down the model lagrangian containing the vector and scalar fields, and the fields of the electron and electron neutrino as representatives of fermions. At the end of the chapter we shall discuss the addition of other leptons and quarks. We pay maximum attention to obtaining a relationship between the masses of the intermediate bosons m_W and m_Z, the fine structure constant α, and the Fermi constant G.

The first papers presenting what is now called the standard variant of the model appeared in the 60's. Since then, hundreds and possibly even thousands of papers were devoted to the study and development of this model. Several dozens of modified versions were suggested but the initial structure proved to be in better agreement with nature than the numerous sophisticated modifications.

We shall see in the next chapter that the standard model fits well the experimental data on neutral currents. We must emphasize that the model could easily be altered by the introduction of additional intermediate bosons and/or new leptons (besides τ and ν_τ), all within the same fundamental ideas. These ideas will be tested by means of the next generation of colliding beams facilities which would refute or confirm the existence of the W- and Z-bosons with masses of the order of 80 to 100 GeV, as well as that of massive scalar bosons.

21.1. Main features of the model

It is impossible to construct a renormalizable theory of weak interactions without including the photon. This occurs because in the general case the electromagnetic interaction of massive vector bosons is not renormalizable. To make it renormalizable, one should introduce the photon and intermediate bosons on an equal basis, as gauge fields, and then give masses spontaneously to intermediate bosons leaving the photon massless (the latter means strict conservation of electric charge). Accordingly, the standard model is based on local SU(2) × U(1) symmetry corresponding to four gauge fields, two of them charged and two neutral. Let us denote the three fields corresponding to non-abelian SU(2) symmetry by $A = \{A_1, A_2, A_3\}$ and the field corresponding to abelian U(1) symmetry by B. We shall see later that the photon and Z-boson are two orthogonal superpositions of the fields A_3 and B, and the bosons W^+ and W^- of the fields A_1 and A_2.

We shall include first only the lightest leptons, that is the electron and electron neutrino. The remaining leptons and quarks will be included only after we establish the properties of the simplified model. Two aspects are important when fermions are included: first, we have to take into account parity non-conservation, that is introduce coupling of gauge fields both to vector and to axial currents; second, conservation of currents coupled to gauge fields must not be violated. The two requirements can be satisfied by assuming that the lagrangian includes massless fermions with completely independent left-handed and right-handed components, $\psi_{L,R} = \frac{1}{2}(1 \pm \gamma_5)\psi$. Assume now that ν_L and e_L form an isotopic doublet

$$L = \begin{pmatrix} \nu_L \\ e_L \end{pmatrix},$$

while ν_R and e_R are isotopic singlets. We have put the left-handed fermions in the isotopic doublet so that they could emit (owing to their non-zero isospin) $W^\pm$ bosons. We have thus satisfied the conditions forced on us by experiment: the weak charged current contains left-handed spinors. Right-handed components of spinors will be put into an isosinglet, in order to avoid right-handed charged currents. The isotopic invariance of the lagrangian now demands the fermions to be massless: indeed, the mass term $m\bar{e}e = m(\bar{e}_L e_R + \bar{e}_R e_L)$ cannot be inserted "by hand" since this would violate isotopic invariance.

Finally, we introduce into the lagrangian an isotopic doublet of scalar fields,

$$\varphi = \begin{pmatrix} \varphi^+ \\ \varphi^0 \end{pmatrix}$$

(together with antiparticles, this makes four fields: φ^+, φ^0, φ^-, $\bar{\varphi}^0$). After developing a non-zero vacuum-expectation value, the field $(\varphi^0 + \bar{\varphi}^0)/\sqrt{2}$ will give masses to the W- and Z-bosons and the electron.

Returning to the gauge fields, we must add that the field B interacts with the hypercharge Y of the particles, defined as twice the mean charge of the multiplet, $Y = 2\langle Q \rangle$. It is easy to see that

$Y = -1$ for ν_L and e_L,
$Y = -2$ for e_R,
$Y = 0$ for ν_R,
$Y = 1$ for φ^+ and φ^0.

We must emphasize that the above isospin and hypercharge have nothing in common with the isospin and hypercharge of the usual hadrons. In a certain sense, the isotopic group discussed in this chapter has a more profound physical meaning than the usual isotopic invariance of strong interactions, since weak isotopic charges are sources of vector particles. At the same time, from the viewpoint of the quark-gluon theory, the usual isotopic invariance merely reflects the "accidental" smallness of the u- and d-quark masses as compared to their kinetic energies in hadrons.

Having made these preliminary remarks, we shall explicitly write down the lagrangian.

21.2. Nine terms of the lagrangian

The full expression for the lagrangian of our system has nine terms:

$$\mathcal{L} = -\tfrac{1}{4}G_{\mu\nu}G^{\mu\nu} - \tfrac{1}{4}F_{\mu\nu}F^{\mu\nu} + i\bar{L}\hat{D}L + i\bar{e}_R\hat{D}e_R$$

$$+ i\bar{\nu}_R\hat{D}\nu_R + |D_\mu\varphi|^2 - \tfrac{1}{2}\lambda^2\big(|\varphi|^2 - \tfrac{1}{2}\eta^2\big)^2$$

$$- f_e\big(\bar{L}e_R\varphi + \bar{e}_R L\varphi^\dagger\big) - f_{\nu_e}\big(\bar{L}\nu_R\varphi_c + \bar{\nu}_R L\varphi_c^\dagger\big).$$

The notation and the meaning of each term will now be clarified. We begin with the covariant derivative

$$D_\mu = \partial_\mu - ig\boldsymbol{T}\cdot\boldsymbol{A}_\mu - ig\tfrac{1}{2}YB_\mu.$$

The values of Y for the fields included in the expression for D_μ were listed above. The second term in D_μ vanishes in the case of isoscalar fields with $T = 0$ (e_R and ν_R). For isospinor fields with $T = \tfrac{1}{2}$ (φ and L), $\boldsymbol{T} = \tfrac{1}{2}\boldsymbol{\tau}$.

The field strengths of the gauge fields are

$$G_{\mu\nu} = \partial_\mu A_\nu - \partial_\nu A_\mu - ig\big[A_\mu A_\nu - A_\nu A_\mu\big] = \boldsymbol{G}_{\mu\nu}\boldsymbol{T},$$

where $A_\mu = \boldsymbol{A}_\mu\cdot\boldsymbol{T}$; $F_{\mu\nu} = \partial_\mu B_\nu - \partial_\nu B_\mu$. The term $i\bar{L}\hat{D}L$ describes both the free motion of the left-handed fermions and their interaction with gauge fields:

$$\bar{L}\hat{D}L = (\bar{\nu}_L \bar{e}_L)\hat{D}\begin{pmatrix}\nu_L \\ e_L\end{pmatrix},$$

where $\hat{D} = D_\mu \gamma^\mu = \hat{\partial} - \tfrac{1}{2}ig\boldsymbol{\tau}\cdot\hat{A} + \tfrac{1}{2}ig'\hat{B}$.

For the right-handed electron we obtain

$$\bar{e}_R \hat{D} e_R = \bar{e}_R(\hat{\partial} + ig'\hat{B})e_R,$$

and for the right-handed neutrino,

$$\bar{\nu}_R \hat{D} \nu_R = \bar{\nu}_R \hat{\partial} \nu_R.$$

The right-handed neutrino has zero isospin and zero hypercharge, and therefore does not interact with the gauge fields. The quantity $D_\mu \varphi$ in the sixth term has the form

$$D_\mu \varphi = \left(\partial_\mu - \tfrac{1}{2}ig\boldsymbol{\tau}\cdot A_\mu - \tfrac{1}{2}ig'B_\mu\right)\varphi.$$

The expression for $|D_\mu \varphi|^2$ represents an isotopic scalar:

$$|D_\mu \varphi|^2 = (D_\mu \varphi)_i^* (D^\mu \varphi)^i.$$

This term describes both the free motion of scalar fields and their interaction with the gauge fields A_μ and B_μ. When the field φ acquires a vacuum-expectation value equal to $\eta/\sqrt{2}$ (see the seventh term), the sixth term gives masses to intermediate bosons, in analogy to what has been described in chapter 20. The eighth and ninth terms describe the interaction of fermions with scalar fields. After the field φ acquires a non-zero vacuum-expectation value, these terms give non-zero masses to fermions: the term with f_e to the electron, and that with f_{ν_e} to the neutrino.

The isospinor φ_c included in the term containing f_{ν_e} is the charge conjugate of the isospinor φ, and has the form:

$$\varphi_c = \begin{pmatrix}\bar{\varphi}^0 \\ -\varphi^-\end{pmatrix}.$$

The minus sign in the isospinor appears because of the charge conjugation matrix in isotopic space:

$$(\varphi_c)_i = \varphi^{*k}\varepsilon_{ik}.$$

If we want the neutrino mass to be zero, we must set $f_{\nu_e} = 0$ (the experimental upper bound is $m_{\nu_e} \leq 30$ eV). If $m_{\nu_e} = 0$, the right-handed neutrino is completely isolated from other particles and does not take part either in strong, electromagnetic or weak interactions. However, ν_R is allowed to participate in gravitational interactions, with gravitons transforming into $\nu_R \bar{\nu}_R$ pairs. Produced by this mechanism, ν_R and $\bar{\nu}_R$ would be absolutely "sterile" but would be deflected by gravitational fields. Another possibility is the absence of right-handed neutrinos in nature.

It is easy to show that all nine terms of the lagrangian obey local isotopic invariance and local hypercharge conservation. The lagrangian contains no constants with dangerous dimension (m to negative powers), and already in the pioneer papers it was conjectured that it is renormalizable. It proved to be far from elementary, however, to demonstrate the validity of this hypothesis. The proof was given by 't Hooft (1971). A discussion of this proof goes beyond the scope of this book.

Making use of the gauge invariance of the lagrangian (see chapter 20), let us apply the phase transformation $\exp(\frac{1}{2}i\tau\theta(x))$ to the isospinor φ, giving it the form

$$\varphi = \frac{1}{\sqrt{2}} \begin{pmatrix} 0 \\ \eta + \chi(x) \end{pmatrix},$$

where η is a real constant entering the lagrangian, and χ is a real scalar field. After going to a new vacuum, the field $\chi(x)$ represents excitations over this vacuum, namely massive scalar mesons with mass $\lambda\eta$. The other three components of the field φ are now "serving" as longitudinal components of massive W^+, W^- and Z-bosons.

21.3. Masses of W- and Z-bosons

In order to clarify how the masses of these bosons appear in the model, let us consider the terms proportional to η, in the expression for $D_\mu \varphi$:

$$D_\mu \varphi = \left(\partial_\mu - \tfrac{1}{2} ig\tau \cdot A_\mu - \tfrac{1}{2} ig' B_\mu \right) \begin{pmatrix} 0 \\ (\eta + \chi)/\sqrt{2} \end{pmatrix}$$

$$\Rightarrow \tfrac{1}{4}\sqrt{2}\, i\left(gA_{3\mu} - g'B_\mu \right) \begin{pmatrix} 0 \\ \eta \end{pmatrix} - \tfrac{1}{2} ig A_\mu^- \begin{pmatrix} \eta \\ 0 \end{pmatrix}.$$

Then the term $|D_\mu \varphi|^2$ yields the mass terms of the intermediate bosons

$$\tfrac{1}{8}\eta^2 \left(gA_{3\mu} - g'B_\mu \right)^2 + \tfrac{1}{4}\eta^2 g^2 A^- A^+.$$

The following notation will be convenient for further analysis:

$$\bar{g} = \sqrt{g^2 + g'^2}, \qquad g/\bar{g} = \cos\theta_W, \qquad g'/\bar{g} = \sin\theta_W.$$

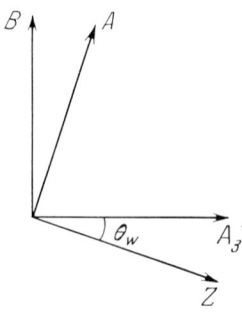

Fig. 21.1.

The parameter θ_W is called the Weinberg angle. Let us introduce the mutually orthogonal fields of the intermediate boson Z and the photon A (fig. 21.1). Furthermore, we shall use new symbols for charged bosons:

$$W^{\pm} \equiv A^{\pm} = \sqrt{\tfrac{1}{2}}\,(A_1 \pm iA_2).$$

With new notation, the mass terms of the intermediate bosons are rewritten in the form:

$$\tfrac{1}{8}\bar{g}^2\eta^2 Z^2 + \tfrac{1}{4}g^2\eta^2 |W_\mu|^2.$$

A comparison with standard expressions for mass terms in the lagrangian,

$$\tfrac{1}{2}m_Z^2 Z^\mu Z_\mu, \qquad m_W^2 W^\mu W_\mu^*,$$

yields

$$m_Z = \tfrac{1}{2}\bar{g}\eta, \qquad m_W = \tfrac{1}{2}g\eta, \qquad \frac{m_W}{m_Z} = \cos\theta_W.$$

As for the photon A, its mass remains zero.

If the Higgs sector of the lagrangian is even more complicated, the formulas for m_W^2 and m_Z^2 must be altered. It is obvious, however, that even in the case of several Higgs multiplets the parameter $\rho = \bar{g}^2 m_W^2 / g^2 m_Z^2 = m_W^2 / m_Z^2 \cos^2\theta_W$ which characterizes the strength of the four-fermion interaction of the neutral axial currents (see chapter 2) would remain equal to unity if all Higgs multiplets were doublets. The relation $\rho = 1$ would be invalid if some of the scalar multiplets have isospin larger than $\tfrac{1}{2}$.

We should emphasize that expressions for the photon and Z-boson,

$$A = B\cos\theta_W + A_3 \sin\theta_W,$$

$$Z = -B\sin\theta_W + A_3 \cos\theta_W,$$

where $\tan\theta_W = g'/g$, are not related to the structure of the Higgs sector and are determined only by the structure of gauge fields. This is so because the photon representing a linear combination of fields B and A_3 does not interact, and this is normal for the photon, with neutral particles (and in particular, with φ^0 and $\bar\varphi^0$), while a combination orthogonal to this and representing the Z-boson is coupled both to charged and neutral particles. Let us discuss the couplings of the photon and of the Z-boson in more detail.

21.4. Relation between electric charge and the constants g and g'

As we can see from the expression for D_μ, the interaction of neutral gauge fields with hypercharge Y and isospin projection T_3 has the form

$$-ig'\tfrac{1}{2}YB - igT_3A_3.$$

Making use of the equality

$$Q = T_3 + \tfrac{1}{2}Y,$$

where Q is electric charge in units of e, we transform the above expression to

$$-ig'\tfrac{1}{2}YB - igT_3A_3 = -ig'(\tfrac{1}{2}Y + T_3)B + i(g'B - gA_3)T_3$$

$$= -ig'QB - i\bar{g}T_3Z$$

$$= -i\frac{g'g}{\bar{g}}QA + i\frac{g'^2}{\bar{g}}QZ - i\bar{g}T_3Z$$

$$= -i\bar{g}(T_3 - Q\sin^2\theta_W)Z - i\frac{g'g}{\bar{g}}QA.$$

This expression is universal. In particular, it is valid both for the left-handed and for the right-handed spinors, and of course for scalar particles. The expression shows that a neutral particle (one with $Q = 0$) is coupled only to the Z-boson but not to the photon. Therefore the Z-boson becomes massive when φ^0 acquires a non-zero vacuum-expectation value, and the photon remains massless (see the preceding section). Obviously, the last term in this expression, $-i(g'g/\bar{g})QA$, describes the photon coupling to the electric charge, whence $g'g/\bar{g} = e = \sqrt{4\pi\alpha}$. This relation can be rewritten in the form

$$e = g\sin\theta_W, \qquad e = g'\cos\theta_W, \qquad e = \bar{g}\sin\theta_W\cos\theta_W,$$

or

$$\frac{1}{e^2} = \frac{1}{g^2} + \frac{1}{g'^2}.$$

We conclude that interactions of all gauge fields are determined by the electric charge e and one free parameter, namely the Weinberg angle. In particular, the "charge" characterizing the coupling of the left-handed or right-handed spinor to the Z-boson is equal to

$$-\frac{e}{\sin\theta_W \cos\theta_W}(T_3 - Q\sin^2\theta_W),$$

where T_3 and Q are the particle isospin projection and particle charge.

The fact that both weak and electromagnetic interactions are characterized by the same charge e is the most conclusive demonstration of the standard model being a unified theory of weak and electromagnetic interactions. The theory is not yet completed because it contains a free parameter θ_W which is not calculated or predicted theoretically (we shall see in chapter 22 that experimentally $\theta_W \cong 30°$). The free parameter θ_W appears because the symmetry group of the weak interaction is a direct product of two simple groups, SU(2) and U(1). The arbitrary choice of θ_W could be eliminated if we assumed both groups to be subgroups of a larger group (see Chapter 25).

21.5. Relation between the vacuum-expectation value η and the Fermi constant G

In order to find the relation between η and G, let us consider the interaction of charged currents shown in fig. 21.2. The term $i\overline{L}\hat{D}L$ in the lagrangian yields for the vertices of this graph:

$$-\tfrac{1}{2}ig(\bar{\nu}_L \bar{e}_L)[\tau_- \hat{W}^- + \tau_+ \hat{W}^+]\binom{\nu_L}{e_L}$$

$$= -\tfrac{1}{2}\sqrt{2}\, ig(\bar{e}_L \hat{W}^- \nu_L + \bar{\nu}_L \hat{W}^+ e_L).$$

(recall that $\tau_\pm = \sqrt{2}\begin{pmatrix} 0 & 1 \\ 0 & 0 \end{pmatrix}$). The amplitude corresponding to fig. 21.2 can now be found by means of the standard rules for Feynman graphs:

$$M = \tfrac{1}{2}g^2(\bar{e}_L \gamma^\alpha \nu_L)(\bar{\nu}_L \gamma^\beta e_L)\frac{g_{\alpha\beta} - q_\alpha q_\beta / m_W^2}{m_W^2 - q^2}.$$

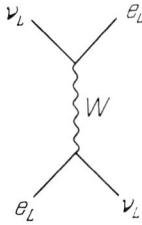

Fig. 21.2.

In the low-energy limit this expression becomes

$$M = \frac{g^2}{2m_W^2}(\bar{e}_L\gamma^\alpha \nu_L)(\bar{\nu}_L\gamma_\alpha e_L),$$

which must be compared to the standard amplitude for νe scattering obtained by multiplying charged currents,

$$M = \sqrt{\tfrac{1}{2}}\, G(\bar{e}\gamma^\alpha(1+\gamma_5)\nu)(\bar{\nu}\gamma_\alpha(1+\gamma_5)e).$$

The comparison yields:

$$\frac{g^2}{8m_W^2} = \frac{G}{\sqrt{2}}.$$

It must be emphasized that this relation is much more general than the specific model in which it was derived. Recalling an expression derived earlier,

$$m_W^2 = \tfrac{1}{4}g^2\eta^2,$$

we find that

$$\frac{1}{\eta^2} = \sqrt{2}\, G, \qquad \frac{1}{\eta} = \left(\sqrt{2}\, G\right)^{1/2},$$

or $\eta = 246$ GeV.

21.6. More about the masses of the W- and Z-bosons

By using the formulas

$$\frac{g^2}{8m_W^2} = \frac{G}{\sqrt{2}} \quad \text{and} \quad g^2 = \frac{4\pi\alpha}{\sin^2\theta_W}$$

we obtain for the W-boson mass

$$m_W = \frac{1}{\sin\theta_W}\left(\frac{\pi\alpha}{\sqrt{2}\,G}\right)^{1/2} = \frac{37.3}{\sin\theta_W}\,\text{GeV}.$$

Note that this relation is independent of the structure of the Higgs sector. Assuming the scalar fields to be isodoublet fields, we obtain the following expression for the Z-boson mass:

$$m_Z = \frac{m_W}{\cos\theta_W} = \frac{37.3}{\sin\theta_W \cos\theta_W} = \frac{74.6}{\sin 2\theta_W}\,\text{GeV}.$$

The minimum masses of the W- and Z-bosons are therefore 37.3 and 74.6 GeV, respectively. No wonder they were not found at the existing accelerators.

21.7. Electron mass

The next term of the lagrangian to be considered is

$$f_e(\bar{L}e_R\varphi + \bar{e}_R L\varphi^+)$$

giving the mass of the electron. Substitution of

$$\varphi = \frac{1}{\sqrt{2}}\begin{pmatrix} 0 \\ \eta + \chi \end{pmatrix}$$

yields

$$\sqrt{\tfrac{1}{2}}\,f_e(\bar{e}_L e_R + \bar{e}_R e_L)(\eta + \chi) = \sqrt{\tfrac{1}{2}}\,f_e \bar{e}e(\eta + \chi).$$

The first term gives the electron a mass which is found to be $m_e = f_e\eta/\sqrt{2}$. There are no predictions about the value of f_e which is a serious shortcoming indicating that the theory is incomplete. The experimental value of f_e is very small,

$$f_e = \frac{\sqrt{2}\,m_e}{\eta} \simeq 3\cdot 10^{-6}.$$

So far we cannot understand how such a small parameter can appear in the theory. The constant f_e characterizes both the electron mass and its interaction with the field χ:

$$\sqrt{\tfrac{1}{2}}\,f_e\bar{e}e(\eta + \chi) = m_e\bar{e}e + (\sqrt{2}\,G)^{1/2} m_e \chi\bar{e}e.$$

We see that Higgs scalar bosons interact with particles the more strongly, the greater the masses of these particles.

Now it will be appropriate to mention how other leptons and quarks are included in the standard model.

21.8. Introduction of other leptons and quarks

All that we have said above concerning the electron and its neutrino, is valid for the muon and its neutrino. The leptonic sector of the lagrangian is expanded by including a doublet $(\nu_{\mu L}, \mu_L)$ and two singlets $\nu_{\mu R}$ and μ_R. Similarly, we add a doublet $(\nu_{\tau L}, \tau_L)$ and two singlets $\nu_{\tau R}$ and τ_R. If $m_{\nu_\mu} = 0$, $m_{\nu_\tau} = 0$, then $f_{\nu_\mu} = 0$ and $f_{\nu_\tau} = 0$ respectively. From what we have said about the electron mass, we obtain that $f_e : f_\mu : f_\tau = m_e : m_\mu : m_\tau$.

Addition of quarks to the standard model is carried out in a similar manner: left-handed particles form isodoublets, and right-handed ones form isosinglets. Three new aspects must be taken into account, however, when quarks are introduced into the model: first, quark hypercharges differ from those of leptons because of the quark fractional charges; second, not only f_u-type constants must be non-zero but f_d-type ones must be non-vanishing as well; third, the Cabibbo-like angles must be included.

The last of these three aspects will now be discussed for the four quarks, u, d, s, c (to cover the b- and t-quarks would only entail algebraic complications, see Chapter 15).

Four left-handed quarks form two isodoublets,

$$\begin{pmatrix} u \\ d' \end{pmatrix}_L, \quad \begin{pmatrix} c \\ s' \end{pmatrix}_L$$

where

$$d' = d \cos \theta_C + s \sin \theta_C,$$
$$s' = -d \sin \theta_C + s \cos \theta_C,$$

and θ_C is the Cabibbo angle ($\theta_C \simeq 15°$). This choice of left-handed quarks is based on the form of the quark charged current known from the experimental data:

$$\bar{u}_L \gamma_\alpha d'_L + \bar{c}_L \gamma_\alpha s'_L.$$

The right-handed quarks are represented by four singlets. Different choices are possible here: u_R, d'_R, s'_R, c_R, or u_R, d_R, s_R, c_R, or any other orthogonal set. This arbitrariness takes place because right-handed quarks enter only the neutral current. We have shown above that the coupling to the Z-boson is completely determined by the quantity $T_3 - Q\sin^2\theta_W$, so that we obtain two combinations,

$$\bar{u}_R \gamma_\alpha u_R + \bar{c}_R \gamma_\alpha c_R, \quad \bar{d}_R \gamma_\alpha d_R + \bar{s}_R \gamma_\alpha s_R,$$

invariant with respect to Cabibbo rotations

$$\bar{d}'_R \gamma_\alpha d'_R + \bar{s}'_R \gamma_\alpha s'_R = \bar{d}_R \gamma_\alpha d_R + \bar{s}_R \gamma_\alpha s_R.$$

This is also valid for neutral currents of left-handed quarks,

$$\bar{u}_L \gamma_\alpha u_L + \bar{c}_L \gamma_\alpha c_L \quad \text{and} \quad \bar{d}'_L \gamma_\alpha d'_L + \bar{s}'_L \gamma_\alpha s'_L.$$

This explains why the neutral currents, in contrast to charged currents, conserve the type (flavor) of quarks. They are diagonal in flavor indices. Then why cannot we choose d' and s' as quarks, instead of d and s, and thus get rid of the Cabibbo angle? The answer to this question is related to quark masses and therefore to their interactions with scalar (Higgs) bosons. Contrary to the leptonic case in which the interactions of the Higgs bosons with e and μ are characterized by two constants f_e and f_μ, the interaction of the Higgs bosons with d' and s' contains three terms:

$$\sqrt{\tfrac{1}{2}} f_{d'}(\bar{d}'_L d'_R + \bar{d}'_R d'_L)(\eta + \chi) = \sqrt{\tfrac{1}{2}} f_{d'} \bar{d}' d'(\eta + \chi),$$

$$\sqrt{\tfrac{1}{2}} f_{s'}(\bar{s}'_L s'_R + \bar{s}'_R s'_L)(\eta + \chi) = \sqrt{\tfrac{1}{2}} f_{s'} \bar{s}' s'(\eta + \chi),$$

$$\sqrt{\tfrac{1}{2}} f_{d's'}(\bar{s}'_L d'_R + \bar{s}'_R d'_L + \bar{d}'_R s'_L + \bar{d}'_L s'_R)(\eta + \chi)$$
$$= \sqrt{\tfrac{1}{2}} f_{d's'}(\bar{s}' d' + \bar{d}' s')(\eta + \chi).$$

Rotation by the Cabibbo angle transforms the expression

$$\sqrt{\tfrac{1}{2}} \left[f_{d'} \bar{d}' d' + f_{s'} \bar{s}' s' + f_{d's'}(\bar{s}' d' + \bar{d}' s') \right] \eta$$

to a diagonal form:

$$m_d \bar{d} d + m_s \bar{s} s,$$

where m_d and m_s are the masses of the d- and s-quarks. Comparison of the last two expressions yields

$$m_d = \sqrt{\tfrac{1}{2}} \left(f_{d'} \cos^2 \theta_C + f_{s'} \sin^2 \theta_C - 2 f_{d's'} \cos \theta_C \sin \theta_C \right) \eta,$$

$$m_s = \sqrt{\tfrac{1}{2}} \left(f_{d'} \sin^2 \theta_C + f_{s'} \cos^2 \theta_C + 2 f_{d's'} \cos \theta_C \sin \theta_C \right) \eta,$$

$$\tan 2\theta_C = \frac{2 f_{d's'}}{f_{s'} - f_{d'}}.$$

We see that the Cabibbo angle appears because $f_{d's'} \neq 0$.

There is a complete degeneracy with respect to flavors, as long as quarks interact only with gluons. Introduction of photons and Z-bosons separates the space of "up" quarks (with $Q = \frac{2}{3}$) from the space of "down" quarks (with $Q = -\frac{1}{3}$). Within each of these two groups degeneracy still persists. Introduction of the interaction with the Higgs bosons (H-bosons) removes this degeneracy and fixes a certain direction in each of these spaces. The interaction with the W-bosons singles out some other direction in the space of the d- and s-quarks; the difference in these two directions (H and W) is characterized by the Cabibbo angle.

A question is in order: why rotate only the d- and s-quarks, but not the u- and c-quarks? The answer is: it is sufficient to rotate either d and s, or u and c, while simultaneous rotation of all of them yields no new observable phenomena. Indeed, the expression for the charged current is a function of only the difference in these rotation angles. Let us prove this statement. Let us assume the u-, d-, s-, c-quarks to have definite masses. Introduce the superpositions

$$d' = d\cos\alpha + s\sin\alpha,$$
$$s' = -d\sin\alpha + s\cos\alpha,$$
$$u' = u\cos\beta + c\sin\beta,$$
$$c' = -u\sin\beta + c\cos\beta,$$

and consider a charged current of the type $\bar{u}'\gamma_\alpha d' + \bar{c}'\gamma_\alpha s'$. After elementary manipulations we obtain

$$\bar{u}'_L\gamma_\alpha d'_L + \bar{c}'_L\gamma_\alpha s'_L = \bar{u}_L\gamma_\alpha d_L \cos\theta_C + \bar{u}_L\gamma_\alpha s_L \sin\theta_C$$
$$+ \bar{c}_L\gamma_\alpha s_L \cos\theta_C - \bar{c}_L\gamma_\alpha d_L \sin\theta_C,$$

where $\theta_C = \alpha - \beta$. This result is a special case of a more general analysis given in Chapter 15, where we considered a unitary $n \times n$ matrix in flavor space, describing the weak charged current.

If the theory in question is correct, the key point of the "flavor" physics is the interaction of Higgs fields. Specific properties of different types of hadrons stem from the properties of the quark mass matrix, that is from the properties of the constants f. Small values of f for the u- and d-quarks make their masses small and thus result in the standard isotopic invariance. It is essential that the masses m_u and m_d may be very different. It will be sufficient for both to be much smaller than μ_c, where $1/\mu_c$ is the size of the hadron determined by the confinement radius. Comparatively large masses of the c-quarks result in striking peculiarities in the behavior of charmonium and charmed particles. Were it not for the differences in interactions of various quarks with Higgs bosons, we would find in nature SU(N) invariance of strong interactions (where N is the total number of distinct quark "flavors"), slightly violated by virtual photons (and intermediate bosons). So far the spectrum of quark and lepton masses, that is the pattern of the constants f, is not understood at all. There is no doubt that the experimental discovery of Higgs bosons would be an important step to solving this key problem of elementary particle physics.

CHAPTER 22

Neutral currents

The prediction of neutral currents constituted the main contribution of the unified electroweak model to the domain of weak interactions at low energies, that is energies of the existing accelerators. Neutral currents in neutrino-nucleon interactions were discovered in 1973. Since then total cross sections of these processes, cross sections of some inelastic channels, and elastic cross sections were measured both for ν_μ and for $\bar{\nu}_\mu$. All these data agree with predictions of the standard model for $\theta_W \simeq 30°$.

The subject of the present chapter is a detailed discussion of the neutrino-electron and neutrino-nucleon interactions. In addition, we shall discuss the effects due to the interference of the weak and electromagnetic interaction in neutrinoless processes: $e^+ e^- \to \mu^+ \mu^-$ annihilation, scattering of electrons on nucleons, parity violation in atoms. At the end of the chapter we shall touch on the problem of parity violation in atomic nuclei.

22.1. Scattering of ν_e and $\bar{\nu}_e$ on the electron

It has been established in Chapter 21 that the "charge" characterizing the coupling of the "particle" to the Z-boson is $-\bar{g}(T_3 - \xi Q)$. Here we introduce the notation $\xi = \sin^2 \theta_W$, with a view to shortening the subsequent formulas. The word "particle" is in quotation marks because in the case of leptons and quarks we consider separately the left-handed and right-handed components as having different values of T and T_3. The coupling to the Z-boson of the (ν_L, e_L) doublet has the form

$$i\bar{g}\tfrac{1}{2}\bar{\nu}_L \hat{Z} \nu_L + i\bar{g}(-\tfrac{1}{2} + \xi)\bar{e}_L \hat{Z} e_L,$$

and that of the singlet e_R:

$$i\bar{g}\xi \bar{e}_R \hat{Z} e_R.$$

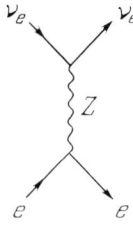

Fig. 22.1.

On the whole, the electron vertex has the form:

$$i\bar{g}\left[\left(-\tfrac{1}{2}+\xi\right)\bar{e}_L\gamma_\alpha e_L + \xi\bar{e}_R\gamma_\alpha e_R\right]Z^\alpha = i\bar{g}\left[-\tfrac{1}{2}\bar{e}_L\gamma_\alpha e_L + \xi\bar{e}\gamma_\alpha e\right]Z^\alpha.$$

The effective lagrangian of $\nu_e e$ scattering due to the exchange of the Z-boson is represented by the graph of fig. 22.1; at energies much smaller than the Z-boson mass, it takes the form:

$$2\sqrt{2}\,G(\bar{\nu}_{eL}\gamma^\alpha\nu_{eL})\left[\left(-\tfrac{1}{2}+\xi\right)\bar{e}_L\gamma_\alpha e_L + \xi\bar{e}_R\gamma_\alpha e_R\right].$$

(Here we have taken into account that $\bar{g}^2/m_Z^2 = \rho g^2/m_W^2 = 4\sqrt{2}\,G\rho$ and that in the standard model $\rho = 1$ (see p. 190). Recalling that the squared total neutral current includes twice the product of the neutrino and electron currents, it is then easy to reconstruct the normalization of the four-fermion lagrangian $\mathcal{L}''$ given on p. 9). The contribution of this lagrangian to the total amplitude of $\nu_e e$ scattering is summed up with that of the effective lagrangian due to exchange of the W-boson (fig. 22.2). This second contribution is

$$\frac{g^2}{m_W^2}\tfrac{1}{2}(\bar{e}_L\gamma_\alpha\nu_L)(\bar{\nu}_L\gamma^\alpha e_L) = 2\sqrt{2}\,G(\bar{e}_L\gamma_\alpha e_L)(\bar{\nu}_L\gamma^\alpha\nu_L).$$

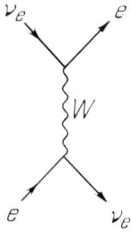

Fig. 22.2.

Here the factor $\frac{1}{2}$ appears because the vertex of W emission has the form

$$\sqrt{\tfrac{1}{2}}\, ig\bar{e}_L \hat{W} \nu_{eL}.$$

A point of special interest is the sign of the expression obtained by means of the Fierz transformation. Recall that this transformation changes the sign of the $V-A$ amplitude. However, the Fierz transformation does not change the sign of the effective $V-A$ lagrangian, since anticommutation of spinor operators supplies an additional minus sign.

The total contribution of the graphs of figs. 22.1 and 22.2 is then

$$2\sqrt{2}\, G(\bar{\nu}_{eL}\gamma^\alpha \nu_{eL})\left[(\tfrac{1}{2}+\xi)\bar{e}_L\gamma_\alpha e_L + \xi \bar{e}_R\gamma_\alpha e_R\right].$$

Standard calculations (see chapter 16) demonstrate that this interaction corresponds to the cross sections

$$\sigma_{\nu_e e} = \frac{G^2 s}{\pi}\left[(\tfrac{1}{2}+\xi)^2 + \tfrac{1}{3}\xi^2\right],$$

$$\sigma_{\bar{\nu}_e e} = \frac{G^2 s}{\pi}\left[\tfrac{1}{3}(\tfrac{1}{2}+\xi)^2 + \xi^2\right].$$

The last expression does not contradict the data obtained in experiments with reactor-produced antineutrinos. No data are available yet on $\nu_e e$ scattering.

22.2. Scattering of ν_μ and $\bar{\nu}_\mu$ on the electron

In the theory without neutral currents, the muon neutrino is not scattered by the electron in the first order in G because there is no $\bar{e}\nu_\mu W$ vertex and therefore there is no graph similar to that of fig. 22.2. The theory with neutral currents predicts a process (see the graph in fig. 22.3) which goes via the exchange of the Z-boson.

We have already discussed the invariance of all intermediate boson couplings with respect to the $\nu_e \leftrightarrow \nu_\mu$ and $e \leftrightarrow \mu$ substitution. Consequently, the effective lagrangian corresponding to the graph of fig. 22.3 has the same form as that for fig. 22.1, namely:

$$2\sqrt{2}\, G(\bar{\nu}_{\mu L}\gamma^\alpha \nu_{\mu L})\left[(-\tfrac{1}{2}+\xi)\bar{e}_L\gamma_\alpha e_L + \xi \bar{e}_R\gamma_\alpha e_R\right].$$

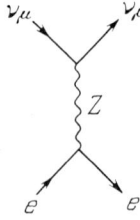

Fig. 22.3.

This interaction corresponds to the cross sections

$$\sigma_{\nu_\mu e} = \frac{G^2 s}{\pi} \left[\left(-\tfrac{1}{2} + \xi\right)^2 + \tfrac{1}{3}\xi^2 \right],$$

$$\sigma_{\bar\nu_\mu e} = \frac{G^2 s}{\pi} \left[\tfrac{1}{3}\left(-\tfrac{1}{2} + \xi\right)^2 + \xi^2 \right].$$

Within experimental error, the data on $\nu_\mu e$ scattering are in agreement with these expressions for $\xi \simeq \tfrac{1}{4}$.

22.3. Annihilation $e^+ e^- \to \mu^+ \mu^-$

Annihilation may proceed in lower order of perturbation theory both via a virtual photon and via a virtual Z-boson (figs. 22.4 and 22.5). The graph in fig. 22.4 corresponds to production of a $\mu^+ \mu^-$ pair with negative C-parity (since the photon is C-odd). The graph in fig. 22.5 represents production of $\mu^+ \mu^-$ in a superposition of states with $C = +1$ and $C = -1$, because the Z-boson vertex contains both the axial and the vector interaction. Interference of states with $C = \pm 1$ must result in charge asymmetry in the distribution of μ^+ and μ^- (with respect to the e^+ momentum in the center-of-mass frame of the colliding e^+ and e^-). Let us calculate this effect. The amplitude corresponding to the sum of graphs in figs. 22.4 and 22.5, is

$$M = \frac{e^2}{q^2} [\bar\mu(p)\gamma_\alpha \mu(-p')][\bar e(-k')\gamma^\alpha e(k)]$$

$$+ \frac{g^2}{q^2 - m_Z^2} \left[\left(-\tfrac{1}{2} + \xi\right)\bar\mu_L(p)\gamma_\alpha \mu_L(-p') + \xi \bar\mu_R(p)\gamma_\alpha \mu_R(-p')\right]$$

$$\times \left[\left(-\tfrac{1}{2} + \xi\right)\bar e_L(-k')\gamma^\alpha e_L(k) + \xi \bar e_R(-k')\gamma^\alpha e_R(k)\right].$$

§3] Annihilation $e^+e^- \to \mu^+\mu^-$

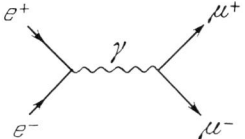

Fig. 22.4.

Here $\bar{\mu}(p)$ is the wave function of the created μ^- with 4-momentum p; $\mu(-p')$ is the wave function of the created μ^+ with 4-momentum p'; $e(k)$ is the wave function of the annihilated e^- with 4-momentum k; $\bar{e}(-k')$ is the wave function of the annihilated e^+ with 4-momentum k'. Minus signs in front of k' and p' appear because of crossing. Neglecting the e and μ masses as compared to their energies, we have dropped the term

$$-\frac{q_\alpha q_\beta}{m_Z^2(q^2 - m_Z^2)}$$

in the Z-boson propagator.

Let us introduce the notation

$$x = \frac{\bar{g}^2}{e^2}\frac{q^2}{q^2 - m_Z^2} \quad \left(x = -\frac{2\sqrt{2}\,G}{\pi\alpha}q^2 \quad \text{for } q^2 \ll m_Z^2\right).$$

Omitting the arguments of the wave functions, M can now be written in the form:

$$M = \frac{4\pi\alpha}{q^2}\left\{(\bar{\mu}_L\gamma_\alpha\mu_L)(\bar{e}_L\gamma^\alpha e_L)\left[1 + x\left(-\tfrac{1}{2} + \xi\right)^2\right] + (\bar{\mu}_R\gamma_\alpha\mu_R)(\bar{e}_R\gamma^\alpha e_R)\right.$$

$$\times\left[1 + x\xi^2\right] + \left[(\bar{\mu}_L\gamma_\alpha\mu_L)(\bar{e}_R\gamma^\alpha e_R) + (\bar{\mu}_R\gamma_\alpha\mu_R)(\bar{e}_L\gamma^\alpha e_L)\right]$$

$$\left.\times\left[1 + x\xi\left(-\tfrac{1}{2} + \xi\right)\right]\right\}.$$

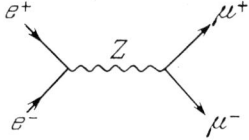

Fig. 22.5.

Let us introduce the new variables:

$$s = (k' + k)^2 = (p' + p)^2 = q^2,$$
$$t = (p - k)^2 = -\tfrac{1}{2}s(1 - \cos\theta) = -2pk = -2p'k',$$
$$u = (p' - k)^2 = -\tfrac{1}{2}s(1 + \cos\theta) = -2p'k = -2pk'$$

where θ is the angle between the momenta of the e^- and μ^- in the center-of-mass frame, and calculate $\overline{|M|^2}$, with the bar over $|M|^2$ denoting summation over the spins of e^+, e^-, μ^+, and μ^-. Note that in this calculation the four terms contained in M do not interfere, since they correspond to different helicities of the initial and (or) final particles.

The Fierz transformation of the term $(\bar{\mu}_L \gamma_\alpha \mu_L)(\bar{e}_R \gamma^\alpha e_R)$ yields:

$$(\bar{\mu}_L \gamma_\alpha \mu_L)(\bar{e}_R \gamma^\alpha e_R) = 2(\bar{\mu}_L e_R)(\bar{e}_R \mu_L)$$
$$= \tfrac{1}{2}(\bar{\mu}(1 - \gamma_5)e)(\bar{e}(1 + \gamma_5)\mu),$$

whence

$$|(\bar{\mu}_L \gamma_\alpha \mu_L)(\bar{e}_R \gamma^\alpha e_R)|^2 \Rightarrow \tfrac{1}{4} \operatorname{Tr} \hat{p}'(1 - \gamma_5)\hat{k}(1 + \gamma_5) \operatorname{Tr} \hat{k}'(1 + \gamma_5)\hat{p}'(1 - \gamma_5)$$
$$= 4(pk)4(p'k') = 4t^2.$$

Likewise,

$$|(\bar{\mu}_R \gamma_\alpha \mu_R)(\bar{e}_L \gamma^\alpha e_L)|^2 \Rightarrow 4t^2.$$

It is not difficult to obtain, by using the Fierz transformation, the relation

$$|(\bar{\mu}_L \gamma_\alpha \mu_L)(\bar{e}_L \gamma^\alpha e_L)|^2 = |(\bar{\mu}_R \gamma_\alpha \mu_R)(\bar{e}_R \gamma^\alpha e_R)|^2$$
$$= 4(kp')4(pk') = 4u^2.$$

These results are quite similar to those derived for νe scattering (see Chapter 16). They can be written down immediately if we consider the Feynman graph in the μe scattering channel (fig. 22.6), and take into account that the

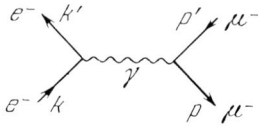

Fig. 22.6.

final result for a collision of two left-handed particles is proportional to the product $(kp')(k'p)$.

Now let us calculate the cross section:

$$\frac{d\sigma}{d\cos\theta} = \frac{\pi\alpha^2}{s^3}[au^2 + bt^2]$$

$$= \frac{\pi\alpha^2}{4s}\left[(a+b)(1+\cos^2\theta) + (a-b)2\cos\theta\right].$$

Here

$$a = \tfrac{1}{2}\left\{\left[1 + x\left(-\tfrac{1}{2} + \xi\right)^2\right]^2 + \left[1 + x\xi^2\right]^2\right\},$$

$$b = \left[1 + x\xi\left(-\tfrac{1}{2} + \xi\right)\right]^2.$$

If $x = 0$, then $a = 1$ and $b = 1$, and our cross section is equal to the well-known cross section of the electromagnetic annihilation $e^+e^- \to \mu^+\mu^-$:

$$\frac{d\sigma}{d\cos\theta} = \frac{\pi\alpha^2}{2s}(1 + \cos^2\theta) \Rightarrow \sigma = \frac{4\pi\alpha^2}{3s}.$$

If $x \neq 0$, then $a - b = \tfrac{1}{4}[x + \tfrac{1}{2}x^2(\tfrac{1}{2} - 2\xi)^2]$. If $s \ll m_Z^2$, then $x < 0$ and $a - b \simeq \tfrac{1}{4}x$, and therefore negative muons must be emitted predominantly in the direction of the initial positron.

Charge asymmetry appears even if Z bosons are ignored, because of the interference of graphs shown in figs. 22.6 and 22.7; it will also appear as a result of the interference of graphs in figs. 22.8 and 22.9, in which the $\mu^+\mu^-$ pairs have different values of C-parity. However, the purely electromagnetic charge asymmetry is a slowly varying function of the energy of the colliding beams, and reaches a maximum for small angles θ. This fact makes it possible to extract the effect of neutral currents.

Experiments of this type are to be run in the colliding e^+e^- beam facilities PETRA 2 × 19 GeV (Hamburg) and PEP 2 × 15 GeV (Stanford). The expected event rate may be several events per hour for storage ring luminosity $L = 10^{31}$ event/cm²·s, and the cross section $\sigma = 4\pi\alpha^2/3s \simeq 0.8 \cdot 10^{-34}$cm² (for $s = 900$ GeV²).

In principle, the contribution of neutral currents can be detected through longitudinal muon polarization due to parity non-conservation. It appears, however, that this would be a more difficult experiment.

Now we shall turn to neutral current effects in lepton-hadron interactions.

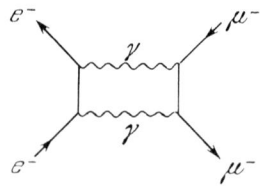

Fig. 22.7.

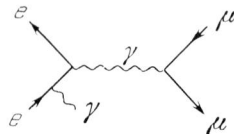

Fig. 22.8.

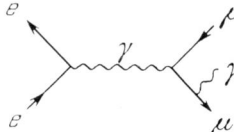

Fig. 22.9.

22.4. Neutral currents and neutrino–nucleon interaction

The interaction of a neutrino with a nucleon could be reduced to its interaction with quarks. Starting with the "Z-charge" equal to $-\bar{g}(T_3 - \xi Q)$, we can write the couplings of the u- and d-quarks to the Z-bosons as

$$i\bar{g}\left[(\tfrac{1}{2} - \tfrac{2}{3}\xi)\bar{u}_L \hat{Z} u_L - \tfrac{2}{3}\xi \bar{u}_R \hat{Z} u_R\right],$$

$$i\bar{g}\left[(-\tfrac{1}{2} + \tfrac{1}{3}\xi)\bar{d}_L \hat{Z} d_L + \tfrac{1}{3}\xi \bar{d}_R \hat{Z} d_R\right].$$

Hereafter we consider only "small" momentum transfer, $|q^2| \ll m_Z^2$. In this limit, there is the effective four-fermion interaction

$$2\sqrt{2}\, G(\bar{\nu}_L \gamma_\alpha \nu_L)\left[(\tfrac{1}{2} - \tfrac{2}{3}\xi)\bar{u}_L \gamma^\alpha u_L - \tfrac{2}{3}\xi \bar{u}_R \gamma^\alpha u_R \right.$$
$$\left. + (-\tfrac{1}{2} + \tfrac{1}{3}\xi)\bar{d}_L \gamma^\alpha d_L + \tfrac{1}{3}\xi \bar{d}_R \gamma^\alpha d_R\right]$$

(we recall that $\bar{g}^2/2m_Z^2 = 2\sqrt{2}\, G$, see Chapter 21).

Making use of this equation, we can derive an expression for the cross section for the deep-inelastic interaction. We shall calculate the cross section per "mean nucleon", that is $\frac{1}{2}(\sigma_p + \sigma_n)$. As before, $u(x)/x$ and $d(x)/x$ denote u- and d-quark distribution densities in the proton (in order to simplify formulas we shall neglect, as a crude approximation, the contribution of s, $\bar{s}$, $\bar{u}$, and $\bar{d}$). The distribution densities in the "mean nucleon" are identical for the u- and d-quark, and are equal to $(1/2x)[u(x) + d(x)]$. Recall that x stands for the fraction of the total nucleon momentum carried by a quark. The cross sections are given in the parton model in terms of mean quantities (see Chapter 17),

$$U = \int_0^1 u(x)\,dx \quad \text{and} \quad D = \int_0^1 d(x)\,dx,$$

and are equal to

$$\sigma_\nu^Z = \frac{G^2 s}{2\pi}(U + D)\left[\left(\tfrac{1}{2} - \tfrac{2}{3}\xi\right)^2 + \tfrac{1}{3}\left(\tfrac{2}{3}\xi\right)^2 \right.$$

$$\left. + \left(-\tfrac{1}{2} + \tfrac{1}{3}\xi\right)^2 + \tfrac{1}{3}\left(\tfrac{1}{3}\xi\right)^2\right],$$

$$\sigma_{\bar{\nu}}^Z = \frac{G^2 s}{2\pi}(U + D)\left[\tfrac{1}{3}\left(\tfrac{1}{2} - \tfrac{2}{3}\xi\right)^2 + \left(\tfrac{2}{3}\xi\right)^2 \right.$$

$$\left. + \tfrac{1}{3}\left(-\tfrac{1}{2} + \tfrac{1}{3}\xi\right)^2 + \left(\tfrac{1}{3}\xi\right)^2\right].$$

These expressions take into account that cross sections for fermions with the same helicities are larger by a factor of three than those for fermions with opposite helicities. It is convenient to consider the ratios of neutral current cross sections, NC, to charged current cross sections, CC. We recall that in the latter case, the cross sections in the approximation under discussion are

$$\sigma_\nu^W = \frac{G^2 s}{2\pi}(U + D), \qquad \sigma_{\bar{\nu}}^W = \frac{1}{3}\frac{G^2 s}{2\pi}(U + D).$$

The above-mentioned ratios are:

$$R_\nu = \frac{\sigma_\nu^Z}{\sigma_\nu^W} = \tfrac{1}{2} - \xi + \tfrac{20}{27}\xi^2,$$

$$R_{\bar{\nu}} = \frac{\sigma_{\bar{\nu}}^Z}{\sigma_{\bar{\nu}}^W} = \tfrac{1}{2} - \xi + \tfrac{20}{9}\xi^2.$$

Initially the experimental values of R_ν and $R_{\bar\nu}$ were very different ($R_\nu \cong 0.15$, $R_\nu \cong 0.45$). A trend of convergence has appeared with time, however, and according to the most recent results, $R_\nu = 0.29 \pm 0.01$, $R_{\bar\nu} = 0.35 \pm 0.05$. This immediately gives

$$\xi = \tfrac{1}{2}(1 + R_{\bar\nu} - 3R_\nu) \cong \tfrac{1}{4}.$$

Recall that $\xi = \sin^2 \theta_W$, whence $\theta_W \cong \pm 30°$.

Until now we were discussing only inclusive reactions on heavy nuclei. Reactions on hydrogen were also studied experimentally: elastic, $\nu p \to \nu p$ and $\bar\nu p \to \bar\nu p$; inelastic, $\nu p \to \nu X$ and $\bar\nu p \to \bar\nu X$; and exclusive, $\nu p \to \nu p \pi^0$, $\nu p \to \nu n \pi^+$, and $\nu n \to \nu p \pi^-$. The data for all these reactions are in agreement with the value $\xi \cong \tfrac{1}{4}$. Note that the experiments show the elastic cross section to be different for the neutrino and antineutrino:

$$\frac{\sigma(\bar\nu p \to \bar\nu p)}{\sigma(\nu p \to \nu p)} \cong 0.4 \pm 0.2.$$

This fact is of special importance. The ratio would be exactly unity for the pure vector or pure axial interaction. Hence, experiment points to the interference of these interactions. This means for any model with a single Z-boson that the Z coupling to quarks does not conserve parity. (With two Z-bosons, one of them could be coupled only to the vector current and the other to the axial one. In this case the cross sections for ν and $\bar\nu$ would differ because of the interference of the vector and axial amplitudes even if parity were conserved).

22.5. Isotopic properties of the neutral current

Isotopic properties of the weak neutral current play an important role in the analysis of weak interactions involving nucleons. This holds both for neutrino reactions and for some other processes: P-odd electron-nucleon interactions and P-odd nuclear forces.

The quark neutral current has the form

$$\sum_q (g_L^q \bar q_L \gamma_\alpha q_L - g_R^q \bar q_R \gamma_\alpha q_R),$$

where summation covers all quark flavors $q = $ u, d, s, ... In the standard model,

$$g_L^q = T_3^w - Q\xi,$$

$$g_R^q = -Q\xi,$$

where T_3^w is the weak isospin of the quark in the left-handed isodoublet, and Q is the quark charge. The values of T_3^w and those of T_3 (projection of the usual isospin characterizing hadrons) are identical for the u- and d- quarks which play the decisive role in weak interactions of nucleons:

$$T_3^w(u) = T_3(u) = +\tfrac{1}{2},$$
$$T_3^w(d) = T_3(d) = -\tfrac{1}{2}.$$

This enables us to determine the properties of weak neutral currents in the usual isospace: the right-handed current is an isoscalar (term $Q\xi$), while the left-handed current is a superposition of an isoscalar (the term $Q\xi$, and terms T_3^w for the s-, c-, and b-quarks) and a component of an isovector (terms T_3^w for the u- and d-quarks). Sometimes it is convenient to change from the left-handed and right-handed currents to the vector and axial ones:

$$g_V^q = g_L^q + g_R^q, \quad g_A^q = g_L^q - g_R^q.$$

This gives

$$g_V^q = T_3^w - 2Q\xi,$$
$$g_A^q = T_3^w.$$

The axial neutral current of the u- and d-quarks is therefore a pure isovector. This current is in one triplet with the axial charged currents coupled to the W^+ and W^- bosons. The isoscalar component of the vector current of the u- and d-quarks is proportional to $\sin^2\theta_W$. We recall that this component results from the mixing of the isovector and isoscalar intermediate bosons (see chapter 21).

22.6. Parity non-conservation in scattering of electrons by nucleons

Neutral currents must result in breaking of mirror symmetry in electron-nucleon (or muon-nucleon) interactions. The effective four-fermion interaction responsible for parity violation is

$$4\sqrt{2}\, G[g_L^e \bar{e}_L \gamma_\alpha e_L + g_R^e \bar{e}_R \gamma_\alpha e_R] \sum_q [g_L^q \bar{q}_L \gamma^\alpha q_L + g_R^q \bar{q}_R \gamma^\alpha q_R]$$

$$= -\tfrac{1}{4}\sqrt{2}\, G[(1 - 4\xi)\bar{e}\gamma_\alpha e + \bar{e}\gamma_\alpha\gamma_5 e]$$

$$\times \sum_q [(2T_3 - 4Q\xi)\bar{q}\gamma^\alpha q + 2T_3 \bar{q}\gamma^\alpha\gamma_5 q].$$

The summation is taken over all quark flavors. (We recall that $g_L^e = -\frac{1}{2} + \xi$, $g_R^e = \xi$). Terms of two types violate conservation of parity: (i) the product of the axial quark current with the vector electron current, and (ii) the product of the vector quark current with the axial electron current.

The electron-nucleon P-odd interaction was first observed in bismuth atoms (see the next section). Soon the same phenomenon was found in deep-inelastic scattering of polarized electrons when the asymmetry

$$A = \frac{d\sigma_R - d\sigma_L}{d\sigma_R + d\sigma_L}$$

was measured at the Stanford accelerator (R and L denote the left-handed and right-handed polarization, respectively, of the incident electron beam). The result obtained was $A = (-9.5 \pm 1.6) \cdot 10^{-5}(Q^2/\text{GeV}^2)$ for scattering by deuterium, ed $\to$ eX with the momentum transfer squared $Q^2 \simeq 1.6$ GeV^2. It is easy to derive a theoretical expression for the asymmetry A if one takes into account that asymmetry is produced by the interference of the large P-even amplitude corresponding to photon exchange (fig. 22.10) and the small P-odd amplitude (Fig. 22.11). By restricting the analysis to the contribution of the valence u- and d-quarks, we obtain by means of standard calculations:

$$A = -\frac{GQ^2}{2\sqrt{2}\,\pi\alpha} \frac{9}{10}\left[\left(1 - \frac{20}{9}\xi\right) + (1 - 4\xi)K(y)\right]$$

$$= -1.62 \cdot 10^{-4}\left[\left(1 - \frac{20}{9}\xi\right) + (1 - 4\xi)K(y)\right]\left(\frac{Q}{\text{GeV}}\right)^2.$$

Here

$$K(y) = \frac{1 - (1-y)^2}{1 + (1-y)^2},$$

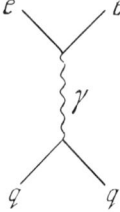

Fig. 22.10.

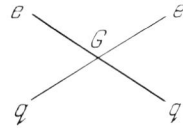

Fig. 22.11.

where $y = 1 - E'/E_0$, and E_0 and E' are the electron energy prior to and after the collision, respectively. (Note that the term $(1 - \frac{20}{9}\xi)$ appears because of the product of the axial electron current with the vector quark current, while the product of the vector electron current with the axial quark current is responsible for the term $(1 - 4\xi)K(y)$.) In the Stanford experiment, $y \simeq 0.21$. A comparison of the theoretical expression for A with its experimental value gives $\xi = 0.20 \pm 0.03$.

22.7. Parity non-conservation in atoms

The product of the axial electron current with the vector quark current is responsible for the appearance of the main P-odd term in the interaction between atomic electrons and nucleons of the nucleus. The point is, the vector "Z-charge" of the nucleus is the the sum of vector "Z-charges" of nucleons, and thus may be quite high in heavy nuclei. By using the vector charges of the proton and neutron, and the electron axial charge, we obtain the interaction

$$\sqrt{2}\, G\big(-\tfrac{1}{2}\bar{e}\gamma_\alpha\gamma_5 e\big)\big[-\tfrac{1}{2}\bar{n}\gamma^\alpha n + \big(\tfrac{1}{2} - 2\xi\big)\bar{p}\gamma^\alpha p\big].$$

In the coordinate representation, this interaction is described by a potential

$$V(r) = \frac{G\kappa}{2\sqrt{2}\, m_e}\big[\boldsymbol{\sigma}\cdot\mathbf{p}\delta(r) + \delta(r)\boldsymbol{\sigma}\cdot\mathbf{p}\big],$$

where $\mathbf{p} = -i\nabla$ is the momentum operator, m_e is the electron mass,

$\kappa = -\tfrac{1}{2}$ for the neutron, and

$\kappa = \tfrac{1}{2}(1 - 4\xi)$ for the proton.

It was established above that $\xi \simeq \tfrac{1}{4}$. Therefore, we expect that the coefficient κ for the proton in the standard model is much smaller than κ for the neutron, and that for the nucleus, $\kappa \simeq -\tfrac{1}{2}(A - Z)$. The potential V

may produce mixing of atomic levels with different parity. As a result, a P-even level acquires a small P-odd admixture. The amplitude F of this admixture may be estimated as follows:

$$F = \frac{\langle V \rangle}{\Delta E},$$

where $\langle V \rangle$ is the matrix element of the potential V between two mixed levels, and ΔE is the spacing between these levels. In hydrogen $\langle V \rangle$ is typically of the order of $Gva^{-3} \simeq Gm_e^2\alpha^4$, since the electron velocity $v \simeq \alpha$ and the Bohr radius $a \simeq 1/m_e$; finally, typical level spacing is $\Delta E \simeq \alpha^2 m_e$. Therefore, $F \simeq \alpha^2 Gm_e^2 \simeq 10^{-15}$. However, this negligibly small quantity may in some cases be enhanced by specific factors. For instance, let us consider the transition between the hydrogen levels $2S_{1/2}$ and $1S_{1/2}$. In the absence of P-odd effects, this is a magnetic dipole transition with a small amplitude $A(M1) \simeq \alpha^2\mu$, where $\mu = e/2m_e$ is the Bohr magneton. (The transition is forbidden in the non-relativistic approximation, because of the orthogonality of the wave functions of the ground and excited S-states).

The P-odd admixture must lead to emission, due to the "impure" $2S_{1/2}$ level, of the electric dipole photon as well. The amplitude of this transition is $FA(E1)$, where $A(E1) \simeq ea$ is the amplitude of the $2P_{1/2} \to 1S_{1/2}$ transition, and F is the mixing amplitude of the levels $2P_{1/2}$ and $2S_{1/2}$. These levels are known to be split only by the Lamb shift which is of the order of $\alpha^5 m_e$. As a result, in this case $F \simeq Gm_e^2/\alpha \simeq 10^{-9}$. The interference of the electric and magnetic transitions must produce circular polarization of photons:

$$P = 2F\frac{A(E1)}{A(M1)} \simeq \frac{Gm_e^2}{\alpha}\left(\frac{ea}{\alpha^2\mu}\right) \simeq \frac{Gm_e^2}{\alpha^4} \simeq 10^{-3}.$$

Unfortunately, observation of this effect in hydrogen is hindered by the same reasons that render it strong: long lifetime of the $2S_{1/2}$ level, and the background of the adjacent $2P_{1/2}$ level. The effect may possibly be detected in beams of hydrogen atoms excited to a metastable level. In heavy atoms (Tl, Pb, Bi) the effect is smaller, viz. $P \simeq Gm_e^2\alpha^2 Z^3(ea/\mu) \simeq 10^{-7}$, but easier to observe.

Parity non-conservation in atomic transitions was first observed early in 1978 by Barkov and Zolotorev (Novosibirsk). They measured the rotation angle of the polarization plane $((-3 \pm 0.5) \cdot 10^{-8}$ rad) in a linearly polarized laser light after transmission through atomic bismuth vapor. Both the value and the sign of the observed optical activity of bismuth vapor are in agreement with the predictions of the standard electroweak model for $\sin^2 \theta_W \simeq \frac{1}{4}$.

Another parity non-conservation effect was found at the end of 1978 in Berkeley when a circularly polarized laser beam was passed through thallium vapor; the absorption cross section for right-hand polarized photons was found to be higher than that for left-hand polarized photons, which again is in agreement with the standard electroweak model. While the effect in bismuth is due to parity non-conservation in the real part of the refractive index, in thallium it was measured in the imaginary part.

Parity non-conservation effects are also expected for μ-atoms, but so far experimental accuracy is too low to enable detection.

22.8. *P*-odd nuclear forces

We shall not discuss in detail the contribution of neutral currents to the nucleon-nucleon weak interaction. The experimentally observed parity non-conservation in nuclear forces may be caused both by charged and by neutral currents. The effect of neutral currents could be ascertained by measuring the isovector component of *P*-odd nuclear forces. The reason for this is that this term appears with only a small factor, $\sin^2 \theta_C$, when occurring in the product of charged currents, but is not suppressed when occurring in the product of neutral currents. In particular, the term with $\Delta T = 1$ must appear in the S-wave *P*-odd isovector vertex $fe_{ikl}\bar{N}\tau_i N \pi_k$ (similar vertices with $\Delta T = 0, 2$ are forbidden by *CP* conservation).

An analysis of a comparatively large effect of parity non-conservation in the nuclei ^{19}F, ^{41}K, ^{175}Lu and ^{181}Ta, observed in a number of experiments, indicates that f is higher by about an order of magnitude than could be expected in the case of charged currents only. If this interpretation is correct, one must expect considerable, $\cong 0.5\%$, circular polarization of photons emitted in transitions from the excited level 0^+ to the ground level 1^+ in the nucleus ^{18}F. (The effect is anticipated to be high, because of the presence of a 0^- level with $T = 1$ close to the level 0^+ with $T = 0$.) Unfortunately, this prediction could be very much suppressed by purely nuclear effects, whose magnitudes are not yet known.

CHAPTER 23

Properties of intermediate bosons

In this chapter we shall calculate partial and full widths of the W- and Z-bosons, and discuss projects for experiments designed to produce these particles.

The masses of intermediate bosons are unambiguously predicted by the standard model of the electroweak interaction provided the Weinberg angle is known:

$$m_W = \left(\frac{\pi\alpha}{\sqrt{2}\,G}\right)^{1/2} \frac{1}{\sin\theta_W} = \frac{37.3}{\sin\theta_W}\text{ GeV},$$

$$m_Z = \frac{m_W}{\cos\theta_W} = \frac{74.6}{\sin 2\theta_W}\text{ GeV}.$$

The neutrino experiments discussed in Chapter 22 yield $\sin\theta_W \cong \frac{1}{2}$. This means that $m_W \cong 75$ GeV, $m_Z \cong 87$ GeV. If the model is extended by introducing additional bosons, some of them may be much lighter. In this chapter, however, we shall not leave the framework of the standard model.

23.1. Decays of W-bosons

We begin with calculating the width of the $W^\pm \to e^\pm \nu$ decay. We remind the reader that the coupling of the W-boson to the $\bar{e}\nu$ current is fixed by gauge invariance. It immediately follows from the term

$$i\bar{L}\hat{D}L = i\bar{L}(\hat{\partial} - \tfrac{1}{2}ig\boldsymbol{\tau}\cdot\hat{\boldsymbol{A}} - \tfrac{1}{2}ig'Y\hat{B})L,$$

where

$$L = \begin{pmatrix} \nu_{eL} \\ e_L \end{pmatrix}, \qquad \boldsymbol{\tau}\cdot\boldsymbol{A} = \tau_3 A_3 + \tau_+ W^- + \tau_- W^+,$$

that the Weν vertex has the form

$$\tfrac{1}{2}\sqrt{2}\,g(\bar{\nu}_L \hat{W} e_L + \bar{e}_L \hat{W}^* \nu_L).$$

We make use of the standard formula (see appendix, Chapter 29, sect. 4):

$$\Gamma = \frac{\overline{|M|^2}}{3 \cdot 2 m_W \cdot 8\pi}.$$

Here the factor $\tfrac{1}{3}$ appears as a result of averaging over spin states of the W-boson, and the bar denotes summation over spin states of the leptons and the W-boson.

$$\overline{|M|^2} = \tfrac{1}{2}g^2 \left(-g^{\alpha\beta} + \frac{k^\alpha k^\beta}{m_W^2}\right) \tfrac{1}{4} \operatorname{Tr} \hat{p}_e \gamma_\alpha \hat{p}_\nu \gamma_\beta (1+\gamma_5)^2$$

$$= \tfrac{1}{2}g^2 4(p_e p_\nu) = g^2 m_W^2.$$

The term $k^\alpha k^\beta / m_W^2$ in the W-boson density matrix gives zero contribution since the electron mass was neglected. Then we immediately obtain:

$$\Gamma(W \to e\nu) = \frac{g^2 m_W}{48\pi} = \frac{G m_W^3}{6\sqrt{2}\,\pi} \simeq 180 \text{ MeV}, \quad \text{if } m_W \simeq 75 \text{ GeV}$$

(we have made use of the equality $g^2/2m_W^2 = 4G/\sqrt{2}$).

Compared to the W-boson mass, we can neglect not only the electron mass but also the muon, and τ-lepton masses; consequently,

$$\Gamma(W \to e\nu_e) = \Gamma(W \to \mu\nu_\mu) = \Gamma(W \to \tau\nu_\tau).$$

Consider hadronic decays of the W-boson. According to the parton model, the total probability of the W-boson decay to a hadronic state with the quantum numbers of the $u\bar{d}$ pair is equal to the probability of decay into this pair calculated with strong interactions switched off. The same is true for the $u\bar{s}$, $c\bar{s}$, and $c\bar{d}$ pairs. Taking into account the Cabibbo angle θ_C and the color factor 3, and neglecting quark masses, we obtain:

$$\Gamma(W \to u\bar{d}) = \Gamma(W \to c\bar{s}) = 3\Gamma(W \to e\nu)\cos^2\theta_C,$$

$$\Gamma(W \to u\bar{s}) = \Gamma(W \to c\bar{d}) = 3\Gamma(W \to e\nu)\sin^2\theta_C.$$

So far we know very little about b- and t-quarks. As an estimate, assume

$$\Gamma(W \to t\bar{b}) = 3\Gamma(W \to e\nu).$$

Finally, the total width of the W-boson is (provided no heavier quarks exist)

$$\Gamma_{\text{tot}}^W \cong 12\Gamma(W \to e\nu) \cong 2.2 \text{ GeV}.$$

23.2. Decays of Z-bosons

It will be convenient to begin our analysis of Z decays by considering the vertex

$$\tfrac{1}{2}\bar{g}\bar{\nu}_{eL}\hat{Z}\nu_{eL}.$$

The corresponding width is

$$\Gamma(Z \to \nu_e\bar{\nu}_e) = \frac{\bar{g}^2 m_Z}{96\pi} = \frac{G m_Z^3}{12\sqrt{2}\,\pi} \simeq 140 \text{ MeV} \quad \text{for } m_Z \simeq 87 \text{ MeV}.$$

(we recall that $\bar{g}/g = m_Z/m_W = 1/\cos\theta_W$ so that

$$\frac{\Gamma(Z \to \nu_e\bar{\nu}_e)}{\Gamma(W \to e\nu_e)} = \frac{1}{2\cos^3\theta_W} \simeq 0.77 \text{ for } \theta_W = 30°).$$

Obviously, $\Gamma(\nu_e\bar{\nu}_e) = \Gamma(\nu_\mu\bar{\nu}_\mu) = \Gamma(\nu_\tau\bar{\nu}_\tau)$. It is not difficult, using the expression for the "Z-charge", $-\bar{g}(T_3 - \xi Q)$, to write expressions for Z-boson couplings to other leptons and quarks (recall that $\xi = \sin^2\theta_W$):

$$\bar{g}\left[(-\tfrac{1}{2} + \xi)\bar{e}_L\hat{Z}e_L + \xi e_R\hat{Z}e_R\right] \quad \text{(the same for } \mu \text{ and } \tau);$$

$$\bar{g}\left[(\tfrac{1}{2} - \tfrac{2}{3}\xi)\bar{u}_L\hat{Z}u_L - \tfrac{2}{3}\xi\bar{u}_R\hat{Z}u_R\right] \quad \text{(the same for c and t);}$$

$$\bar{g}\left[(-\tfrac{1}{2} + \tfrac{1}{3}\xi)\bar{d}_L\hat{Z}d_L + \tfrac{1}{3}\xi\bar{d}_R\hat{Z}d_R\right] \quad \text{(the same for s and b).}$$

From these we obtain:

$$\Gamma(Z \to e\bar{e}) \simeq \Gamma(Z \to \mu\bar{\mu}) \simeq \Gamma(Z \to \tau\bar{\tau})$$
$$= \left[(-1 + 2\xi)^2 + 4\xi^2\right]\Gamma(Z \to \nu_e\bar{\nu}_e) \simeq 0.5\Gamma(Z \to \nu_e\bar{\nu}_e);$$

$$\Gamma(Z \to u\bar{u}) \simeq \Gamma(Z \to c\bar{c}) \simeq \Gamma(Z \to t\bar{t})$$
$$= 3\left[(1 - \tfrac{4}{3}\xi)^2 + (\tfrac{4}{3}\xi)^2\right]\Gamma(Z \to \nu_e\bar{\nu}_e) \simeq \tfrac{5}{3}\Gamma(Z \to \nu_e\bar{\nu}_e);$$

$$\Gamma(Z \to d\bar{d}) \simeq \Gamma(Z \to s\bar{s}) \simeq \Gamma(Z \to b\bar{b})$$
$$= 3\left[(-1 + \tfrac{2}{3}\xi)^2 + (\tfrac{2}{3}\xi)^2\right]\Gamma(Z \to \nu_e\bar{\nu}_e) \simeq \tfrac{13}{6}\Gamma(Z \to \nu_e\bar{\nu}_e).$$

The total width of the Z-boson for $\xi = \tfrac{1}{2}$ is

$$\Gamma_{\text{tot}}^Z \simeq 8(3 - 6\xi + 8\xi^2)\Gamma(Z \to \nu_e\bar{\nu}_e) \simeq 16\Gamma(Z \to \nu_e\bar{\nu}_e) \simeq 2.2 \text{ GeV}.$$

By measuring Γ_{tot}^Z one can find whether the products of Z-boson decays include, in addition to ν_e, ν_μ, ν_τ, neutrinos of other types.

23.3. Production of Z-bosons

An optimal method of creation of a Z-particle is to use colliding e^+e^- beams:

$$e^+e^- \to Z \to \text{decay products}.$$

Production of a Z-boson must appear as a resonance at $E = \sqrt{s} = m_Z$, with width $\Gamma \cong 2$ GeV. In a resonance, the cross section for transition to the final state f is given by the Breit-Wigner formula:

$$\sigma = \frac{4\pi(2J+1)}{2 \cdot 2 m_Z^2} \frac{\Gamma_i \Gamma_f}{(E-m_Z)^2 + \tfrac{1}{4}\Gamma^2}.$$

Here $J = 1$ is the spin of the Z-boson. The factor $2 \cdot 2$ in the denominator comes from averaging over the electron and positron spins, and

$$\Gamma_i = \Gamma(Z \to e^+e^-), \quad \Gamma_f = \Gamma(Z \to f), \quad \Gamma = \Gamma(Z \to \text{all}).$$

Assume that the resonance is realized, that is $(E - m_Z)^2 \ll \tfrac{1}{4}\Gamma^2$, and that we wish to find the total cross section, that is $\Gamma_f = \Gamma$. We have:

$$\sigma = \frac{12\pi}{m_Z^2} \frac{\Gamma(Z \to e\bar{e})}{\Gamma} \cong \frac{0.5}{16} \frac{12\pi}{m_Z^2} \cong 4 \cdot 10^{-32} \text{ cm}^2.$$

It means that colliding electron-positron beams with energy of about 2×45 GeV and luminosity $L = 10^{32}$ event/s·cm² would produce several Z-bosons per second. This would be a veritable Z-boson factory. It should be instructive to compare the resonance cross section of the process

$$e^+e^- \to Z \to \mu^+\mu^-,$$

with that of electromagnetic production $e^+e^- \to \gamma \to \mu^+\mu^-$ at $\sqrt{s} = m_Z$. The former cross section is equal to $(12\pi/m_Z^2)(\Gamma_{ee}/\Gamma)^2$, and the latter is $\tfrac{4}{3}\pi(\alpha/m_Z)^2$. Their ratio is

$$\frac{9}{\alpha^2}\left(\frac{\Gamma_{ee}}{\Gamma}\right)^2 \simeq 160.$$

Note that for $\sqrt{s} = m_Z$, the expected energy resolution in colliding beams is of the order of 100 MeV, that is much less than the Z-boson width. Therefore radiative corrections to the above expressions are fairly small, of the order of α, and without large logarithmic terms (as in the case of production of very narrow resonances, such as the J/ψ meson).

23.4. Production of W-bosons

Only non-resonance production of pairs of charged W-bosons is possible in colliding e^+e^- beams (fig. 23.1). The pair production cross section of W-bosons is smaller than the resonance production cross section of Z-bosons by approximately 3–4 orders of magnitude. Besides, this process requires even higher energies for the colliding electrons and positrons. It seems more feasible, therefore, to realize production of single W-bosons in colliding $p\bar{p}$ or pp beams with energies of several hundreds of GeV or more in each beam, i.e., the inclusive reactions

$$p + \bar{p} \rightarrow W + X,$$
$$p + p \rightarrow W + X,$$

where X stands for any hadronic state.

These reactions are based on resonance production of W-bosons in quark-antiquark collisions. We have already mentioned in chapter 17 that a fast proton can be treated as an ensemble or partons, that is quarks, antiquarks, and gluons which carry approximately 44, 7, and 49% of its momentum, respectively. The cross section of W production in a collision of a u-quark (from the proton) and a $\bar{d}$-quark (from the colliding antiproton) is equal to

$$\sigma = \frac{12\pi}{2 \cdot 2 \cdot 3 \cdot 3 m_W^2} \frac{\Gamma_i \Gamma}{(E - m_W)^2 + \frac{1}{4}\Gamma^2}.$$

This formula is quite similar to that describing production of Z-bosons in e^+e^- collisions, and differs by a factor $3 \cdot 3$ in the denominator. This factor appears as a result of averaging over color states of the colliding quarks and antiquarks. (With other conditions equal, the width of the W-decay into colored quarks is three times that of the decay into colorless leptons since the sum over three colors is included. However, the cross section of W production from colored quarks is smaller by a factor of three than that of

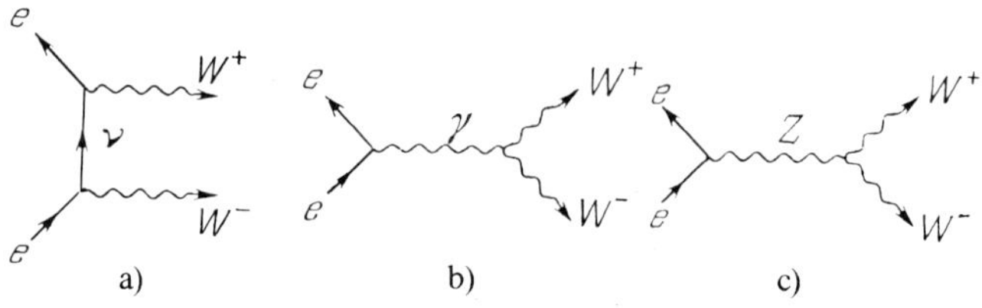

Fig. 23.1.

§4] Production of W-bosons

production from colorless leptons, since a quark of a given color can annihilate into W only with an antiquark of the same color. The reduction factor is three and not nine because one of the factors of three in the denominator is cancelled by a factor of three in the numerator.) Hence, the cross section of resonance W production would be equal to

$$\sigma_{res}(u + \bar{d} \to W) = \frac{12\pi}{9m_W^2} \frac{\Gamma(W \to u\bar{d})}{\Gamma} \simeq \frac{12\pi}{9m_W^2} \frac{3}{12}$$

$$= \frac{\pi}{3m_W^2} \simeq 6 \cdot 10^{-32} \text{ cm}^2.$$

Both the proton and antiproton can be viewed as bunches of quarks and antiquarks whose energies form a continuous spectrum. It is therefore convenient to write out the cross section of W production by quarks in the limit in which the quark energy spread is much wider than the W-boson width. In this limit it has the following form:

$$\sigma(u + \bar{d} \to W \to f) = \frac{\pi}{3m_W^2} \frac{\Gamma_i \Gamma_f}{(E - m_W)^2 + \tfrac{1}{4}\Gamma^2}$$

$$= \frac{4\pi}{3} \frac{\Gamma_i \Gamma_f}{\left(s_{u\bar{d}} - m_W^2\right)^2 + m_W^2 \Gamma^2}$$

$$\Rightarrow \frac{4\pi}{3} \frac{\Gamma_i \Gamma_f}{m_W \Gamma} \pi \delta(s_{u\bar{d}} - m_W^2) = \frac{4\pi^2}{3} \frac{\Gamma_i \Gamma_f}{m_W \Gamma s} \delta(x_1 x_2 - \tau),$$

where $\Gamma_i = \Gamma(W \to u\bar{d})$, $s = (p_p + p_{\bar{p}})^2$, $\tau = m_W^2/s$, the u-quark 4-momentum is equal to $x_1 p_p$, and that of the $\bar{d}$-quark is $x_2 p_{\bar{p}}$; p_p and $p_{\bar{p}}$ are the proton and antiproton 4-momenta, respectively. Therefore,

$$s_{u\bar{d}} = (p_u + p_{\bar{d}})^2 \simeq 2 p_u p_{\bar{d}} = x_1 x_2 \cdot 2 p_p p_{\bar{p}}$$

$$\simeq x_1 x_2 (p_p + p_{\bar{p}})^2 = x_1 x_2 s.$$

Averaging over the u-quark and $\bar{d}$-quark spectra yields the following cross section of W production in channel i (followed by decay in the channel f):

$$\sigma = \frac{4\pi^2}{3} \frac{\Gamma B_i B_f}{m_W^3} \int u(x_1) d(x_2) \delta(x_1 x_2 - \tau) \, dx_1 \, dx_2.$$

Here $B_i = \Gamma_i/\Gamma \equiv \Gamma(W \to u\bar{d})/\Gamma \simeq \tfrac{1}{4}$. Let us choose the channel $W \to \mu\nu$ as an example of channel f. Then $B_f \Gamma = \Gamma(W \to \mu\nu) = Gm_W^3/6\pi\sqrt{2}$. If the integrand included no δ-function, we would have

$$\sigma \simeq \frac{\pi GUD}{28\sqrt{2}} \simeq 2.4 \cdot 10^{-35} \text{ cm}^2.$$

Recall that $U = \int u(x)\,dx \simeq 0.28$, $D = \int d(x)\,dx \simeq 0.15$ (distribution of d-quarks in the antiproton is the same as that of d-quarks in the proton), so that $UD \simeq 0.04$.

Let us change variables from x_1, x_2 to $\tau' = x_1 x_2$ and $\xi = x_1 - x_2$. Then $x_{1,2} = \tfrac{1}{2}\xi \pm \sqrt{\tfrac{1}{4}\xi^2 + \tau'}$ and

$$\int u(x_1)\,d(x_2)\delta(x_1 x_2 - \tau)\,dx_1\,dx_2 = \int \frac{u(x_1)\,d(x_2)}{\sqrt{\xi^2 + 4\tau'}}\,d\xi.$$

The variable ξ is the longitudinal momentum of the W-boson in units of the maximum possible momentum, $\tfrac{1}{2}\sqrt{s}$. It will be convenient to introduce a normalized quantity K, whose values are obtained by numerical integration for various values of τ:

$$K = \frac{1}{UD}\int \frac{u(x_1)\,d(x_2)}{\sqrt{\xi^2 + 4\tau}}\,d\xi; \qquad K(\tau = 10^{-2}) \simeq 10,$$

$$K(\tau = 10^{-1}) \simeq 4.$$

The above estimates of the cross section based on the parton model constitute a distant extrapolation from the available experimental data. Quantum chromodynamics predicts deviations from the parton model. In particular, emission of bremsstrahlung gluons must increase the transverse momenta of the annihilating quark and antiquark, and hence the transverse momenta of the W-bosons produced. Nevertheless, QCD calculations show that qualitatively the parton model predictions hold.

In order to have abundant collisions of sufficiently fast quarks, the energy $\sqrt{s}$ of the colliding p and p̄ must be several times larger than the W-boson mass, that is the parameter τ must be $\lesssim \tfrac{1}{10}$. If two proton beams collide, their energy $\sqrt{s}$ must be still higher since the "sea" antiquarks in the proton are much softer than valence antiquarks in the antiproton.

23.5. Colliding beam projects

At present there is no doubt that clarification of the dynamics of weak interactions calls for colliding beam experiments at energies above the Z- and W-boson thresholds.

Projects of colliding lepton beams with energies of the order of 50 to 100 GeV in each beam have been the subject of discussion since the beginning of the late 60's. For instance, the Yerevan project of colliding e^+e^- beams for 2×100 GeV and Budker's idea of colliding $\mu^+\mu^-$ beams for 2×250 GeV were debated at the 1969 Yerevan International Conference on Accelerators. In 1975 Richter published a detailed analysis of colliding e^+e^- beams with energy 2×100 GeV. By Richter's estimates, a machine of this type will cost around 10^9 dollars. Such a machine would be an extremely powerful generator of synchrotron radiation whose industrial application would pay back at least partly the cost of construction. Detailed plans for the LEP project of colliding electron-positron beams, with energy of about 80 GeV in each beam, were discussed at CERN. LEP is a ring about 10 km in diameter, with luminosity of the order of $10^{31} - 10^{32}$ cm^{-2}s^{-1} and energy spread of the order of 0.1%.

The pp and p̄p colliders will be built in the following laboratories. An accelerating and storage complex (UNK, for Uskoritel'no Nakopitel'ny Kompleks) with energies of colliding beams equal to 3 TeV is being considered for construction at Serpukhov. The proton-proton storage ring ISABELLE (2×400 GeV) is under construction at Brookhaven. The construction of two p̄p colliders, at CERN (2×270 GeV) and FNAL (2×1000 GeV), will be completed in the eighties.

23.6. DUMAND

It would be interesting to observe creation of W-bosons in a purely leptonic collision. The neutrino energy required to produce a W-boson in a collision $\bar{\nu}e^- \rightarrow W^-$ is

$$E_0 = \frac{m_W^2}{2m_e} \cong 5 \cdot 10^{15} \text{ eV}.$$

Such neutrinos can only be found in cosmic rays. The process $e\nu \rightarrow W$ could yield unique information on the intensity of the flux of cosmic neutrinos.

The project DUMAND (for Deep Underwater Muon and Neutrino Detector), which is now actively discussed, could in principle allow observation of this reaction. The project envisages the construction deep in the ocean of a gigantic cubic lattice of detectors, with the length of the cube edge of about one kilometer. The lattice period is to be about 10 m if the detectors are optical, and about a hundred meters if they are acoustic. The project would make it possible to work with a water target whose mass exceeds 10^{15} g, and to observe neutrino reactions at very high energies.

Assume the differential flux of neutrinos with energy E to be $j(E)$ neutrino/cm²·s·sr. Let us find the number of $\bar{\nu} + e^- \to W^-$ events per second, n, in a target containing N_e electrons:

$$n = 2\pi \int N_e j(E) \frac{dE}{dE'} dE' \sigma(E').$$

Here E is the laboratory energy of the neutrino, and E' is the total energy in the center-of-mass frame of the electron and neutrino:

$$E'^2 = 2m_e E \quad \text{and} \quad \frac{dE}{dE'} = \frac{E'}{m_e} = \frac{m_W}{m_e}.$$

The factor 2π appears because resonance-energy neutrinos do not enter the detector from the lower hemisphere, being absorbed within the earth. This angular asymmetry is an indication of W-boson production. We now take into account that

$$\sigma(E') = \frac{4\pi(2J+1)}{2m_W^2} \frac{\Gamma_i \Gamma}{(E' - m_W)^2 + \tfrac{1}{4}\Gamma^2}$$

$$\simeq \frac{12\pi^2}{m_W^2} \Gamma(W \to e\nu) \delta(E' - m_W).$$

Then

$$n = 4\sqrt{2}\,\pi^2 j(E_0) E_0 G N_e.$$

Therefore, the expected number of events is determined by a single variable, namely by the neutrino flux $j(E_0)$ at the neutrino energy E_0 corresponding to the W-boson mass: $E_0 = m_W^2/2m_e$. (In the literature the integral flux $\Phi(E)$, obeying a power law, is used:

$$\Phi(E) = \int_E^\infty j(E)\,dE = A E^{-\gamma},$$

Hence, in this case $E_0 j(E_0) = \gamma \Phi(E_0)$.)

CHAPTER 24

Properties of Higgs bosons

This chapter will discuss in more detail the role played by Higgs bosons (H-bosons) in weak interactions at high energies. We shall also consider the possible mechanisms of creation and decay of these particles.

We have established earlier, when discussing how the electron mass appears in the standard model of the electroweak interaction, that the coupling of the electron to a scalar Higgs boson represented by the real field χ is determined by the value of the electron mass:

$$\frac{m_e}{\eta}(\eta + \chi)\bar{e}e = m_e\bar{e}e + \left(\sqrt{2}\,G\right)^{1/2} m_e \chi \bar{e}e$$

(recall that $\eta = \left(\sqrt{2}\,G\right)^{-1/2} = 246$ GeV). Interaction of H-bosons with other fermions is similar: the coupling is the stronger the heavier the fermion is. H-bosons are coupled weakly to light leptons and light quarks. This conclusion is valid if the theory contains only one isodoublet of scalar mesons, as the standard electroweak model does. With more than one doublet, some of the bosons may have anomalously strong coupling to light particles.

Before considering the reactions involving H-bosons, a few remarks must be made about the expected masses of these particles.

24.1. On the H-boson mass

In contrast to the W- and Z-boson masses, the mass of the H-boson is not fixed by the Weinberg angle and represents a free parameter of the model. The H-boson mass is determined by the potential

$$V(\varphi) = \tfrac{1}{2}\lambda^2\left(|\varphi|^2 - \tfrac{1}{2}\eta^2\right)^2$$

$$= \tfrac{1}{2}\lambda^2\left(|\varphi| - \sqrt{\tfrac{1}{2}}\,\eta\right)^2\left(|\varphi| + \sqrt{\tfrac{1}{2}}\,\eta\right)^2 = \tfrac{1}{8}\lambda^2(2\eta + \chi)^2\chi^2$$

and equals

$$m_H = \lambda \eta.$$

Nothing was said so far about the dimensionless constant λ. The smaller λ, the lighter H-bosons are. However, λ cannot be very small. Indeed, λ characterizes the nonlinear self-coupling of the field χ which cannot be very small because H-bosons must be coupled, due to exchange by pairs of Z- and W-bosons (fig. 24.1a). Hence, even with a small bare constant λ^2, the effective value of λ^2 (taking into account quantum corrections) will not be smaller than a quantity of the order of $\lambda^2_{\min} = e^4 = \alpha^2$, that is $\lambda \geq \alpha$. Consequently, $m_H > \alpha \eta$.

To be more precise, the argument is based not only on the graphs of fig. 24.1a but also on those of figs. 24.1b, c and other more complicated graphs with a large number of Higgs "whiskers". It can be shown that the sum of these graphs yields an effective potential $\tilde{V}(\varphi)$ of the following type:

$$\tilde{V}(\varphi) = \mu^2 |\varphi|^2 + \gamma |\varphi|^4 \ln(|\varphi|^2/m^2).$$

Here $\gamma = 3(m_Z^4 + 2m_W^4)/16\pi^2 \eta^4$ (in the general case when the vector, scalar, and fermion loops are taken into account,

$$\gamma = \frac{3\sum_v m_v^4 + \sum_s m_s^4 - 4\sum_f m_f^4}{16\pi^2 \eta^4};$$

however, hereafter we shall neglect the contribution of scalar and fermion loops). The parameters m and μ are renormalization constants and must be chosen in such a manner that $\tilde{V}(|\varphi|^2)$ has a minimum at $|\varphi|^2 = \eta^2$ (recall that $|\varphi|^2 = \frac{1}{2}(\eta + \chi)^2$). The condition $\tilde{V}'(|\varphi|)|_{|\varphi| = \eta/\sqrt{2}} = 0$ gives a relation between μ^2 and m^2:

$$\gamma \ln \frac{\eta^2}{2m^2} = -\frac{1}{2}\left(\frac{2\mu^2}{\eta^2} + \gamma\right).$$

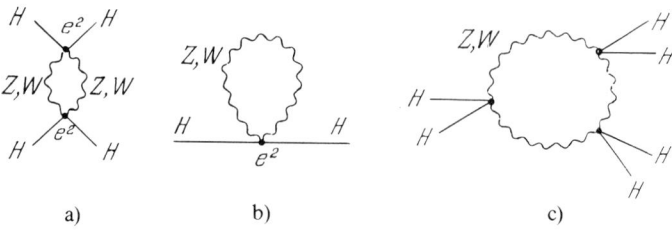

Fig. 24.1.

We easily see that for $\mu^2 > 0$, the potential $\tilde{V}(|\varphi|)$ has another minimum, at $\varphi = 0$. If we demand this minimum to be unstable and the one at $|\varphi|^2 = \frac{1}{2}\eta^2$ to be stable, it is necessary that $\tilde{V}(\frac{1}{2}\eta^2) < 0$, whence

$$\frac{2\mu^2}{\eta^2} + \gamma \ln \frac{\eta^2}{2m^2} < 0$$

or

$$2\mu^2 < \gamma\eta^2.$$

Let us now find the restriction on the H-boson mass imposed by this inequality:

$$m_H^2 = \frac{d^2\tilde{V}}{d\chi^2} = \frac{1}{2}\frac{d^2\tilde{V}}{d|\varphi|^2} = \mu^2 + 3\gamma\eta^2 \ln \frac{\eta^2}{2m^2} + \tfrac{7}{2}\gamma\eta^2$$

$$= 2\gamma\eta^2 - 2\mu^2 = \gamma\eta^2 + (\gamma\eta^2 - 2\mu^2) \geq \gamma\eta^2.$$

Therefore,

$$m_H \geq \sqrt{\frac{3(m_Z^4 + 2m_W^4)}{16\pi^2\eta^2}} = \frac{\alpha\eta}{4\sin^2\theta_W}\sqrt{3\left(2 + \frac{1}{\cos^4\theta_W}\right)} \cong 7.3 \text{ GeV}.$$

(This value is obtained for $\sin^2\theta_W = 0.2$). Were this inequality violated in the model with a single Higgs doublet, our physical vacuum (at $|\varphi| = \eta/\sqrt{2}$) would be unstable and sooner or later would spontaneously explode, switching to the stable vacuum with $\varphi = 0$. It must be emphasized, however, that the probability of explosion is exponentially small for $m_H \gtrsim 0.6$ GeV.

24.2. The role of H-bosons at high energies

We cannot indicate the upper bound for the H-boson mass. If, however, it is much larger than those of the W- and Z-bosons, the coupling in the sector of H-, W-, and Z-particles becomes strong, so that perturbation theory fails at $m_H \gtrsim 1$ TeV. This is easy to understand qualitatively: we recall that λ^2 characterizes not only the self-coupling of the H-boson field (the terms $\lambda^2\chi^4$ and $\lambda^2\eta\chi^3$) but also the interaction between longitudinal components of the $W^\pm$ and Z-boson fields and between them and H-bosons. Indeed, all these four fields (W_3^+, W_3^-, Z_3^0, and H) are contained in the isodoublet φ whose self-coupling is described by the potential $V(\varphi)$.

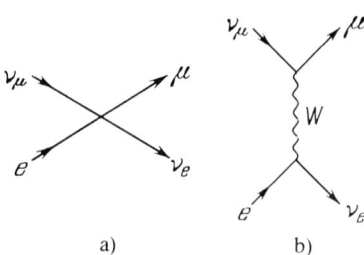

Fig. 24.2.

If $m_H \gtrsim 1$ TeV, then the strong interaction of the H-, Z-, and W-bosons at energies of the order of 1 TeV may resemble, by the diversity of its manifestations, the resonance region of the usual hadrons in the range around 1 GeV. If, however, $m_H \lesssim m_W$, then perturbation theory must be valid at high energies, and all amplitudes growing with energy will cancel out. The H-bosons play an important role in this compensation. We recall (see chapter 18) that the linear increase of the cross section ($\sigma \propto G^2 s$), corresponding to the graph of fig. 24.2a, is eliminated when the W-boson is included (fig. 24.2b).

However, the processes involving W-bosons continue to grow as s increases. This is illustrated by the cross section corresponding to the graph of fig. 24.3a. For the vector electron current the growth is suppressed when compensation of graphs given in figs. 24.3a and b is taken into account. Note, however, that this compensation is achieved only for the vector electron current but not for the axial one. In the latter case

$$\left.\frac{d\sigma}{dt}\right|_{|t|\approx s} \propto \frac{m_e^2 s}{m_Z^4} \frac{1}{s^2},$$

while unitarity requires

$$\left.\frac{d\sigma}{dt}\right|_{|t|\approx s} \propto \frac{1}{s^2}.$$

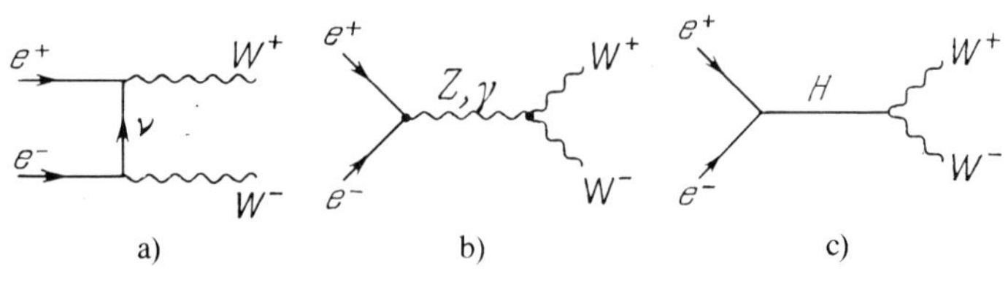

Fig. 24.3.

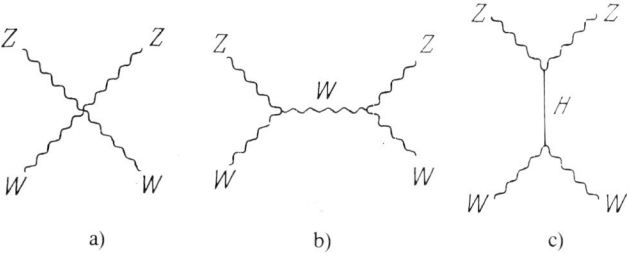

Fig. 24.4.

Addition of a graph with an H-boson (fig. 24.3c) eliminates the term in the amplitude proportional to the electron mass, and results in the desired behavior of the cross section.

Non-renormalizability of the theory without H-bosons manifests itself even better in the scattering of intermediate bosons on one another. It is not difficult to show (see Chapter 19, sect. 7) that the cross section corresponding to the sum of graphs in figs. 24.4a and b grows linearly with s. This growth is suppressed by addition of the H-boson because the graphs in figs. 24.4a, b and c cancel each other out. Therefore, for a very large H-boson mass, the scattering of intermediate bosons may exceed the unitarity limit before the graph 24.4c becomes effective. This will mean the strong interaction of intermediate bosons.

24.3. Coupling of H-bosons to heavy quarks

An H-boson with mass below that of the Υ-meson could be produced in decays of the Υ-meson or of heavier mesons composed of heavier quarks. Indeed, the quark–H-boson coupling is the stronger the heavier the quark is. Let us compare the decay rates for $\Upsilon \to H + \gamma$ (fig. 25.5a) and $\Upsilon \to \mu^+\mu^-$

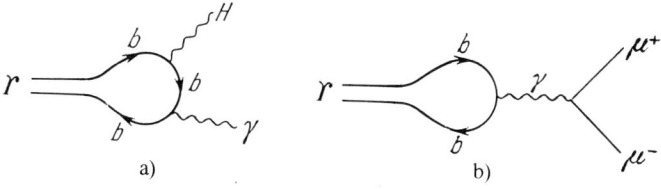

Fig. 24.5.

(fig. 25.5b). It can easily be shown that

$$\frac{\Gamma(\Upsilon \to H\gamma)}{\Gamma(\Upsilon \to \mu^+\mu^-)} = \frac{Gm_b^2}{\sqrt{2}\,\pi\alpha}\left(1 - \frac{m_H^2}{m_\Upsilon^2}\right) = 0.008\left(1 - \frac{m_H^2}{m_\Upsilon^2}\right),$$

(the numerical coefficient 0.008 corresponds to $m_b = 4.75$ GeV) which is quite measurable. The main decays of the H-boson must be $H \to \tau^+\tau^-$ and $H \to c\bar{c}$, provided its mass is in the range from 7 to 9 GeV. The latter channel implies decays into pairs of charmed particles: $D\bar{D}$, $F\bar{F}$, $D^*\bar{D}^*$, and so on. Taking into account the color factor 3, we obtain:

$$\frac{\Gamma(H \to c\bar{c})}{\Gamma(H \to \tau^+\tau^-)} = \frac{3m_c^2}{m_\tau^2} = \frac{3(1.3)^2}{1.8^2} \cong 1.6.$$

In its turn,

$$\Gamma(H \to \tau^+\tau^-) = \frac{1}{8\pi}\sqrt{2}\,Gm_H m_\tau^2 v^3, \quad \text{where } v = \left(1 - \frac{4m_\tau^2}{m_H^2}\right)^{1/2}.$$

24.4. Coupling of H-bosons to gluons

The gluon mass being zero, there is no direct coupling of H-bosons to gluons. However, there must be an indirect coupling due to quark loops (fig. 24.6). The decay amplitude is

$$M(H \to 2g) = N_H \frac{\alpha_s(m_H)}{6\pi}(\sqrt{2}\,G)^{1/2}\chi G_{\mu\nu}^{1a}G_{2a}^{\mu\nu},$$

where $G_{\mu\nu}^{ia} = k_{i\mu}B_{i\nu}^a - k_{i\nu}B_{i\mu}^a$, $i = 1,2$, $a = 1,2,\ldots,8$, k_1 and k_2 are 4-momenta of two gluons, B_1 and B_2 are their wave functions, χ is the H-boson wave function, $\alpha_s = g_s^2/4\pi$ is the chromodynamic running coupling constant (the $H \to 2g$ decay amplitude is determined by the value of $\alpha(Q)$ for $Q \simeq m_H$), and N_H is the number of flavors of heavy quarks. Here we refer to quarks as heavy if their masses satisfy the inequality $m_q > \frac{1}{2}m_H$ where m_H is the H-boson mass. It is significant that the contribution of

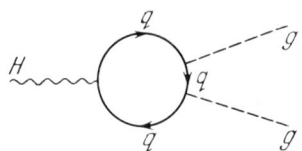

Fig. 24.6.

Coupling of H-bosons to gluons

heavy quarks is independent of the masses of these quarks. Hence, even a comparatively light H-boson is something of a "cash register" of heavy quarks (including quarks so heavy that we cannot hope to produce them even using future accelerators). If N_H is sufficiently high, then the decays H → 2g → hadrons become predominant. It is readily obtained that

$$\Gamma(H \to 2g) = N_H^2 \left(\frac{\alpha_s(m_H)}{\pi}\right)^2 \frac{\sqrt{2}\,G}{72\pi} m_H^3$$

and therefore,

$$\frac{\Gamma(H \to 2g \to \text{hadrons})}{\Gamma(H \to \tau^+\tau^-)} = N_H^2 \left(\frac{\alpha_s(m_H)}{3\pi}\right)^2 \left(\frac{m_H}{m_\tau}\right)^2 \left(1 - \frac{4m_\tau}{m_H^2}\right)^{-3/2}.$$

Note that the H-boson decay into two gluons via the light quark loops is suppressed by an additional factor of the order of $(2m_q/m_H)^4$.

The above amplitude $M(H \to 2g)$ can easily be derived on the basis of the expression for the quark loop contribution (fig. 24.7) to the color charge renormalization. This contribution is equal (see Chapter 7) to

$$G_{\mu\nu}^{1a} G_{2a}^{\mu\nu} \frac{\alpha_s}{12\pi} \ln\frac{\Lambda^2}{m_q^2},$$

where Λ is the ultraviolet cut-off constant, and m_q is the quark mass. From this we find the transition amplitude of an arbitrary number of H-bosons into two gluons by using the substitution

$$m_q^2 \to m_q^2\left(1 + \frac{\chi}{\eta}\right)^2$$

(this substitution takes care of the Higgs-boson origin of the quark mass; we assume that $m_H \ll m_q$). The amplitude for H → 2g is obtained by taking the expansion of the logarithm and retaining only the term linear in χ:

$$M(H \to 2g) = G_{\mu\nu}^{1a} G_{2a}^{\mu\nu} \frac{\alpha_s}{6\pi} \frac{\chi}{\eta}.$$

This amplitude corresponds to one quark loop.

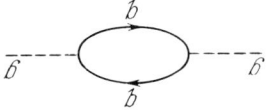

Fig. 24.7.

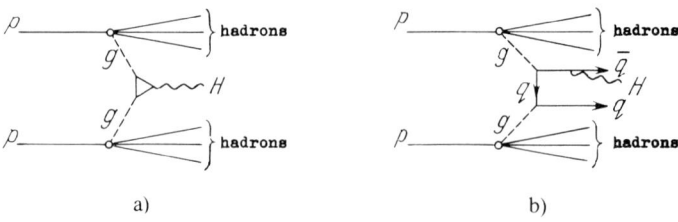

Fig. 24.8.

H-bosons can be produced in high-energy pp and p̄p collisions via a reverse reaction: gluon + gluon → H-boson (fig. 24.8a). The expected cross section for this process is of the order of 10^{-35} cm². The process of associated production of a pair of heavy quarks and one H-boson is easier to observe but it has a smaller cross section (fig. 24.8b).

24.5. "Higgs charge" of nucleons

It will be very instructive to discuss the effective constant M which has the dimension of mass and characterizes the H-boson–nucleon coupling vertex, $(M/\eta)\chi\bar{\psi}_N\psi_N$, for $q_H^2 = 0$. At first glance, $M = m_N$, as in the case of quarks. However, a nucleon is not an elementary particle; it is composite, so that our guess $M = m_N$ is wrong. At second glance, it would seem that $M = 2m_u + m_d$ for the proton and $M = 2m_d + m_u$ for the neutron, that is M is negligibly small on the hadronic mass scale. Alas, we are wrong again. In reality, the H-boson–nucleon interaction vertex is mostly determined by the interaction of H-bosons with gluons, the respective lagrangian being

$$\mathcal{L}_{Hgg} = N_H \frac{\alpha_s}{12\pi} \frac{\chi}{\eta} G^a_{\mu\nu} G_a^{\mu\nu}.$$

(This effective lagrangian differs from a similar expression for the amplitude $M(H \to 2g)$, corresponding to fig. 24.6, by a factor of $\frac{1}{2}$, taking into account that the gluons are identical). On the other hand, the nucleon mass in the chiral limit is again determined by the contribution of gluons:

$$m_N \bar{\psi}_N \psi_N = \langle N|\theta^\mu_\mu|N\rangle = \langle N|\tilde{\theta}^\mu_\mu|N\rangle.$$

Here $\tilde{\theta}^\mu_\mu$ is the trace of that part of the operator of the total energy-momentum tensor θ^μ_ν, which is due to the contribution of light quarks and gluons:

$$\tilde{\theta}^\mu_\mu = -\frac{\tilde{b}\alpha_s}{8\pi} G^a_{\mu\nu} G_a^{\mu\nu},$$

where $\tilde{b} = 11 - \frac{2}{3} \cdot 3 = 9$. Taking into account that

$$\frac{M}{\eta}\chi\bar{\psi}_N\psi_N = \langle N|\mathcal{L}_{Hgg}|N\rangle = \frac{N_H}{\eta}\frac{2}{3}\langle N|\frac{\alpha_s}{8\pi}G^a_{\mu\nu}G^{\mu\nu}_a|N\rangle$$

$$= -\frac{N_H}{\eta}\frac{2m_N}{3\tilde{b}}\bar{\psi}_N\psi_N,$$

we obtain that $M = -(2m_N/27)N_H \cong -70$ MeV$\cdot N_H$. Recall that N_H stands for the number of flavors of heavy quarks.

24.6. Digression on the trace of the energy-momentum operator

The result obtained above stems from a spectacular property of the energy-momentum tensor. We shall now discuss this property. The trace of the total energy-momentum operator of gluons and quarks is written in the form

$$\theta^\mu_\mu = \sum_l m_{q_l}\bar{q}_l q_l + \sum_h m_{q_h}\bar{q}_h q_h - \frac{b\alpha_s}{8\pi}G^a_{\mu\nu}G^{\mu\nu}_a,$$

where $b = 11 - \frac{2}{3}N_l - \frac{2}{3}N_h \equiv \tilde{b} - \frac{2}{3}N_h$, and indices l and h refer to light and heavy quarks, respectively. In contrast to the preceding sections of the present chapter, we define here the u-, d-, and s-quarks whose masses are below μ_c ($1/\mu_c$ is the confinement radius), as light quarks, and all other quarks as heavy. (In the general case, therefore, $N_h \geq N_H$; the equality $N_H = N_h$ then holds only if $m_H \simeq \mu_c$).

The last term in the expression for θ^μ_μ is the contribution of the so-called triangle anomaly. A pertinent image is that of the contribution due to the "shadow cabinet" of gluons and quarks with infinitely high masses, that is the so-called Pauli-Willars particles which regularize (cut off) divergent graphs. These particles have entered the energy-momentum tensor from the lagrangian where they are introduced in order to regularize the theory. The triangle anomalies in question are represented in the graphs of fig. 24.9 (recall that gravitons interact with the energy-momentum tensor). The graphs (a) and (c) of this figure contain physical particles, while the graphs (b) and (d) contain regularizing particles in loops. The graphs (c) and (d) cancel each other out, and heavy quarks drop out if the virtualities of external gluons are small compared to quark masses in the triangle.

Graph 24.9a gives zero contribution to the trace of the energy-momentum tensor since the gluon mass is zero. For light quarks, the graph (c) contains a small factor m_q/μ_c, where $1/\mu_c$ is the confinement radius. Therefore, the regularizing contribution to θ^μ_μ is not cancelled out for light quarks and gluons.

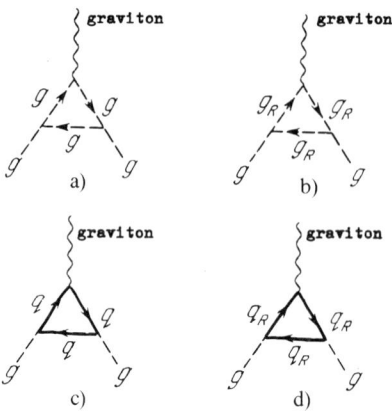

Fig. 24.9.

In the case of the nucleon the matrix elements

$$\langle N|m_{q_h}\bar{q}_h q_h|N\rangle \quad \text{and} \quad \langle N|\tfrac{2}{3}N_h \frac{\alpha_s}{8\pi} G^a_{\mu\nu} G^{\mu\nu}_a|N\rangle$$

cancel to an accuracy of terms of the order of $(\mu_c/m_{q_h})^2$. If we also use the fact that in the chiral limit $m_{q_l} \to 0$, we come to the conclusion that in this limit the nucleon mass is determined by the contribution of $\tilde{\theta}^\mu_\mu$.

24.7. Coupling of H-bosons to W- and Z-bosons

Obviously, coupling of H-bosons to W- and Z-bosons must be especially strong because of the large masses of the intermediate bosons; this coupling is characterized by the constant α. Consequently, creation of W- and Z-bosons must be accompanied by a rather intensive internal bremsstrahlung of Higgs particles. For example, if $m_H \cong 10$ GeV, $p\bar{p}$ collisions at $\sqrt{s} = 540$ GeV must produce one Z + H pair per approximately each 300 singly produced Z-bosons.

At the total energy $\sqrt{s} = m_Z + \sqrt{2}m_H$, the e^+e^- beams could be an intensive source of H-bosons, i.e. we mean the process $e^+e^- \to Z + H$ represented by the graph of fig. 24.10. The cross section for this process at $m_H \cong 10$ GeV is approximately three times the standard electromagnetic

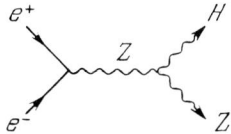

Fig. 24.10.

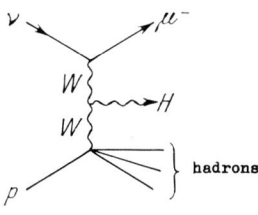

Fig. 24.11.

cross section for $e^+e^- \to \gamma \to \mu^+\mu^-$, equal to $4\pi\alpha^2/3s$, that is it comes to approximately $3 \cdot 10^{-35}$ cm^2.

Production of H-bosons in neutrino experiments must also proceed mostly owing to their strong ($\sim e$) coupling to the W-bosons (fig. 24.11). The cross section for this process must come to approximately $10^{-3}Gs$ of the cross section for the basic process $\nu + p \to \mu^- +$ hadrons. The factor Gs is easily derived on the basis of dimensional arguments (see the graph of fig. 24.11).

24.8. Coupling of H-bosons to photons

In the case of a single Higgs doublet the decay $H \to 2\gamma$ must have a small branching ratio. If, however, the number of Higgs particles is considerable, this decay may become important and in principle, even predominant. The decay $H \to 2\gamma$ proceeds via triangular graphs with virtual charged particles (fig. 24.12). We can write the decay amplitude in the form

$$M(H \to 2\gamma) = \frac{\alpha F}{4\pi}(\sqrt{2}\,G)^{1/2}\chi F^1_{\mu\nu}F^{\mu\nu}_2 = \frac{\alpha F}{4\pi}(\sqrt{2}\,G)^{1/2}m_H^2\chi e_1 e_2.$$

Here χ, e_1, and e_2 are wave functions of the H-boson and photons, and m_H is the H-boson mass. The constant F is dimensionless, its value being determined by the contribution of the graph of fig. 24.12. In cases when the virtual particles forming the triangle in fig. 24.12 have unit charges and

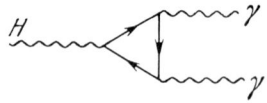

Fig. 24.12.

spins 0, $\tfrac{1}{2}$ and 1, respectively, the expressions for F are:

$$F_0 = \beta(1 - \beta x^2),$$

$$F_{1/2} = -2\beta[(1-\beta)x^2 + 1],$$

$$F_1 = [2 + 3\beta + 3\beta(2-\beta)x^2].$$

Here $\beta = 4m^2/m_H^2$, m is the virtual particle mass,

$$x = \arctan \frac{1}{\sqrt{\beta - 1}} \quad \text{for } \beta > 1,$$

$$x = \frac{1}{2}\left(\pi + i\ln\frac{1+\sqrt{1-\beta}}{1-\sqrt{1-\beta}}\right) \quad \text{for } \beta < 1.$$

For $\beta \to 0$: $F_0, F_{1/2} \to 0$, $F_1 \to 2$.
For $\beta \to \infty$: $F_0 \to -\tfrac{1}{3}$, $F_{1/2} \to -\tfrac{4}{3}$, $F_1 \to 7$.

Consequently, the contribution of virtual heavy particles to the amplitude $M(H \to 2\gamma)$ does not tend to zero as their masses increase. The above expressions for F are easily derived if we first calculate the contribution of the scalar, fermion and vector loops to the photonic vacuum polarization, and then carry out a substitution $m \to m(1 + \chi(\sqrt{2}\,G)^{1/2})$ (recall a similar discussion of the H-boson–gluon interactions). In the general case F is the sum $F = \Sigma_i Q_i^2 F_i$, where Q_i is the charge of the ith virtual particle, and summation is carried out over all virtual particles. With two or more Higgs doublets, the coefficients in this sum are not unity any more, since now the mass of the particle does not fix its coupling to the H-boson. As a result, the amplitude $M(H \to 2\gamma)$ may be considerably enhanced. By using the expression for $M(H \to 2\gamma)$ we readily obtain $\Gamma(H \to 2\gamma) = (\alpha F/4\pi)^2 \times (Gm_H^3/8\pi\sqrt{2})$.

24.9. General remarks on Higgs bosons

The Higgs mechanism of generation of mass and elimination of divergences is a characteristic element of the electroweak model. This makes the experimental search for Higgs bosons very interesting. Some theorists regard the Higgs mechanism as artificial and are not inclined to believe that it is realized in nature. Of course, a number of important elements are lacking in the present-day picture; in particular, we do not understand what fixes the spectrum of lepton and quark masses, as well as the values of the Cabibbo and Weinberg angles. However, the very idea of the existence of scalar elementary particles does not appear so artificial after we have accepted the existence of elementary particles with spins of $\frac{1}{2}$, 1 (and 2). Supersymmetry schemes developed in recent years include all these particles (and those with spin $\frac{3}{2}$) into one supermultiplet (see Chapter 26), with interactions determined by a single dimensionless constant. It seems natural from this point of view that the number of Higgs particles is not minimal, and that their mass spectrum is very broad.

The so-called grand synthesis (or grand unification) models, which will be considered in Chapter 25, also favor a large number of Higgs bosons.

Two remarks must be made concerning charged H-bosons. First, they must be effectively produced in colliding electron-positron beams. Above the threshold we have:

$$R_H = \frac{\sigma(e^+e^- \to \gamma \to H^+H^-)}{\sigma(e^+e^- \to \gamma \to \mu^+\mu^-)} = \frac{1}{4}.$$

Second, semiweak decays of heavy quarks with emission of H-bosons could be possible. If, for instance, the H-boson were lighter than the difference in masses of the b- and c-quarks, we would have

$$\frac{\Gamma(b \to H^-c)}{\Gamma(b \to c\bar{u}d)} \cong \frac{6\pi^2}{G_F m_b^2} \cong 10^5.$$

This chapter was already completed when active discussion began concerning the so-called "technicolor" model. According to this model, Higgs bosons are composite particles built of so-called "techniquarks". "Techniquarks" interact via special "technigluons". The confinement radius for the technicolor interaction is of the order of 1 TeV^{-1}.

CHAPTER 25

Grand unification

Despite their apparent pomposity, the words "Grand Synthesis", or "Grand Unification" are physical terms, not mere expressions of delight. They represent theoretical models which unify the weak, electromagnetic and strong interactions. This unification becomes possible because the effective constants of these interactions, which depend on momentum transfer owing to polarization of the vacuum, tend to the same limiting value when the momentum transfer increases. According to the guiding idea of grand unification, a single coupling constant characterizes all three interactions at small distances. The grand synthesis models explain why quarks have fractional charges, predict the value of $\sin^2\theta_W$ and, what is especially interesting, predict a qualitatively new phenomenon: instability of the proton. Mostly, we shall discuss a model based on SU(5) symmetry.

25.1. Three generations of fermions

It would seem the most consistent to put all leptons and quarks into a single fundamental multiplet of some group, gauge the corresponding symmetry, and then spontaneously violate this symmetry. This program, however, immediately runs into such unimaginably vast multiplets of intermediate and Higgs bosons that one is forced to begin with a piecemeal approach.

The first step in the outlined direction is the unification on the basis of SU(5) symmetry. The SU(5) model classifies all known fermions into three generations. The scheme starts with the first generation; the subsequent generations are included in an analogous manner. Each generation contains 15 fermion states. Let us discuss the first generation in some detail. It contains three doublets of left-handed quarks and six singlets of right-handed quarks:

$$\begin{pmatrix} u_i \\ d_i \end{pmatrix}_L, \quad u_{iR}, \quad d_{iR}$$

(where $i = r, y, b$ are color indices). It also contains a left-handed lepton doublet and a right-handed lepton singlet,

$$\begin{pmatrix} \nu_e \\ e^- \end{pmatrix}_L \quad \text{and} \quad e_R^-$$

(the right-handed neutrino is not included since this particle does not participate even in weak interactions). The second generation contains c, s, ν_μ, and μ, and the third contains t, b, ν_τ, and τ. With the exception of the neutrino, all particles become heavier with each subsequent generation.

Sometimes the upper quarks (u, c, t) are called anoquarks and denoted by α or a (for the Greek $\alpha\nu\omega$ meaning top) while the lower quarks (d, s, b) are called catoquarks and denoted by κ or c (for the Greek $\kappa\alpha\tau\omega$ meaning bottom). In this notation, the generation is

$$\begin{pmatrix} \alpha_i \\ \kappa_i \end{pmatrix}_L, \quad \alpha_{iR}, \quad \kappa_{iR}, \quad \begin{pmatrix} \nu_\ell \\ \ell^- \end{pmatrix}_L, \quad \ell_R^-.$$

Instead of right-handed particles, it will be convenient to operate with the CP conjugate left-handed antiparticles. All fifteen particles of the generation are then left-handed:

$$\begin{pmatrix} \alpha_i \\ \kappa_i \end{pmatrix}_L, \tilde{\alpha}_{iL}, \tilde{\kappa}_{iL}, \begin{pmatrix} \nu_\ell \\ \ell^- \end{pmatrix}_L, \ell_L^+,$$

where $i = r, y, b$ are color indices. We must emphasize that ℓ_L^+ is an antiparticle with respect to ℓ_R^- and not to ℓ_L^-, with a similar statement valid for $\tilde{\alpha}_{iL}$ and $\tilde{\kappa}_{iL}$ (the tilde sign here is short and does not cover the index L; the tilde indicates C-conjugation).

25.2. Quintet and decuplet in SU(5)

A spinor with 15 components is still too large: it corresponds to the group SU(15) and to 224 vector bosons. Therefore the 15 left-handed particles are placed in two multiplets of SU(5):

$$15 = \bar{5} + 10.$$

For the first generation, for example, the quintet is chosen as follows:

$$\bar{Q}_L^a = \left(\bar{d}_r, \bar{d}_y, \bar{d}_b, e^-, \nu_e \right)_L, \quad a = 1, 2, \ldots, 5,$$

or

$$Q_{aR} = (d_r, d_y, d_b, e^+, \bar{\nu}_e)_R.$$

As for the decuplet, it transforms as an antisymmetrized product

$$D_{ab} \sim \sqrt{\tfrac{1}{2}}(Q_a Q_b - Q_b Q_a).$$

(In analogy to SU(3) where $3 \times 3 = 6 + \bar{3}$, in SU(5) we write $5 \times 5 = 10 + 15$). The antisymmetric matrix describing our decuplet has the form

$$D_{ab} = \frac{1}{\sqrt{2}} \begin{pmatrix} 0 & \bar{u}^b & -\bar{u}^y & -u_r & -d_r \\ -\bar{u}^b & 0 & \bar{u}^r & -u_y & -d_y \\ \bar{u}^y & -\bar{u}^r & 0 & -u_b & -d_b \\ u_r & u_y & u_b & 0 & -e^+ \\ d_r & d_y & d_b & e^+ & 0 \end{pmatrix}_L.$$

A useful mnemonic picture can be constructed: each particle of the decuplet consists of two right-handed particles of the quintet, "quintons", and of a gravitino, that is a massless neutral particle with spin $\tfrac{3}{2}$. (The existence of the gravitino is predicted in the framework of so-called supergravitation, (see Chapter 26.) The spins of the gravitino and two quintons add up as shown in fig. 25.1. Thin arrows in this figure denote particle momenta, and open arrows denote their spins. We see that our composite fermions are left-handed. It is easily verified that colors and charges of these composite fermions are those of the particles in the above decuplet matrix D_{ab}. For instance,

$$D_{12} = \frac{1}{\sqrt{2}} \bar{u}^b \sim \frac{d_r d_y - d_y d_r}{\sqrt{2}} = \frac{d_i d_k \varepsilon^{ikb}}{\sqrt{2}}.$$

It is not impossible that fig. 1 is not purely mnemonic, but reflects some important physical message, some composite model with dynamics which at

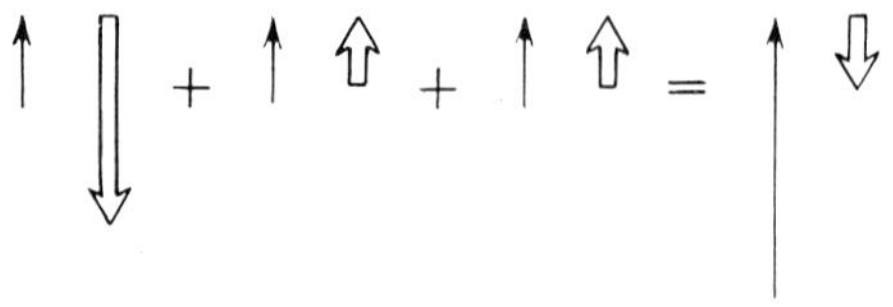

Fig. 25.1.

present is absolutely unclear. However, a more natural explanation of the fact that fundamental particles are grouped into two SU(5) multiplets, is given by SO(10) symmetry (see below).

25.3. 24 vector bosons

The SU(5) group has 24 generators ($5 \times \bar{5} = 24 + 1$). Let us consider a theory in which the SU(5) symmetry is local. Then each of these generators corresponds to a vector boson.

One of the generators of the group is the electric charge. We know that the generator trace is zero (recall the matrices τ in SU(2) and λ in SU(3)). This means that the sum of the charges of particles in an SU(5) multiplet must be zero. Hence,

$$Q_{d_r} + Q_{d_y} + Q_{d_b} = Q_e,$$

and in the case of the exact color symmetry $Q_d = -\frac{1}{3}$. Therefore, SU(5) symmetry explains why quarks have fractional charges.

Let us consider now the properties of the 24 vector bosons. In the limit of exact SU(5) symmetry all of them are massless. It is convenient to discuss these bosons looking at the quintet.

The eight bosons giving transitions between three colored quarks are gluons, and they correspond to the subgroup $SU(3)_c$. The three bosons giving transitions in the lepton sector of the quintet are W^+, W^-, W^0, corresponding to the group $SU(2)_L$. Finally, there is the twelfth boson, B^0, with particle hypercharges as its source; this boson corresponds to the group U(1). The photon and Z^0 boson are (as in the standard electroweak model which is part of the SU(5) model) linear mutually orthogonal superpositions of W^0 and B^0. It is readily shown, when decuplet particles are taken into account, that gluons interact not only with left-handed currents $\bar{d}_{Li}\gamma_\mu d_L^k$, but also with right-handed currents $\bar{d}_{Ri}\gamma_\mu d_R^k$, and not only with d-quarks but with u-quarks as well. It is also easy to verify that W-bosons interact both with leptons and with quarks.

The remaining twelve vector bosons are much more unusual. They form two charged color triplets:

$$X^i_{+4/3}, \bar{X}^i_{-4/3}, Y^i_{+1/3}, \bar{Y}^i_{-1/3},$$

where i = r, y, b are color indices, and subscripts indicate electric charge.

The bosons X and $\bar{X}$ are emitted in transitions $\bar{d} \leftrightarrow e^-$, and Y and $\bar{Y}$ in transitions $\bar{d} \leftrightarrow \nu$. We shall see later, considering transitions in the decuplet in addition to those in the quintet, that the exchange of X- or Y-bosons results in instability of protons. The X- and Y-boson masses must be very high ($\gtrsim 10^{15}$ GeV) in order to make this instability acceptably small. Judging by the enormous spread in particle masses, SU(5) is very badly broken. We shall see, however, that some important features of the symmetry are retained despite this extremely strong symmetry breaking. Under the ruins of the symmetry we shall find the surviving pattern of the coupling constants.

25.4. Running coupling constants

At very short distances when momentum transfer is much larger than the masses of the heaviest vector bosons, SU(5) symmetry is restored and all interactions of vector bosons with other particles (fermions and Higgs bosons) and with one another are determined by a single coupling constant, that is by a single charge g. The term "constant" for this quantity is kept for historical reasons only. In reality it is a logarithmic function of the momentum transfer. At $q \gg 10^{15}$ GeV all interactions—weak, strong, and electromagnetic—are of the same magnitude. As the momentum transfer diminishes, the constants characterizing these interactions diverge from one another. However, broken SU(5) is still valuable because this divergence is logarithmic and hence, not large. Let us consider the charges g_3, g_2, g_1 which correspond to the groups SU(3), SU(2), U(1), respectively. The corresponding vertices are:

$$\text{SU(3):} \qquad g_3 G_\mu \sum_q \bar{q} \gamma^\mu \tfrac{1}{2} \lambda q,$$

where G_μ is the octet of gluons, and the sum is taken over all quarks;

$$\text{SU(2):} \qquad g_2 W_\mu \left(\sum_q \bar{q}_L \gamma^\mu \tfrac{1}{2} \tau q_L + \sum_\ell \bar{\ell}_L \gamma^\mu \tfrac{1}{2} \tau \ell_L \right),$$

where W_μ is the triplet of W-bosons, and the sums are taken over all doublets of left-handed quarks and left-handed leptons; and

$$\text{U(1):} \qquad g_1 B_\mu \sum_f \bar{f} C \tfrac{1}{2} Y \gamma^\mu f,$$

where the sum is taken over all particles: quarks and leptons, singlets and doublets. The coefficient C in the last expression serves to normalize the U(1) current in the same manner as the SU(3) and SU(2) currents. The point is, both matrices $\tfrac{1}{2}\lambda$ and $\tfrac{1}{2}\tau$ are generators of the group SU(5), while the diagonal matrix $\tfrac{1}{2}Y$ is only proportional but not equal to the SU(5) generator. The coefficient C was introduced to normalize $\frac{1}{2}CY$ as $\frac{1}{2}\lambda$ and $\frac{1}{2}\tau$

$$\mathrm{Tr}\left(\tfrac{1}{2}\tau_i\right)^2 = \tfrac{1}{2}, \qquad \mathrm{Tr}\left(\tfrac{1}{2}\lambda_a\right)^2 = \tfrac{1}{2}$$

for any $i = 1, 2, 3$ and any $a = 1, 2, \ldots, 8$. We now demand

$$\mathrm{Tr}\, C^2 \left(\tfrac{1}{2}Y\right)^2 = \tfrac{1}{2}.$$

Taking the quintet as an example and using the relation $Q = T_3 + \frac{1}{2}Y$ we find that $\frac{1}{2}Y(\bar{d}_L) = +\frac{1}{3}$, $\frac{1}{2}Y(e_L^-) = \frac{1}{2}Y(\nu_L) = -\frac{1}{2}$ and therefore

$$\mathrm{Tr}\left(\tfrac{1}{2}Y\right)^2 = \tfrac{1}{9}\cdot 3 + \tfrac{1}{4}\cdot 2 = \tfrac{5}{6};$$

hence,

$$C^2 = \tfrac{3}{5}.$$

Recall that the standard electroweak model contains the constants g and g' (see Chapter 21) and that

$$g'/g = \tan\theta_W, \qquad g = e/\sin\theta_W,$$

where θ_W is the Weinberg angle. Experiment yields

$$\sin^2\theta_W \simeq 0.2 - 0.25.$$

It is easy to see that in our new notation, $g = g_2$, $g' = g_1 C = g_1\sqrt{\tfrac{3}{5}}$. Therefore, for $q \gg 10^{15}$ GeV, when $g_1 = g_2$, $\tan^2\theta_W^0 = \tfrac{3}{5}$ and $\sin^2\theta_W = \tfrac{3}{8} \simeq 0.38$, which is almost twice the observed value of $\sin^2\theta_W$ for small transferred momenta. We know that for such momenta ($q \ll 100$ GeV)

$$\alpha_3 \simeq \tfrac{1}{5}, \qquad \alpha_2 \simeq \frac{\alpha}{\sin^2\theta_W} \simeq \frac{1}{26}, \qquad \alpha_1 = \frac{\alpha}{\cos^2\theta_W}\frac{5}{3} \simeq \frac{1}{67},$$

where we have denoted

$$g_i^2/4\pi = \alpha_i, \qquad i = 1, 2, 3.$$

Let us find how the α_i change as functions of momentum transfer. These changes are mainly brought about by loops of virtual particles, namely vector bosons (fig. 25.2) and fermions (fig. 25.3) (we neglect Higgs loops). Summation of these loops gives

$$\frac{1}{\alpha_i(M)} - \frac{1}{\alpha_i(\mu)} = \frac{b_i}{2\pi}\ln\frac{M}{\mu},$$

where M and μ are two values of the momentum transfer, both much larger than the masses of the particles whose contribution is taken into account in the calculation. The quantities b_i are determined by the order of the (local)

Fig. 25.2.

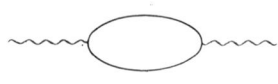

Fig. 25.3.

group, i.e. by the number of vector bosons, and by the number of fermion flavors. For gluons in $SU(3)_L$ we have $b_3 = 11 - \frac{2}{3}N_f$ where N_f is the number of quark flavors ($b_3 = 7$ if $N_f = 6$). Here the first term corresponds to fig. 25.2 and the second one to fig. 25.3. In the limit of strict SU(5), $b_2 = b_3$. However, as a result of the breaking of SU(5) to $SU(3) \times SU(2) \times U(1)$, the loops of fig. 25.2 at $q \ll m_X, m_Y$ give for W-bosons $\frac{22}{3}$ instead of 11 (the corresponding term equals $\frac{11}{3}n$ for the $SU(n)$ group). Consequently, $b_3 - b_2 = \frac{11}{3}$. The boson B_μ interacts with hypercharge while neither gluons nor W-bosons have hypercharge; hence, the loops of fig. 25.2 give no contribution to α_1 as a function of q, so that $b_3 - b_1 = 11$. We see that b_3 and b_2 are positive while b_1 is negative. This means that α_3 and α_2 decrease as the momentum increases (asymptotic freedom) while α_1 increases. Neglecting the threshold effects near m_t, m_W and m_X, it is convenient to plot $1/\alpha_i$

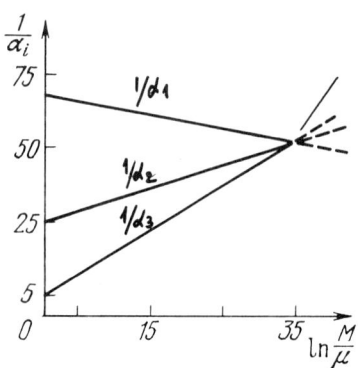

Fig. 25.4.

instead of α_i (see fig. 25.4). Making use of the three equations

$$\frac{1}{\alpha_i(M)} - \frac{1}{\alpha_i(\mu)} = \frac{b_i}{2\pi} \ln \frac{M}{\mu},$$

where

$$\frac{1}{\alpha_1(\mu)} = \frac{3\cos^2\theta_W}{5\alpha}, \quad \frac{1}{\alpha_2(\mu)} = \frac{\sin^2\theta_W}{\alpha}, \quad \frac{1}{\alpha_3(\mu)} = \frac{1}{\alpha_s},$$

$$b_1 = -4, \quad b_2 = 3.3, \quad b_3 = 7,$$

and assuming unbroken SU(5) symmetry at the point M, that is,

$$\alpha_1(M) = \alpha_2(M) = \alpha_3(M) = \bar{\alpha}(M),$$

we easily find three unknowns: $\ln(M/\mu)$, $\bar{\alpha}(M)$, and $\sin^2\theta_W$. For $\ln(M/\mu)$ we obtain

$$\ln \frac{M}{\mu} = \frac{\pi}{11} \left[\frac{1}{\alpha(\mu)} - \frac{8}{3\alpha_s(\mu)} \right].$$

If we start at $\mu \sim m_W$ and take into account that $\alpha(m_W) \cong \frac{1}{129}$, $\alpha_s(m_W) \cong 0.1$, then $\ln(M/m_W) \cong 29$ and $M/m_W \cong 4 \cdot 10^{12}$. It is important to emphasize that this result is independent of $\sin^2\theta_W$. We can find $\bar{\alpha}(M)$ just as easily; it is close to 0.02 (see fig. 25.4). The following relation is obtained for $\sin^2\theta_W$:

$$\sin^2\theta_W = \frac{1}{6} + \frac{5}{9} \frac{\alpha}{\alpha_s} \cong 0.20.$$

This is strikingly close to the experimental value ($\cong 0.23$). However, a slight discrepancy between theory and the experiment may prove dangerous for the theory if the predicted value is found to lie beyond the experimental errors.

25.5. Unstable proton

Let us turn again to the X- and Y-bosons. The coupling of X-bosons to fermions is written in the form:

$$\sqrt{\tfrac{1}{2}} g X_{i\mu} \left(\bar{d}^i \gamma^\mu e^+ + \bar{\tilde{u}}_{Lm} \gamma^\mu u_{Lk} \varepsilon^{ikm} \right).$$

Note that both $\bar{\tilde{u}}$ and u are covariant color spinors (with lower indices); the term $\bar{d}^i \gamma^\mu e^+$ is a sum of two terms: $\bar{d}_L \gamma^\mu e_L^+$ from the quintet and $\overline{e_L^- \gamma^\mu \tilde{d}_L^i}$

from the decuplet, with the last term being equivalent to $\bar{d}_R^i \gamma^\mu e_R^+$. The coefficient $\sqrt{\frac{1}{2}}$ in the above expression for the coupling is introduced to normalize the current, responsible for the emission of X-bosons in the same way as the current responsible for the emission of W-bosons. In the last case the factor $\sqrt{\frac{1}{2}}$ is implicitly contained in the isotopic vector $\bar{e}\frac{1}{2}\tau\nu$:

$$\tfrac{1}{2}\tau = \left\{ \tfrac{1}{2}\begin{pmatrix} 1 & 0 \\ 0 & -1 \end{pmatrix}, \sqrt{\tfrac{1}{2}}\begin{pmatrix} 0 & 1 \\ 0 & 0 \end{pmatrix}, \sqrt{\tfrac{1}{2}}\begin{pmatrix} 0 & 0 \\ 1 & 0 \end{pmatrix} \right\}.$$

The Y-boson interaction vertex is

$$\sqrt{\tfrac{1}{2}}\, g Y_{i\mu} \left(\bar{d}_L^i \gamma^\mu \nu_L + \bar{d}_{Lm} \gamma^\mu \bar{u}_{Lk} \varepsilon^{ikm} + \overline{e_L^+} \gamma^\mu u_L^i \right).$$

Exchange of X- and Y-bosons obviously results in non-conservation of the baryonic and leptonic charges, due to elementary processes of the type of

$$uu \xrightarrow{X} e^+ \bar{d}, \quad ud \xrightarrow{Y} \bar{d}\bar{\nu}, \quad ud \xrightarrow{Y} e^+ \bar{u}.$$

These processes make the proton unstable:

$$p = uud \to (e^+ \bar{d}d \text{ or } e^+ \bar{u}u) \Rightarrow e^+ \pi^0 (\eta^0, \rho^0, \omega^0),$$

$$p = uud \to u\bar{d}\bar{\nu} \Rightarrow \bar{\nu}\pi^+ (\rho^+).$$

Neutrons must undergo similar decays (in nuclei which usually are considered stable):

$$n = ddu \to d\bar{d}\bar{\nu} \Rightarrow \bar{\nu}\pi^0 (\eta^0, \rho^0, \omega^0).$$

We can estimate the proton decay rate by noting that the matrix element is proportional to $g^2/m_X^2 \cong \alpha/m_X^2$. A dimensional estimate then gives

$$\frac{1}{\tau_p} \cong \frac{\alpha^2 m_p^5}{m_X^4}.$$

A quantitative calculation confirms this estimate. The coupling of X- and Y-bosons to the second-generation fermions makes the processes shown in fig. 25.5 possible. The process 25.5c does not contribute to the proton decay because of the large mass of the c-quark. The processes of figs. 25.5a and b give the following decays:

$$p = uud \to \mu^+ \bar{s}d \Rightarrow \mu^+ K^0,$$

$$p = uud \to \bar{\nu}_\mu \bar{s}u \Rightarrow \bar{\nu}_\mu K^+,$$

$$n = ddu \to \bar{\nu}_\mu \bar{s}d \Rightarrow \bar{\nu}_\mu K^0.$$

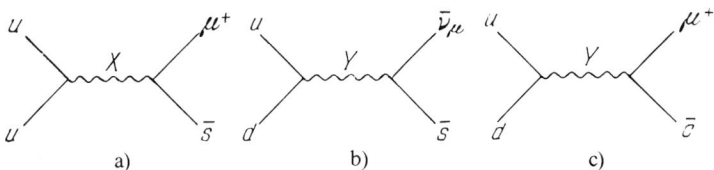

Fig. 25.5.

The probability of the decay with emission of μ^+ must be several times lower than that of the decay with emission of e^+. First, the virtual Y-boson does not contribute to the muonic decay, and second, the phase-space volume is reduced (this reduction is caused by creation of comparatively heavy strange particles). It can be conjectured that muons must be produced in approximately 10% of all proton decays.

The experimental lower bound on the proton lifetime is approximately 10^{30} years. From the relation

$$\tau_p \cong \left(\frac{m_X}{m_p}\right)^4 \frac{1}{\alpha^2 m_p} \gtrsim 10^{37} \text{ s,}$$

we derive that

$$\frac{m_X}{m_p} \gtrsim 10^{14}$$

(here we take into account that $1/\alpha^2 m_p \simeq 10^4 \cdot 10^{-24}$ s). The same estimate is true for Y-bosons. Summarizing, we obtain

$$m_X \simeq m_Y \gtrsim 10^{14} \text{ GeV.}$$

We see that this quantity is of the same order of magnitude as the energy $\overline{M}$ at which all three constants α_1, α_2 and α_3, become equal. In order to prevent divergence of these constants at $M > \overline{M}$, the masses of the X- and Y-bosons should not exceed $\overline{M}$. In this case the behavior of all the constants will be determined at $M \gg \overline{M} \cong m_X$ by the coefficient

$$\bar{b} = \tfrac{11}{3} \cdot 5 - \tfrac{2}{3} N_f = 14.3.$$

The trend of the unified charge corresponding to this value of $\bar{b}$ is traced by a solid line in the upper right-hand corner of fig. 25.4.

Detailed calculations give $10^{15} \text{GeV} \gtrsim m_X \gtrsim 10^{14} \text{GeV}$ and $\tau_p \simeq 10^{30 \pm 3}$ years. If additional Higgs fields are introduced into the SU(5) model, the expected value of τ_p may exceed 10^{34} years.

In order to prove that $\tau_p > 10^{30}$ years, it is necessary to show that not one of the 10^{30} nucleons in 1.6 tons of matter decay during one year. Experiments of this type are conducted deep underground where the cosmic ray background is sufficiently low. Significant progress will be achieved if the mass will be increased to about $10^3 - 10^5$ tons. The discovery of proton decay would be of crucial importance because it would confirm unambiguously the principal ideas of grand unification.

During the first moments of the big bang 10^{10} years ago, the processes of baryon charge non-conservation were much more abundant, possibly playing a very important role at temperatures of the order of $10^{17} - 10^{10}$ GeV. It is tempting to interpret the existing charge asymmetry of the universe by non-conservation of baryonic charge and violation of CP invariance. A little more will be said about this in Chapter 27.

25.6. Grand Higgs bosons

The SU(5) model is constructed as a spontaneously broken symmetry. This guarantees its renormalizability. At least two multiplets of scalar bosons are necessary for the spontaneous breaking of symmetry. The first of them, a 24-plet, gives masses to the X- and Y-bosons, and the second, the quintet, gives masses to the W- and Z-bosons and fermions. The scheme of symmetry breaking looks as follows:

$$SU(5) \xrightarrow{24} SU(3) \times SU(2) \times U(1) \xrightarrow{5} SU(3) \times U(1).$$

As a result, we are left with only nine massless fields: eight gluons and the photon. The coupling of vector fields to scalar fields is gauge invariant, and is characterized by a universal charge g. The matrix of vacuum-expectation values, which appear due to spontaneous breaking of SU(5) in the Higgs 24-plet, has the form

$$v_{24} \cdot \begin{pmatrix} 2 & & & & \\ & 2 & & & \\ & & 2 & & \\ & & & -3 & \\ & & & & -3 \end{pmatrix}.$$

Hence, SU(3) and SU(2) symmetries remain unbroken. The X- and Y-boson masses are then of the order of gv_{24}, so that v_{24} must be very large, of the order of 10^{15} GeV. At the second stage, the spontaneous symmetry breaking in the Higgs quintet

$$\left(H_r^{-1/3}, H_y^{-1/3}, H_b^{-1/3}, H^+, H^0 \right)$$

§6] *Grand Higgs bosons* 247

takes place, and the neutral component $\sqrt{\frac{1}{2}}(H^0 + \overline{H}^0)$ gains a non-zero vacuum expectation value $v_5 \simeq 10^2$ GeV. At the same time, H^+, H^-, and $\sqrt{\frac{1}{2}}(H^0 - \overline{H}^0)$ transform into the third components of W^+, W^-, and Z^0 bosons, just as in the standard model of electroweak interactions.

Unfortunately, serious problems appear in the Higgs sector of the SU(5) model. The point is that the two multiplets of scalar particles are not isolated but connected by loops of vector fields (fig. 25.6). As a result, there is a threat that a large quantity, αv_{24}, will be added to a small one, v_5. The required hierarchy of vacuum expectation values cannot be preserved without compensations which today appear very artificial.

Let us look now into the masses of fermions. These appear owing to the coupling of the scalar quintet to fermions. It is easy to show that

$$5 \times 10 = 45 + \bar{5},$$

$$10 \times 10 = 50 + 45 + \bar{5}.$$

Hence, the Higgs quintet could give masses both to the α-fermions in the 10, and to the κ-fermions whose left-handed components belong to the 10 and the right-handed ones to the $\bar{5}$. No doubt, this mechanism does not give a mass to the neutrino. The coupling constants $f_{5,10}$ and $f_{10,10}$ characterizing the interaction of Higgs and fermion multiplets are not fixed by the gauge invariance of the lagrangian; they can be chosen "by hand" so as to provide the best fit to the experimentally observed masses. These constants are different for different fermion generations, with the masses of α-quarks unrelated to those of κ-quarks and leptons. It is easy to show that κ-quarks and leptons have identical masses:

$$m_d = m_e, \quad m_s = m_\mu, \quad m_b = m_\tau.$$

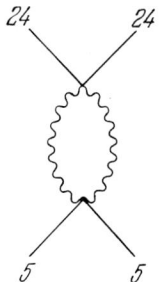

Fig. 25.6.

These relations must be satisfied at momenta of the order of m_X. Taking into account the dependence of the masses on the momentum, we can show that

$$\frac{m_\kappa(\mu)}{m_\ell(\mu)} = \left(\frac{\alpha_s(\mu)}{\bar{\alpha}}\right)^{4/b_3} \left(\frac{\bar{\alpha}}{\alpha_1(\mu)}\right)^{1/b_1},$$

where $b_3 = 11 - \frac{2}{3}N_f$, $b_1 = -\frac{2}{3}N_f$. With μ chosen higher than all thresholds, say $\mu \cong 10$ GeV, we conclude that

$$\frac{m_d}{m_e} = \frac{m_s}{m_\mu} = \frac{m_b}{m_\tau} \cong 2 - 3.$$

The ratio m_b/m_τ is indeed close to the predicted value, but the agreement for the d- and s-quarks is very poor if the standard values, 7 MeV and 150 MeV, are taken.

The above-mentioned contradictions can be removed by introducing into the SU(5) model additional Higgs fields; this reduces, however, the predictive power of the model.

Among other attempts of grand unification we shall mention some models based on orthogonal and exceptional groups.

25.7. The SO(10) group and other orthogonal groups

SO(10) is a group of orthogonal rotations in a ten-dimensional space (S indicates that the determinant of transformation matrices equals unity). In this case vector particles belong to a 45-plet, with fermions naturally placed in the spinor representation. In the case of SO(10) the spinor comprises sixteen components so that in addition to fifteen left-handed fermions we deal with in the SU(5) model, each generation includes the sixteenth fermion, namely the left-handed antineutrino.

The group SO(10) has SU(5) as a subgroup. The 16-plet of fermions consists of three SU(5) multiplets: $16 = 10 + \bar{5} + 1$. This can be regarded as a natural explanation of the fact that in the framework of SU(5), fermions belong to the reducible representation.

Along with a chain of spontaneous symmetry breakings SO(10) → SU(5) → SU(3) × SU(2) × U(1) → SU(3) × U(1), chains of other types are also possible. Some of them make it possible to obtain, among other things, the value of $\sin^2\theta_W$ which is closer to the experimental value of this paramenter than that found in the SU(5) model.

The SO(10) model predicts, just as SU(5), the proton instability. Some variants of this model yield the proton lifetime exceeding 10^{34} years. The experiments planned for the first half of the 80's would be unable to detect such a slow decay.

In addition to the proton decay, another process not conserving the baryonic quantum number ($\Delta B \neq 0$) is discussed in the framework of the SO(10) model, namely neutron-antineutron transitions in vacuum, similar to K^0-$\overline{K}^0$ oscillations but essentially much slower. A crude dimensional estimate relates the period of oscillations τ_{osc} to the deuteron lifetime τ_{dec} for the decay into pions (in both processes $|\Delta B| = 2$):

$$\tau_{dec} \sim \tau_{osc}^2 m_p.$$

This relation shows that the deuteron lifetime of $\sim 10^{30}$ years corresponds to the neutron oscillation period of about one year. If such oscillations really occur, they could be found in neutron beams of nuclear reactors.

The SO(10) symmetry naturally gives non-zero mass to the neutrino. This occurs because the fermion 16-plet comprises both the left-handed and right-handed components of the neutrino wave function. The estimates of the expected neutrino mass, found in the current literature, lie mostly within the interval $10^{-1} - 10^{-6}$ eV.

A large number of papers are devoted to unified gauge theories based on higher-rank orthogonal groups. For instance, the fundamental 64-component spinor of the group SO(14) comprises four generations of leptons and quarks. The group SO(22) contains, in addition to conventional quarks, the techniquarks mentioned in Chapter 24.

25.8. The exceptional groups

In addition to four series of regular groups

$$A_l = SU(l+1), \quad B_l = SO(2l+1), \quad C_l = Sp(2l), \quad D_l = SO(2l),$$

whose rank l can be arbitrarily high, there are five exceptional Cartan groups: G_2, F_4, E_6, E_7, E_8 (the subscript giving the group rank). Recall that the rank of the group is the number of mutually commuting generators. The groups G_2 and F_4 are not broad enough to include all known flavors and colors of quarks, so that discussions in the literature are mostly restricted to the groups E_6, E_7, and E_8.

The group E_6 has SO(10) as its subgroup. The E_6 model comprises 78 gauge bosons; fermions form a 27-plet. Two alternative E_6 models are found in the literature. The first E_6 model includes several generations of fermion 27-plets each of which has the following SO(10) composition: $27 = 16 + 10 + 1$. Each generation thus has 11 new states in addition to the known 16 fermion states: a singlet quark with $Q = -\frac{1}{3}$, a singlet Majorana lepton N^0 and a lepton doublet L^+, L^0. All of them are assumed to be superheavy.

The second E_6 model is more compact: all the known fermions in this model enter, together with several additonal particles, one 27-plet comprising two quarks with $Q = +\frac{2}{3}$ (u, c), four quarks with $Q = -\frac{1}{3}$ (d, s, b, h), four leptons with $Q = -1$ (e, μ, τ, λ), and five leptons with $Q = 0$ (ν_e, ν_μ, ν_τ, ν_λ, ν_ρ). We thus find 18 quarks (color taken into account) and nine leptons. The model therefore predicts four so far unobserved particles. What is essential is that there is no t-quark among them! In this model it is difficult to avoid undesirable neutral currents changing quark flavors. Most of the versions of this E_6 model discussed in the literature did not include weak decays of a b-quark into a c-quark (see footnote on p. 5).

The model based on the group E_7 comprises 133 vector bosons, and fermions belong to a 56-plet containing six color quarks, six color antiquarks, and ten leptons (four charged and six neutral). This model does not include the t-quark if all fermions are assumed to belong to a single 56-plet. The fatal shortcoming of the model is that the value of $\sin^2\theta_W$ it predicts is at least twice as large as the experimental value.

No detailed investigation of the E_8 model has been undertaken. One of its interesting features is that the dimensions of the fundamental and the adjoint representations are identical: 248 fermions and 248 gauge bosons. The number of scalar particles runs into thousands, which is a frightening prospect.

It should be clear from the contents of this chapter that so far it is impossible to select the best among the competing grand models. The program of grand unification runs into difficulties. The most serious difficulty is the incomplete state of the mechanism of spontaneous symmetry breaking and the related problem of the hierarchies of the vacuum expectation values. A serious shortcoming of the grand unification models is the fact that these models ignore the gravitational interaction. The program of uniting all known types of interaction, including gravity, was given the name of superunification. It will be briefly outlined in the next chapter.

CHAPTER 26

Superunification

26.1. Supersymmetry

Until now we have been discussing multiplets containing either fermions or bosons. By definition, supersymmetry is a symmetry which relates particles with integral spin to those with half-integral spin. For instance, the parameters of the isospin group transformations $\exp(i\alpha \cdot \tau)$ are scalars. On the other hand, there are spinors among the parameters of supersymmetric transformations. These transformations turn fermions into bosons and bosons into fermions. Sometimes supersymmetry is referred to as the Fermi-Bose symmetry. When a supersymmetric theory is constructed, spinor generators Q_α are added to the Poincaré group generators P_μ and $M_{\mu\nu}$. The simplest supersymmetric lagrangian is given by the sum of free Maxwell and Majorana lagrangians which describe the photons and the truly neutral neutrinos. More complicated supersymmetric lagrangians comprising fields with higher spins ($J > 1$) were also considered. It was found that a number of obstacles which appear when these fields are considered phenomenologically, are eliminated by treating the fields as components of some supersymmetric multiplet.

A local supersymmetry in which the parameters of transformations are functions of coordinates and time, is of special interest. Recall that the local group U(1), related to conservation of electric charge, can be realized only via introduction of massless photons. The local color group SU(3) is realized via an octet of massless gluons. Localization of the Poincaré group is realized via gravitons, in the framework of general relativity theory. (If Lorentz rotations are different at different points, this means that these points have accelerations with respect to one another. Invariance with respect to such transformations is possible only by taking the gravitational field into account. This invariance manifests itself, in particular, in that the observer in an elevator car cannot distinguish between effects due to the elevator acceleration and effects due to a uniform gravitational field.)

Both gravitons and massless particles with spin $\frac{3}{2}$, the so-called gravitino, are necessary to realize supersymmetry. A local supersymmetry is called supergravity. Serious efforts are made at present to find out the properties of higher orders of perturbation theory in supergravity. We have already mentioned that the gravitational constant is dimensional, so that from a formal standpoint quantum gravity is non-renormalizable. Indeed, even one-loop graphs lead to divergences in the theory which contains other particles in addition to the gravitons. At the beginning of the 70's, however, it was demonstrated that one-loop graphs in the theory containing gravitons only, are finite. Supergravity, like quantum gravity theory with gravitons only, has no divergences in the one-loop approximation. Moreover, the two-loop approximation is also convergent. A superparticle (multiplet of particles) plays a role identical to that of the graviton in the usual theory of gravitation, so that there are very rigid relationships between properties and interactions of various particles.

The models of the so-called extended supergravity combining the geometric and internal symmetries are of special interest. In these models the Planck mass $m_p = G_N^{-1/2} \simeq 10^{19}$ GeV (where G_N is the Newton constant) acts as a natural scale unit of the whole physics of elementary particles. One may hope that extended supergravity is a way to develop the unified theory of all particles and all interactions. Versions of extended supergravity have been discussed, involving n gravitinos ($1 \leq n \leq 8$) and having SO(n) invariance, where SO(n) is a group of orthogonal transformations in n-dimensional space. As n increases, the number of particles with different spins in the multiplet rises as well. For example, for $n = 2$, the theory comprises, in addition to the graviton and two gravitinos, only one particle (with $J = 1$), while for $n = 8$ it comprises one particle with $J = 2$, eight particles with $J = \frac{3}{2}$, 28 with $J = 1$, 56 with $J = \frac{1}{2}$, and 70 with $J = 0$. For $n \geq 9$, the theory already has more than one graviton and also particles with $J > 2$.

It is easy to show that even such a large group as SO(8) is still too narrow to contain all the known fermion and gauge multiplets. Several very interesting papers report SO(8) to have a hidden dynamic symmetry SU(8). The group SU(8) already would be sufficiently broad to comprise not only the SU(5) gauge symmetry but also the three fermion generations.

26.2. Sub-quarks?

What lies behind the large number of quarks, leptons, and other particles which today are considered elementary? In principle, two alternative answers can be given.

Answer One: These particles are indeed elementary; their large number does not spoil the beautiful picture since all of them are components of one superparticle; both the number of these components and the character of the interactions between them are determined by one or several fundamental principles.

Answer Two: The particles that at present are considered elementary are in fact built of some more elementary subparticles. Roughly, the main idea of this approach is: universe – galaxies – stars – atoms – nuclei – quarks – subquarks.

It is then natural to assume that subquarks are the blocks of which not only quarks are built but leptons, gluons, and photons as well. At energies now available (at the achieved momentum transfers) all these particles appear to be elementary, and hence they are very small ($\ll 10^{-15}$ cm). Before the advent of quantum chromodynamics, any theorist would have said that only very heavy particles can be localized at such short distances and that as a result the subquark masses cannot be below several tens of GeV. But now when the idea of confinement has gained so much popularity, it is quite acceptable to picture subquarks as light or even massless particles confined at very short distances. However, by virtue of the uncertainty relation, the kinetic energies of these particles must be high (of the order of the inverse dimension of leptons, photons, etc.). Hence, the intervals between energy levels of such systems consisting of subquarks, that is between masses of the conventional elementary particles, must be large. But how could we explain in this case the differences of the order of hundreds of MeV, or even several MeV, between the experimentally observed masses of these particles?

CHAPTER 27

Particles and the universe

Some of the fundamental characteristics of elementary particles can be deduced from astrophysical observational data with much higher accuracy than from laboratory experiments. As examples we can mention the upper bounds on the mass of the photon and of different types of neutrinos, restrictions on the mass and lifetimes of hypothetical neutral leptons, the upper bound on the number of different types of neutrino, and finally the arguments in favor of quark confinement derived on the basis of the estimated residual concentration of relic quarks. Some of these restrictions are discussed in the present chapter on the basis of the hot universe model.

27.1. Hot universe

At first glance, the most abundant particles in the surrounding universe are protons and electrons: on average, the universe contains several protons and as many electrons (the universe is electrically neutral) per cubic meter. This is equivalent to a density of approximately $10^{-29} - 10^{-30}$ g/cm^3. In reality, however, the number of photons in the universe is much greater than that of protons, by a factor of approximately 10^9 (400 photons per cm^3). These photons exist in the form of radiation with a temperature of approximately 2.7 K or $2.5 \cdot 10^{-4}$ eV, and fill the universe. The total energy of this microwave radiation is smaller by four orders of magnitude than the total mass of protons. It means that although the number of relic photons is large, their role at present is negligible in comparison to that of electrons and protons.

There was a time, however (about 10^{10} years ago), when these photons were highly energetic and to a great extent determined the dynamics of the whole universe.

It has been established, on the basis of the red shift of remote galaxies, that the universe is expanding. This expansion is reminiscent of the growth of scale on the surface of a balloon when it is being inflated. The distances between any two points in the universe, and with them the photon wavelengths, increase; the universe expands and cools down. Going back in time, we come to the point $t = 0$, corresponding to infinite density and temperature of the universe. Today it is not clear how close to this point we are justified in extrapolating the increasing temperature and density. The extrapolation is probably valid up to $T \simeq 1$ MeV, and possibly up to $T \simeq 1$ GeV or even up to the Planck temperature (10^{19} GeV). The hot phase of the evolution of the universe has left us a background of cooled relic photons.

Let us find a relation between the age of the universe, that is the time elapsed since the moment $t = 0$, and its temperature. In general, an analysis of the properties of the expanding universe requires the knowledge of the general theory of relativity. However, the relation in question can be found by means of elementary arguments. Consider a uniform infinite flat universe and a sphere of radius R in this universe. Let a probe particle of mass μ lie on this sphere (see fig. 27.1). Expansion of the universe increases the scale and hence our radius R. This means that $dR/dt > 0$. The kinetic energy of the particle in the reference frame connected with the center of the sphere is

$$T_{\text{kin}} = \tfrac{1}{2}\mu\left(\frac{dR}{dt}\right)^2.$$

Consider now the potential energy of the probe particle. It is determined by the attraction of the particles (both massive and massless) inside the sphere (the matter outside the sphere has no effect on the particle). The energy of this attraction is

$$U = -G_N \frac{4\pi R^3 \rho \mu}{3R},$$

where $G_N \simeq 6 \cdot 10^{-39} m_p^{-2}$ is Newton's constant and ρ is the mean energy density (we recall that the source of gravitation is energy, not mass).

Let us find now what critical density $\rho = \rho_c$ corresponds to a steady-state expansion when the sum of kinetic and potential energies of the probe

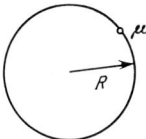

Fig. 27.1.

particle is zero:

$$\frac{\mu}{2}\left(\frac{dR}{dt}\right)^2 = G_N \frac{4\pi R^3 \rho \mu}{3R}.$$

In this expansion mode

$$\left(\frac{1}{R}\frac{dR}{dt}\right)^2 = \tfrac{8}{3}\pi G_N \rho_c.$$

The quantity $H(t) = (1/R)dR/dt$ proves to be independent of R. It is called the Hubble constant and according to observations, $H(t) \simeq 60$ km/s·Megaparsec $\simeq 18$ km/s·light year (the value 60 km/s·Megaparsec is considered the most likely; in fact, 45 km/s·Megaparsec $< H <$ 120 km/s·Megaparsec). The quantity $1/2H \simeq 10^{10}$ years characterizes the age of the Universe. When H is known, it is not difficult to find the critical density ρ_c:

$$\rho_c = \frac{3H^2}{8\pi G_N} \simeq 5 \cdot 10^{-30}\,\text{g/cm}^3 \simeq 3m_p/\text{m}^3.$$

The observed density of matter in the universe is quite close to its critical level. If $\rho > \rho_c$, the expansion will ultimately change to contraction. If $\rho < \rho_c$, then the expansion will continue forever.

Let us turn now to the first moments of the big bang when both T_{kin} and $|U|$ were enormously large compared to their current values, while $T_{\text{kin}} + U$ was the same as it is today. Consider a sufficiently hot phase of the universe for which the particle masses in the expression for the density ρ can be neglected in comparison to their momenta. It is clear on the basis of dimensional arguments that in this case $\rho = a/R^4$ where a is a dimensionless constant. Substitution of this expression into the equation

$$\left(\frac{dR}{dt}\right)^2 = \tfrac{8}{3}\pi G_N \rho R^2$$

yields that

$$\frac{dR}{dt} = \left(\tfrac{8}{3}\pi G_N a\right)^{1/2}\frac{1}{R}.$$

Solution of this equation gives

$$R^2 = \left(\tfrac{32}{3}\pi G_N a\right)^{1/2} t.$$

Returning to the energy density, $\rho = a/R^4$, we obtain

$$\tfrac{32}{3}\pi G_N \rho t^2 = 1.$$

Let us now consider the relation between ρ and the temperature T of the universe:

$$\rho = 4\sigma\kappa T^4,$$

where $\sigma = \frac{1}{60}\pi^2$ is the Stefan-Boltzmann constant (recall that in our system of units $\hbar = c = 1$). The coefficient κ is a function of T: it depends on the number of types of elementary particles contributing to the total energy density at temperature T. In a universe containing photons only, $\kappa = 1$. With other particles taken into account,

$$\kappa = \kappa(T) = 1 + \tfrac{7}{8}N_\nu + \tfrac{7}{4}N_\ell(T) + \ldots,$$

where N_ν is the number of various types of two-component massless neutrino ($\nu_e, \nu_\mu, \nu_\tau, \ldots$), and $N_\ell(T)$ is the number of different four-component leptons for which $2m_\ell \ll T$. The ellipsis in the expression for κ stands for the contributions of quarks, gluons (for $T \gtrsim 1$ GeV), and W- and Z-bosons (for $T \gtrsim 100$ GeV). The factor $\tfrac{7}{8}$ takes into account the difference in densities of the Fermi and Bose distributions for ν and γ, respectively. (Indeed,

$$\int_0^\infty \frac{z^n\,dz}{e^z - 1} - \int_0^\infty \frac{z^n\,dz}{e^z + 1} = 2\int_0^\infty \frac{z^n\,dz}{e^{2z} - 1} = \frac{1}{2^n}\int_0^\infty \frac{y^n\,dy}{e^y - 1},$$

whence

$$\int_0^\infty \frac{z^n\,dz}{e^z + 1} \bigg/ \int_0^\infty \frac{z^n\,dz}{e^z - 1} = 1 - 2^{-n}.$$

This gives $\tfrac{7}{8}$ for $n = 3$).

The contribution of the electron is twice that of the neutrino since the electron has four components. Finally,

$$\tfrac{128}{3}\pi\sigma\kappa G_N T^4 t^2 = 1.$$

This yields the well-known relation of temperature and age of the universe:

$$t\,(\text{seconds}) \simeq \frac{1}{T^2\,(\text{MeV}^2)}.$$

27.2. Upper bound on the neutrino mass

The theory of the hot universe shows, under the assumption of zero total leptonic charge, that the current number of relic neutrinos is of approximately the same order of magnitude as the number of photons. However, the detection of the neutrino background is an incomparably more difficult task which so far seems to be unfeasible. The fact that the theory gives us today's concentration of relic neutrinos makes it possible to find the upper bound on the masses of these particles.

Denote the number of muonic neutrinos (photons) in one cm³ by n_{ν_μ} (n_γ). If $n_{\nu_\mu} \simeq n_\gamma$, then the world contains approximately 10^9 neutrinos per one nucleon. If each neutrino had a mass of, say, 10 eV, the net mass of all neutrinos would be a few times that of all protons in the universe, so that the dynamics of the expansion of the universe would be determined today, as it was several billion years ago, by the gravity effect of these cold and otherwise absolutely passive neutrinos. The neutrino mass cannot be much larger than 10 eV since it would give $\rho \gg \rho_c$. This in turn would mean that the expansion at the earlier stages of the universe was much faster than in the steady-state case. This conclusion would lead to a contradiction since it would mean that the age of the universe is shorter than that of the oldest rocks on earth (which are several billion years old).

In the above paragraph we have made use of the assumption $n_{\nu_\mu} \simeq n_\gamma$. Let us find an exact ratio of these quantities assuming that only left-handed neutrinos participate in the weak interaction. While ν_μ and γ are in equilibrium, $n_{\nu_\mu} = \frac{3}{4} n_\gamma$ (see the above integrals for the Fermi and Bose distributions at $n = 2$). This ratio remained unaltered even when the neutrino ceased to be in equilibrium with electrons, positrons, and photons. However, annihilation of electrons and positrons increased n_γ by a factor of $1 + \frac{7}{4} = \frac{11}{4}$ by virtue of conservation of entropy (the entropy S is proportional to ρ/T) so that from that time until today $n_{\nu_\mu}/n_\gamma = \frac{3}{4} \cdot \frac{4}{11} = \frac{3}{11}$. The same is true for ν_e and ν_τ. As a result, $m_{\nu_e} + m_{\nu_\mu} + m_{\nu_\tau} \lesssim 30$ eV. Note that the cosmological limit is comparable with the laboratory upper limit for ν_e and is better than the laboratory upper bounds on the ν_μ and ν_τ masses by four and seven orders of magnitude, respectively.

27.3. On the number of possible types of neutrinos

The relation between temperature T and the age t of the universe, established above, is

$$t^2 = \frac{3}{32\pi\sigma\kappa G_N T^4},$$

where $\sigma = \frac{1}{15}\pi^2$, $G_N \simeq 6 \cdot 10^{-39} m_p^{-2}$, and $\kappa = \kappa(T)$ is determined by the number of elementary particles whose masses satisfy the condition $T \gg m$.

At $T \simeq 14$ MeV, when the present ratio of abundances of neutrons (mostly contained in the primordial ^{4}He) and protons (hydrogen) had set on, the coefficient κ was

$$\kappa = 1 + \tfrac{7}{4} + \tfrac{7}{8} N_\nu.$$

Here 1 gives the contribution of photons, $\tfrac{7}{4}$ gives that of electron-positron pairs, and $\tfrac{7}{8} N_\nu$ that of neutrinos. The last term is proportional to the number of possible types of neutrinos, N_ν. (Note that the equilibrium between neutrinos of different types and photons, electrons, and positrons is maintained by reactions which proceed via neutral currents, for example, $\nu_\mu + \bar\nu_\mu \to e^+ + e^-$).

We find that at any given temperature, the universe is the younger the greater the number of different types of neutrinos, N_ν, is. In other words, as N_ν is augmented, the time scale contracts, that is the expansion rate of the universe rises:

$$t \to t' = t\sqrt{\kappa/\kappa'}.$$

We shall demonstrate now that the ratio of neutrons to protons is very sensitive to the expansion rate of the universe, so that the observed abundance of ^{4}He is an indicator of the number of different types of neutrinos.

The equilibrium ratio of neutrons to protons at high temperatures is $\exp(-\Delta m/T)$, where $\Delta m = m_n - m_p$. This equilibrium value is maintained by weak reactions of the type $pe^- \to n\nu_e$, $ne^+ \to p\bar\nu_e$, $n \to pe^-\bar\nu_e$. However, when the characteristic time of these weak reactions, t_w, becomes large in comparison with t_c characterizing the expansion of the universe, the ratio of neutron to proton abundance is "quenched" or "frozen", and does not diminish in the course of a further drop in temperature. Dimensional arguments show that

$$\frac{1}{t_w} \simeq G_F^2 T^5,$$

where $G_F = 10^{-5} m_p^{-2}$ is the Fermi constant. By setting t_w equal to t_c we obtain

$$G_F^2 T^5 \simeq (\kappa G_N T^4)^{1/2}.$$

We see that temperature at which neutrons are quenched increases with κ:

$$T_n \simeq \kappa^{1/6} G_N^{1/6} G_F^{-2/3}.$$

As a result, the neutron concentration, $\exp(-\Delta m/T)$, also increases as κ rises.

If we take into account that neutrons are trapped in the primordial ^{4}He, then the abundance (by mass) of the primordial helium was twice the abundance of neutrons. Astrophysical calculations indicate that the abundance of primordial helium was close to 0.24 (this corresponds to $n_n/n_p \simeq 1/7$). This abundance agrees with a cosmological estimate for $N_\nu = 3$. It is not difficult to show that each additional type of neutrino increases helium abundance by about 1.5%. Several years ago astrophysicists considered it possible that the maximum abundance of the primordial helium was above 30%. However, the upper bound on helium abundance as given in the literature gradually diminishes. Assuming, with a number of authors, that the primordial helium abundance did not exceed 25%, we conclude that nature contains not more than one massless neutrino in addition to ν_e, ν_μ, and ν_τ.

27.4. Concentration of relic quarks

As long as we are without a consistent theory of confinement, the main argument against the existence of free quarks with fractional charges is their absence in the surrounding matter. The point is, the expected concentration of relic quarks can be estimated in the framework of the theory of the hot universe, and the estimate is found to be very high, much higher than the upper bound imposed by the experiments on the detection of free quarks. We show below how the estimates of this type are obtained.

Assume that a quark confined within a hadron has a mass much smaller than that of a free quark. (Abdus Salam referred to this phenomenon as the Archimedes effect: a quark dipped into a "hadron bathtub" becomes lighter. Absolute confinement corresponds to infinitely heavy free quarks.) Assume for the purpose of estimation that the free quark mass is of the order of 10 GeV (we shall see that the result is not very dependent on m_q) and consider the universe at $T \simeq 100$ GeV. Quarks at this temperature (it was reached 10^{-10} s after the big bang) are in thermodynamical equilibrium with other fundamental particles, namely photons and leptons. At $T \simeq m_q$ quarks go out of equilibrium: their production stops, and they begin to burn out. This burning goes through reactions of the type of

$q + \bar{q} \rightarrow$ mesons,

$q + q \rightarrow \bar{q} +$ baryon.

These are exothermic reactions: the first releases energy of the order of $2m_q$, and the second of the order m_q. Assume, in analogy to usual exothermic reactions, that σv for these reactions tends, for the quark velocity tending to zero, to a constant limit σ_0. The quark burning rate is characterized then by $n_q \sigma_0$ where n_q is the concentration of free quarks. The process slows down, and then stops, when the rate of burning becomes equal to, and then less than, the rate of the cosmological expansion. The temperature at this stage is still comparable, by order of magnitude, with m_q. Hence,

$$n_q \simeq G_N^{1/2} T^2 / \sigma_0 \simeq G_N^{1/2} m_q^2 / \sigma_0.$$

It will be convenient to consider the ratio of this concentration, n_q, to the concentration of photons $n_\gamma \simeq T^3 \simeq m_q^3$:

$$\frac{n_q}{n_\gamma} \simeq \frac{G_N^{1/2}}{\sigma_0 m_q} \simeq \frac{10^{-18}}{m_p m_q \sigma_0} \simeq 10^{-21} \quad \text{if } \sigma_0 \simeq m_\pi^{-2}.$$

As the universe expands and cools down, the quark and photon concentrations decrease. Their ratio, however, remains constant if the expansion is adiabatic. This ratio, n_q/n_γ, is small because the universe expands slowly, even at $T \simeq 1$ GeV, compared to the time scale typical for elementary particles, $1/\text{GeV}$. A small parameter here is the gravitational constant G_N determining the expansion rate of the universe. A large number of quarks merge into hadrons during this slow expansion.

Now we shall take into account that there are 10^9 photons per one proton. Then the ratio of quark to proton abundances must be equal to

$$n_q/n_p \simeq 10^{-12}.$$

Although this is a very small quantity, it is still unacceptably high. It exceeds the abundance of gold, and would make it possible to "haul cartloads" of quarks. After comparing this result with the experimental limit, $n_q/n_p \lesssim 10^{-27}$, we can hardly avoid the conclusion that there are no free fractionally charged quarks in nature. To lower the quark concentration to an acceptable level, one would have to assign at least atomic (and not nuclear) dimensions to the quark annihilation cross section, σ_0.

27.5. On baryonic asymmetry of the universe

Application of estimates similar to those of the preceding sections, to the annihilation of primordial baryons and antibaryons yields

$$n_p/n_\gamma \simeq G_N^{1/2} m_p \simeq 10^{-18},$$

which is smaller by nine orders of magnitude than the ratio observed in nature. This small value is obtained if one assumes that (i) the initial baryonic charge of the universe is zero, and (ii) there are no interactions violating conservation of baryonic charge. In order to obtain $n_p/n_\gamma \simeq 10^{-9}$ in a world where the baryonic charge is conserved, one has to assume that a small excess of baryonic charge was present from the very beginning, so that $(n_q - n_{\bar{q}})/n_q \simeq 10^{-9}$.

Another point of view appears very attractive. It states that the excess of baryons over antibaryons was produced at the early stages of the evolution of the universe owing to non-conservation of baryonic charge and violation of CP. This idea was first proposed in the middle of the 60's and has been discussed in a number of recent publications in the framework of grand unification models.

CHAPTER 28.

Bibliography

The literature survey given below does not pretend to be complete. It is practically impossible to cover all important publications even if the survey is not too brief.

The literature cited in the survey will help the reader to get a more detailed knowledge of the present state of the problems considered in the book and also, in a number of cases, of the history of these problems. The survey of the literature is divided into the following sections:

28.1. Monographs. Proceedings of Conferences 265
    28.1.1. Monographs on the theory of weak interaction..................265
    28.1.2. Monographs on relativistic quantum theory....................265
    28.1.3. Group theory (monographs and papers) 266
    28.1.4. Conferences, 1960–1982 266
28.2. Decays of leptons and hadrons 267
    28.2.1. On the history of weak interactions, 1896–1958 267
    28.2.2. Muon decay 268
    28.2.3. Strangeness conserving leptonic decays, ...269
    28.2.4. Strangeness changing leptonic decays 269
    28.2.5. Non-leptonic decays of strange particles 270
    28.2.6. Non-conservation of parity in nuclear forces 272
    28.2.7. Neutral K-mesons 272
    28.2.8. Violation of CP invariance 274
    28.2.9. Charmed quark prior to the discovery of the J/ψ meson 275
    28.2.10. Charmed particles after the discovery of the J/ψ meson 276
    28.2.11. Decays of the τ-lepton..............................277
    28.2.12. Properties of the b- and t-quarks 277
28.3. Neutrino reactions 278
    28.3.1. Neutrino-nucleon interactions. Charged currents 278
    28.3.2. Neutrino-nucleon interactions. Neutral currents 279
    28.3.3. Neutrino-electron interaction 279
    28.3.4. Creation of a muon pair by a neutrino in the Coulomb field of a nucleus 280

	28.3.5.	Electron-nucleon interaction 280
	28.3.6.	Phenomenological analysis of data on neutral currents 281
28.4.	Weak interactions at high energies 281	
	28.4.1.	Four-fermion interaction and the unitarity limit 281
	28.4.2.	Gauge symmetry 282
	28.4.3.	Spontaneous breaking of gauge symmetry 283
	28.4.4.	Standard model of the electroweak interaction 284
	28.4.5.	Production and decay of the W- and Z-bosons 285
	28.4.6.	Future facilities designed to search for the W- and Z-bosons 285
	28.4.7.	Properties of Higgs bosons 286
28.5.	Models of grand and super unification 288	
	28.5.1.	Model based on the group SU(4) × SU(2) × SU(2) 288
	28.5.2.	Model based on the group SU(5) 288
	28.5.3.	Model based on the group SO(10) 288
	28.5.4.	Models based on exceptional groups 288
	28.5.5.	Models of simple unification 289
	28.5.6.	Reviews of grand unification models 289
	28.5.7.	Superunification 289
28.6.	Particles and the universe 290	
	28.6.1.	The general theory of relativity and the theory of the hot universe 290
	28.6.2.	The hot universe and leptons 291
	28.6.3.	The hot universe and quarks 292
	28.6.4.	Cosmology and spontaneous symmetry breaking 293
	28.6.5.	Astrophysics and photons 293
	28.6.6.	Possible non-conservation of the baryonic charge, and baryonic asymmetry of the universe 294
	28.6.7.	Possible non-conservation of the leptonic charge 295
	28.6.8.	Concentration of relic monopoles 295
	28.6.9.	On the conservation of the electric charge 296
	28.6.10.	Time independence of the fundamental constants 296
28.7.	Supplementary bibliography (autumn 1980) 296	
	28.7.1.	Electroweak processes. Review papers 296
	28.7.2.	Decays of strange particles 297
	28.7.3.	Violation of CP invariance 297
	28.7.4.	Decays of c- and b-hadrons 298
	28.7.5.	Neutral currents 299
	28.7.6.	Electroweak radiative corrections 300
	28.7.7.	Virtual Higgs bosons 300
	28.7.8.	Dynamical symmetry breaking 300
	28.7.9.	Properties of free neutrinos 301
	28.7.10.	Neutrino masses in grand unification models 302
	28.7.11.	Grand unification models and proton decay 302
	28.7.12.	Various aspects of grand unification 303
	28.7.13.	Superunification 304
	28.7.14.	Cosmology and elementary particles. Review papers 305

28.7.15. Neutrino mass and hidden mass in the universe 305
28.7.16. Calculation of n_B/n_γ 306
28.7.17. Magnetic monopoles and cosmology 306
28.7.18. Search for new particles and interactions 306
28.7.19. Search for deviations from Newton's law 307
28.7.20. Verification of electric charge conservation 307
28.7.21. Future accelerators. 307

28.1. Monographs. Proceedings of conferences

28.1.1. Monographs on the theory of weak interaction

J. M. GAILLARD, M. K. GAILLARD, D. HAIDT, L. JAUNEAU, O. NACHTMAN, H. PIETSCHMANN, F. VANNUCCI, *Weak interactions*, ed. M. K. NIKOLIĆ (Paris, IN2P3, 1977).

D. BAILIN, *Weak interactions* (Sussex University Press, 1977).

J. C. TAYLOR, *Gauge theories of weak interactions* (Cambridge University Press, 1976).

R. E. MARSHAK, RIAZUDDIN, C. P. RYAN, *Theory of weak interactions in particle physics* (Wiley, 1969).

L. B. OKUN, *Weak interactions of elementary particles* (Pergamon, 1965).

The development of weak interaction theory, ed. P. K. KABIR (Gordon and Breach, 1963).
Collection of 39 selected reprints on the theory of weak interactions 1934–1961.

28.1.2. Monographs on relativistic quantum theory

V. B. BERESTETSKY, E. M. LIFSHITZ, L. P. PITAEVSKY, *Relativistic quantum theory*, vol. 4 of Course of theoretical physics, translated from Russian by J. B. SYKS AND J. S. BELL (Pergamon, 1971).

YU. V. NOVOZHILOV, *Vvedeniye v teoriyu elementarnikh chastits* (*Introduction to the theory of elementary particles*) (Nauka, Moscow, 1972) (in Russian).

A. I. AKHIEZER, V. B. BERESTETSKY, *Quantum electrodynamics*, Translated from second Russian edition, ed. G. M. Volkoff (Interscience, 1965).

N. N. BOGOLUBOV, D. V. SHIRKOV, *Introduction to the theory of quantum fields* (Wiley, 1959).

J. D. BJORKEN and S. D. DRELL, *Relativistic quantum mechanics*, vol. I, *Relativistic quantum fields*, vol. II (McGraw-Hill, 1965).

R. P. FEYNMAN, *Theory of fundamental processes* (Benjamin, 1961).

R. P. FEYNMAN, *Quantum electrodynamics* (Benjamin, 1961).

28.1.3. Group theory (monographs and papers)

G. RACAH, *Group theory and spectroscopy*, Preprint of lectures delivered at the Institute of Advanced Studies, Princeton, spring 1951. *Preprint* CERN 61-8, 6 March 1961, Geneva.

F. GÜRSEY, in *Relativity, groups, and topology*, ed. C. DE WITT and B. DE WITT (Gordon and Breach, 1964).

A. O. BARUT and R. RACZKA. *Theory of group representations and applications* (Polish Scientific Publishers, Warszawa, 1977).

K. W. MCVOY, *Symmetry groups in physics*, Rev. Mod. Phys. 37 (1965) 84.

R. E. BEHRENDS, J. DREITLEIN, C. FRONSDAL, W. LEE, *Simple groups and strong interaction symmetries*, Rev. Mod. Phys. 34 (1962) 5.

A. SALAM, *The formalism of Lie groups*, in *Theoretical physics* (IAEA, Vienna, 1963) p. 173.

28.1.4. Conferences, 1960–1982

Rapporteur talks published in the proceedings of regular international conferences are indispensable in following the progress of high-energy physics during the last two decades.

Rochester Conferences on High-Energy Physics. The first seven conferences took place at the University of Rochester, and the subsequent conferences were:
Genève 1958 (VIII), Kiev 1959 (IX), Rochester 1960 (X), Genève 1962 (XI), Dubna 1964 (XII), Berkeley 1966 (XIII), Vienna 1968 (XIV), Kiev 1970 (XV), Batavia 1972 (XVI), London 1974 (XVII), Tbilisi 1976 (XVIII), Tokyo 1978 (XIX), Madison 1980 (XX), Paris (1982) (XXI).

European Conferences on High-Energy Physics
Aix-en-Provence 1961, Sienna 1963, Oxford 1965, Heidelberg 1967, Lund 1969, Amsterdam 1971, Aix-en-Provence 1973, Palermo 1975, Budapest 1977, Genève 1979, Lisbon 1981.

Neutrino Conferences
Moscow 1968, Cortona 1970, Balatonfüred 1972, Philadelphia 1974, Balatonfüred 1975, Aachen 1976, Baksan Valley 1977, West Lafayette 1978, Bergen 1979, Erice 1980, Hawaii, 1981, Balatonfüred, 1982.

Photons and Leptons Conferences
Cambridge 1963, Hamburg 1965, Stanford 1967, Daresbury 1969, Cornwall 1971, Bonn 1973, Stanford 1975, Hamburg 1977, Batavia 1979, Bonn 1981.

28.2. Decays of leptons and hadrons

28.2.1. On the history of weak interactions, 1896–1958

Listed below are some of the papers which played an important role in the genesis of the modern picture of the weak interactions.

A. BECQUEREL, *Compt. Rend.* 122 (1896) 501, 509.
The discovery of radiation of salts of uranium (of β-rays of thorium).

E. RUTHERFORD, *Phil. Mag.*, ser. 6, 21 (1911) 669.
The discovery of the atomic nucleus.

J. CHADWICK, *Nature* 129 (1932) 312.
The discovery of the neutron.

W. PAULI, in: *Noyaux atomique*, VII Conseil de Physique Solvay 1933, Paris, 1934, p. 324.
The hypothesis of the neutrino.

E. FERMI, *Z. Phys.* 88 (1934) 161.
The theory of β-decay (vector variant).

C. D. ANDERSON, S. H. NEDDERMEYER, *Phys. Rev.* 51 (1937) 894.
The discovery of the muon.

C. M. J. LATTES, H. MUIRHEAD, J. P. S. OCCHIALINI, C. F. POWELL, *Nature* 159 (1947) 694.
The discovery of charged pions.

L. Le Prince-Ringuet, M. Lheritier, *Compt. Rend.* 219 (1944) 618.
The first observation of the K-meson.

J. D. Rochester, C. C. Butler, *Nature* 160 (1947) 855.
The first observation of "forks" of decays of neutral strange particles.

B. Pontecorvo, *Phys. Rev.* 72 (1947) 246.
G. Puppi, *Nuovo Cim.* 5 (1948) 505.
O. Klein, *Nature*, 161 (1948) 897.
The hypothesis of the universal weak interaction.

T. D. Lee, C. N. Yang, *Phys. Rev.* 104 (1956) 254.
Formulation of the problem of parity non-conservation in weak interactions.

E. Ambler, R. W. Hayward, D. D. Hoppes, R. R. Hudson, C. S. Wu, *Phys. Rev.* 105 (1957) 1413.
The discovery of the P-odd angular asymmetry of electrons in decays of polarized ^{60}Co nuclei.

R. L. Garwin, L. M. Lederman, M. Weinrich, *Phys. Rev.* 105 (1957) 1415;
J. J. Friedman, V. G. Telegdi, *Phys. Rev.* 105 (1957) 1681.
The discovery of the muon polarization and electron asymmetry in decays $\pi \to \mu \to e$.

R. P. Feynman, M. Gell-Mann, *Phys. Rev.* 109 (1958) 193;
R. E. Marshak, E. C. G. Sudarshan, *Phys. Rev.* 109 (1958) 1860;
J. J. Sakurai, *Nuovo Cim.* 7 (1958) 649.
The $V - A$ theory of weak interactions.

28.2.2. Muon decay

L. Michel, *Proc. Phys. Soc.* A63 (1950) 154, 1371.
Calculation of the spectrum of electrons in muon decay.

A. O. Weissenberg, *Muons* (North-Holland, Amsterdam, 1966);
J. Feinberg, L. Lederman, *Ann. Rev. Nucl. Sci.* 13 (1963).
Discussion of the most important experimental data on muons.

S. M. Berman, A. Sirlin, *Ann. of Phys.* 20 (1962) 20;
A. Sirlin, *Rev. Mod. Phys.* 50 (1978) 573.
Radiative corrections to muon decay.

28.2.3. Strangeness conserving leptonic decays

S. S. GERSHTEIN, YA. B. ZELDOVICH, *ZhETF* 29 (1955) 698 [*JETP* 2(1956) 576];
R. P. FEYNMAN, M. GELL-MANN, *Phys. Rev.* 103 (1958) 193.
Conservation of vector current, non-renormalizability of the vector constant, and prediction of the π_{e3} decay.

M. GELL-MANN, *Phys. Rev.* 111 (1958) 362.
Weak magnetism.

Y. NAMBU, *Phys. Rev. Lett.* 4 (1960) 380;
Y. NAMBU, G. JONA-LASINIO, *Phys. Rev.* 122 (1961) 345,
V. G. VAKS, A. I. LARKIN, *ZhETF* 40 (1961) 282 [*JETP* 13 (1961) 192];
J. GOLDSTONE, *Nuovo Cim.* 19 (1961) 154.
Spontaneous breaking of chiral symmetry and massless bosons.

M. L. GOLDBERGER, S. B. TREIMAN, *Phys. Rev.* 110 (1958) 1178, 111 (1958) 354;
CHOU KUANG-CHAO *ZhETF* 39 (1960) 703 [*JETP* 12 (1961) 492].
Effective pseudoscalar and dispersion relations.

M. GELL-MANN, M. Levy, *Nuovo Cim.* 16 (1960) 705.
The hypothesis of partial conservation of the axial current.

A description of the evolution of the theory of soft pions and partial conservation of the axial current can be found in the review paper:
A. I. VAINSHTEIN AND V. I. ZAKHAROV, in *Usp. Fiz. Nauk*, 100 (1970) 225 [*Uspekhi* 13 (1970) 73];
and in the monographs:
S. L. ADLER, R. F. DASHEN, *Current algebra and applications to particle physics* (Benjamin, 1968);
J. BERNSTEIN, *Elementary particles and their currents* (Freeman, San Francisco, 1968);
S. B. TREIMAN, R. JACKIW, D. J. GROSS, *Lectures on current algebra and its applications* (Princeton University Press, 1972).

28.2.4. Strangeness changing leptonic decays

M. GELL-MANN, *Proc. Rochester Conf.* 8 (1956) 25;
L. B. OKUN, *ZhETF* 34 (1958) 469 [*JETP* 7 (1958) 322].

Selection rules for semileptonic decays of hadrons, $\Delta Q = \Delta S$, $|\Delta S| = 1$, $\Delta T = 1/2$, are formulated on the basis of the model in which all hadrons are built of three fundamental baryons with quantum numbers p, n, Λ (the Sakata model).

N. CABIBBO, *Phys. Rev. Lett.* 10 (1963) 531.
A unified description of leptonic decays of mesons and baryons on the basis of SU(3) symmetry, with a single parameter, θ_C. See also earlier papers:
M. GELL-MANN, M. LEVY, *Nuovo Cim.* 16 (1960) 705;
M. GELL-MANN, *Phys. Rev.* 125 (1962) 1067.
The current $\bar{p}n\cos\theta + \bar{p}\lambda\sin\theta$ is introduced, and the idea of the modified universality of the weak interaction is formulated.

I. YU. KOBZAREV, L. B. OKUN, *ZhETF* 42 (1962) 1400 [*JETP* 15 (1962) 970].
The equality of constants of the strangeness-changing vector and axial currents is demonstrated on the basis of SU(3) symmetry of the strong interaction and by comparing decay rates of $\pi_{\mu 2}$, π_{e3}, $K_{\mu 2}$, $K_{\ell 3}$.

M. ADEMOLLO, R. GATTO, *Phys. Rev. Lett.* 13 (1964) 262.
The theorem on the vanishing of first-order corrections in SU(3) breaking to vector vertices.

All the details concerning SU(3) symmetry can be found in
M. GELL-MANN, Y. NE'EMAN, *The eightfold way* (Benjamin, 1964).

C. G. CALLAN, S. B. TREIMAN, *Phys. Rev. Lett.* 16 (1966) 153.
Relations between amplitudes of $K_{\ell 2}$, $K_{\ell 3}$, and $K_{\ell 4}$ decays.

L. M. CHOUNET, J. M. GAILLARD, M. K. GAILLARD, *Phys. Reports* 4C (1972) 199.
A review of leptonic decays of mesons and hyperons.

W. TANNENBAUM et al., *Phys. Rev. Lett.* 33 (1974) 175;
D. DECAMP et al., *Phys. Lett.* 66B (1977) 295.
Experiments on measuring g_A/g_V for $\Sigma^- \to \text{ne}^- \bar{\nu}_e$ decay.

28.2.5. Non-leptonic decays of strange particles

M. GELL-MANN, A. PAIS, *Proc. Glasgow Conf.*, 1954 (Pergamon, London, 1955) p. 342.
The hypothesis of the selection rule $\Delta T = 1/2$.

B. W. LEE, *Phys. Rev. Lett.* 12 (1964) 83;
H. SUGAWARA, *Progr. Theor. Phys.* 31 (1964) 231;

M. GELL-MANN, *Phys. Rev. Lett.* 12 (1964) 155.
The hypothesis of octet enhancement in the effective non-leptonic strangeness-changing interaction. The Lee-Sugawara relation is derived.

P. K. ELLIAS, C. TAYLOR, *Nuovo Cim.* 44 (1966) 518;
Y. HARA, Y. NAMBU, *Phys. Rev. Lett.* 16 (1966) 875;
M. SUZUKI, *Phys. Rev.* 144 (1966) 1154;
A. D. DOLGOV, V. I. ZAKHAROV, *Yad. Fiz.* 1 (1968) 352 [*Sov. J. Nucl. Phys.* 7 (1968) 232].
The relation between the amplitudes of the decays $K \to 2\pi$ and $K \to 3\pi$ is derived by the soft pion techniques. The last paper demonstrates that the $\Delta T = 1/2$ rule must be strongly broken ($\simeq 50\%$) in slopes of pion spectra in $K \to 3\pi$ decays.

T. J. DEVLIN, J. O. DICKEY, *Rev. Mod. Phys.* 51 (1979) 237.
A review of experimental and theoretical results concerning $K \to 2\pi$ and $K \to 3\pi$ decays.

H. SUGAWARA, *Phys. Rev. Lett.* 15 (1965) 870;
M. SUZUKI, *Phys. Rev. Lett.* 15 (1965) 486;
Y. HARA, Y. NAMBU, T. SCHECHTER, *Phys. Rev. Lett.* 16 (1966) 875;
S. BADIER, C. BOUCHIAT, *Phys. Lett.* 20 (1966) 259.
Application of the soft pion techniques to non-leptonic decays of hyperons.

J. SCHWINGER, *Phys. Rev. Lett.* 12 (1964) 630;
I. YU. KOBZAREV, L. B. OKUN, *Yad. Fiz.* 1 (1965) 1134 [*Sov. J. Nucl. Phys.* 1 (1965) 807].
Phenomenological estimates of the matrix elements of the decays $K^+ \to \pi^+ \pi^0$ and $\Lambda^0 \to p\pi^-$, derived by multiplying hadronic amplitudes for the semi-leptonic decays $K_{\ell 3}$, $\pi_{\mu 2}$, and $\Lambda \to pe^- \bar{\nu}_e$.

K. WILSON, *Phys. Rev.* 179 (1969) 1499.
Operator expansion and its application to the $\Delta T = 1/2$ rule.

M. K. GAILLARD, B. W. LEE, *Phys. Rev. Lett.* 33 (1974) 108;
G. ALTARELLI, L. MAIANI, *Phys. Lett.* 52B (1974) 351.
Hard virtual gluons are shown to enhance amplitudes with $\Delta T = 1/2$ and to suppress those with $\Delta T = 3/2$.

A. I. VAINSHTEIN, V. I. ZAKHAROV, M. A. SHIFMAN, *ZhETF* 72 (1977) 1275 [*JETP* 45 (1977) 670].
A gluon-monopole mechanism is suggested for non-leptonic decays satisfying the rule $\Delta T = 1/2$. The total non-leptonic lagrangian is derived.

J. Finjord, *Phys. Lett.* 76B (1978) 116.
Comparison of the predictions of the gluon-monopole mechanism with the experimental data on decays of the Ω-hyperon.

28.2.6. Non-conservation of parity in nuclear forces

Yu. G. Abov, P. A. Krupchitsky, Yu. A. Oratovsky, *Phys. Lett.* 12 (1964) 25;

Yu. G. Abov, P. A. Krupchitsky, Yu. A. Oratovsky, *Yad. Fiz.* 1 (1965) 479 [*Sov. J. Nucl. Phys.* 1 (1965) 341].
The discovery of the P-odd correlation of the γ-quantum momentum with the neutron spin in the reaction ^{113}Cd$(n\gamma)^{114}$Cd.

V. M. Lobashov, V. A. Nazarenko, L. F. Sayenko, L. M. Smotritsky, G. I. Kharkevich, *Pis'ma ZhETF* 3 (1966) [*JETP Lett.* 3 (1966) 173].
The discovery of the P-odd circular polarization of γ-quanta in decays of the ^{181}Ta nucleus.

Yu. G. Abov, P. A. Krupchitsky, *Usp. Fiz. Nauk* 118 (1976) 141 [*Uspekhi* 19 (1976) 75].
A review of experiments on the study of P-odd nuclear forces.

V. M. Lobashov et al., *Nucl. Phys.* A197 (1972) 241.
Measurements of circular polarization of photons in the reaction np $\to$ dγ. Theoretical discussion of this effect see in:

G. S. Danilov, *Yad. Fiz.* 22 (1975) 776 [*Sov. J. Nucl. Phys.* 22 (1975) 401].

G. V. Danilyan et al., *Yad. Fiz.* 27 (1978) 42 [*Sov. J. Nucl. Phys.* 27 (1978) 21].
Non-conservation of parity in nuclear fission.

R. J. Blin-Stoyle, *Fundamental interactions and the nucleus* (North-Holland, Amsterdam, 1973).

28.2.7. Neutral K-mesons

M. Gell-Mann, A. Pais, *Phys. Rev.* 97 (1955) 1387.
Introduction of K_1^0 and K_2^0 as the C-even and C-odd superpositions of the states K^0 and $\bar{K}^0$.

A. Pais, O. Piccioni, *Phys. Rev.* 100 (1955) 1487.
Theoretical prediction of oscillations in a beam of neutral K-mesons and of the regeneration effect.

K. Lande, E. T. Booth, J. Impeduglia, L. M. Lederman, W. Chinowsky, *Phys. Rev.* 103 (1956) 1901.
Observation of long-lived neutral K-mesons.

B. L. Ioffe, L. B. Okun, A. P. Rudik, *ZhETF* 32 (1957) 396 [*JETP* 5 (1957) 328];
L. D. Landau, *ZhETF* 32 (1957) 405 [*JETP* 5 (1957) 336];
T. D. Lee, C. N. Yang, R. Oehme, *Phys. Rev.* 106 (1957) 340.
An analysis of the situation with K_1^0 and K_2^0 taking into account non-conservation of *C*- and *P*-parities. K_1^0 and K_2^0 are regarded as states with $CP = +1$ and $CP = -1$, respectively.

L. B. Okun, B. Pontecorvo, *ZhETF* 32 (1957) 1587 [*JETP* 5 (1957) 1297].
A small $K_1^0 - K_2^0$ mass difference means that there are no transitions with $|\Delta S| = 2$.

L. B. Okun, in Proc. of 1960 Annual Int. Conf. on high-energy physics at Rochester (University of Rochester, 1960) p. 743.
Cutoff value of the order of 1 GeV is obtained by comparing the contribution of the loop in the second order in the weak interaction with the observed small $K_1^0 - K_2^0$ mass difference.

I. Yu. Kobzarev, L. B. Okun, *ZhETF* 39 (1960) 605 [*JETP* 12 (1961) 426].
Discussion of possible experiments for determining which is heavier, K_1^0 or K_2^0.

M. L. Good, *Phys. Rev.* 106 (1957) 591; 110 (1958) 550.
Description of oscillations of K^0 mesons in a medium.

F. Muller et al., *Phys. Rev. Lett.* 4 (1960) 418.
The first result of measuring the $K_1^0 - K_2^0$ mass difference.
C. Geweniger et al., *Phys. Lett.* 52B (1974) 108. One of the most accurate results for Δm_{LS}.

T. Day, *Phys. Rev.* 121 (1961) 1204.
Quantum-mechanical reduction of a packet at large distances, exemplified by a $K^0 \overline{K}^0$ pair. See also:
A. Einstein, B. Podolsky, N. Rosen, *Phys. Rev.* 47 (1935) 777.

G. Feinberg, *Phys. Rev.* 109 (1958) 1381;
Ya. B. Zeldovich, *ZhETF* 36 (1959) 1381 [*JETP* 9 (1959) 984].
The theory of regeneration $K_2^0 \to K_1^0$ on electrons.
W. R. Molzon et al., *Phys. Rev. Lett.*, 41 (1978) 1213.
Experimental observation of the regeneration $K_2^0 e \to K_1^0 e$.

28.2.8. Violation of CP invariance

J. H. CHRISTENSON, J. W. CRONIN, V. L. FITCH, R. TURLAY, *Phys. Rev. Lett.*, 13 (1964) 138.
The discovery of the decay $K_L^0 \to \pi^+ \pi^-$.

CPT theorem and its application:
G. GRAVERT, G. LÜDERS, G. ROLLNIK, *Fortschr. der Phys.* 7 (1959) 291.
W. PAULI, in: *Niels Bohr and the development of physics*, ed. W. PAULI (Pergamon, London, 1955).

L. WOLFENSTEIN, *Phys. Rev. Lett.* 13 (1964) 562.
The model of superweak violation of *CP* invariance.

T. T. WU, C. N. YANG, *Phys. Rev. Lett.* 13 (1964) 380.
J. S. BELL, J. STEINBERGER, in: *Proc. Int. Conf. on elementary particles, Oxford*, 1965 (Rutherford High Energy Lab., Chilton, 1966) p. 193.
A phenomenological analysis of *CP* violation in decays of K^0-mesons.

For reviews of *CP* violation see:
Review talks presented at the *International Seminar on CP violation*, Moscow, 1968, *Usp. Fiz. Nauk* 95 (1968) No. 3, 4.
L. B. OKUN, C. RUBBIA, in: *Proc. Heidelberg Int. Conf. on elementary particles*, ed. H. FILTHUT (North-Holland, Amsterdam, 1968) p. 299.
K. KLEINKNECHT, in: *Proc. 17th Int. Conf. on high-energy physics, London, 1974* (Rutherford Lab., Chilton, Didcot, 1974) p. 3–23.

The upper bound on the neutron dipole moment:
W. B. DRESS et al., *Phys. Rev.* D15 (1977) 9.
($d_n < e \cdot 3 \cdot 10^{-24}$ cm);
I. S. ALTARYEV et al., *Pis'ma ZhETF* 29 (1979) 794 [*JETP Lett.* 29 (1979) 730]. ($d_n < e \cdot 1.6 \cdot 10^{-24}$ cm).

E. P. SHABALIN, *Yad. Fiz.* 28 (1978) 151 [*Sov. J. Nucl. Phys.* 28 (1978) 75]; preprint ITEP-87 (1979);
J. ELLIS, M. K. GAILLARD, *Nucl. Phys.* B150 (1979) 141.
Estimates of the neutron dipole moment in the six-quark model.

S. WEINBERG, *Phys. Rev. Lett.* 37 (1976) 657;
A. A. ANSELM, D. I. D'YAKONOV, *Nucl. Phys.* B145 (1978) 271.
Breaking of *CP* in a model with two quark doublets and two doublets of Higgs bosons.

A. A. ANSELM, N. G. URAL'TSEV, *Yad. Fiz.* 30 (1979) 465 [*Sov. J. Nucl. Phys.* 30 (1979) 240].
Spontaneous breaking of *CP* invariance in a model with three quark doublets and several multiplets of Higgs fields.

CP-odd gravitational moment of the proton:
I. YU. KOBZAREV, L. B. OKUN, *ZhETF* 43 (1962) 1904 [*JETP* 16 (1963) 1343];
T. LEITNER, S. OKUBO, *Phys. Rev.* 136B (1964) 1542;
B. V. VASILYEV, *Pis'ma ZhETF* 9 (1969) 299 [*JETP Lett.* 9 (1969) 175].

28.2.9. Charmed quark prior to the discovery of the J/ψ meson

A remark stating that the symmetry between leptons and quarks requires the existence of a fourth quark appeared in the paper which introduced three quarks:
M. GELL-MANN, *Phys. Lett.* 8 (1964) 214.

See also:
Y. KATAYAMA, K. MATUMOTO, S. TANAKA, E. YAMADA, *Progr. Theor. Phys.* 28 (1962) 675;
Z. MAKI, M. NAKAGAWA, S. SAKATA, *Progr. Theor. Phys.* 28 (1962) 870.
These papers discuss an extended Sakata model with a fourth fundamental baryon.

Later the four-quark model was discussed by:
Y. HARA, *Phys. Rev.* B134 (1964) 701,
D. AMATI, H. BACRY, J. NUYTS, J. PRENTKI, *Phys. Lett.* 11 (1964) 190.

The term "charm" was introduced in:
J. D. BJORKEN, S. L. GLASHOW, *Phys. Lett.* 11 (1964) 255.

Classification of SU(4) multiplets of mesons and baryons:
V. V. VLADIMIRSKY, *preprints* ITEP, 1964, no. 262 and 299; 1965, no. 353.

A proposal to search for charmed (in the old terminology, supercharged) particles by looking for multilepton events in neutrino experiments:
L. B. OKUN, *Phys. Lett.* 12 (1964) 250.

S. L. GLASHOW, J. ILIOPOULOS, L. MAIANI, *Phys. Rev.* D2 (1970) 1285.
Weak interaction of the (then hypothetical) c-quark and SU(4) symmetry of the strong interaction are used to explain the absence, in experiment, of weak strangeness-changing neutral currents.

(The problem of strange neutral currents became especially acute after the publication of the paper:

B. L. IOFFE, E. P. SHABALIN, *Yad. Fiz.* 6 (1967) 828 [*Sov. J. Nucl. Phys.* 6 (1968) 603],

which demonstrated that the three-quark model predicts a too high value of the mass difference of the K_L^0 and K_S^0 mesons, as well as a too high decay rate for $K_L^0 \to \mu^+ \mu^-$).

Further discussion of this problem:

A. I. VAINSHTEIN, I. B. KHRIPLOVICH, *Pis'ma ZhETF* 18 (1973) 141 [*JETP Lett.* 18 (1973) 83];

M. K. GAILLARD, B. W. LEE, *Phys. Rev.* D10 (1974) 897;

A. I. VAINSHTEIN, V. I. ZAKHAROV, V. A. NOVIKOV, M. A. SHIFMAN, *Yad. Fiz.* 23 (1976) 1024 [*Sov. J. Nucl. Phys.* 23 (1976) 540].

28.2.10. Charmed particles after the discovery of the J/ψ meson

J. J. AUBERT et al., *Phys. Rev. Lett.* 33 (1974) 1404.
An announcement of the discovery of the J-meson by Ting's group.
J. E. AUGUSTIN et al., *Phys. Rev. Lett.*, 33 (1974) 1406.
An announcement of the discovery of the ψ-meson by Richter's group.

B. RICHTER, S. TING, *Nobel Lectures*, December 1976, in: *Les Prix Nobél en 1976*, Stockholm, 1977.

J. ELLIS, M. K. GAILLARD, D. V. NANOPOULOS, *Nucl. Phys.* B100 (1976) 313;

N. CABIBBO, L. MAIANI, *Phys. Lett.* 73B (1978) 418.

M. B. VOLOSHIN, V. I. ZAKHAROV, L. B. OKUN, *Pis'ma ZhETF* 21 (1975) 403 [*JETP Lett.* 21 (1975) 183]; *Yad. Fiz.* 22 (1975) 166 [*Sov. J. Nucl. Phys.* 22 (1975) 81];

M. B. EINHORN, C. QUIGG, *Phys. Rev.* D12 (1975) 2015.

Quark graphs and sextet enhancement for non-leptonic decays of charmed particles.

G. GOLDHABER et al., *Phys. Rev. Lett.* 37 (1976) 255.
I. PERUZZI et al., *Phys. Rev. Lett.* 37 (1976) 569.
The discovery of the D-mesons.

I. PERUZZI et al., *Phys Rev. Lett.* 39 (1977) 1301.

Measurement of $B(D^0 \to K^- \pi^+)$, $B(D^+ \to \overline{K}^0 \pi^+)$, and other partial widths of D-mesons.

28.2.11. Decays of the τ-lepton

First experimental evidence of the existence of the τ lepton was obtained at SLAC:
M. L. PERL et al., *Phys. Rev. Lett.* 35 (1975) 1489;
M. L. PERL et al., *Phys. Lett.* 63B (1976) 466.
Confirmation of the existence of the τ-lepton:
J. BURMESTER et al., *Phys. Lett.* 68B (1977) 297, 301;
R. BRANDELIK et al., *Phys. Lett.* 70B (1977) 125.
Summary of experimental data on the τ-lepton:
G. J. FELDMAN, preprint SLAC-PUB-224 (October 1978); *Proc. of the 19th Int. Conf. on high energy physics, 1978, Tokyo, Japan*, p. 777.
Theoretical papers:
Y. S. TSAI, *Phys. Rev.* D4 (1971) 2821; Relation between the decay width of $\tau \to \nu + 2n\pi$ and the annihilation $e^+ e^- \to 2n\pi$.
YA. I. AZIMOV, L. L. FRANKFURT, V. A. KHOZE, *Usp. Fiz. Nauk* 124 (1978) 459 [*Uspekhi* 21 (1978) 225].

28.2.12. Properties of the b- and t-quarks

S. W. HERB et al., *Phys. Rev. Lett* 76B (1978) 243.
The discovery of the Υ-meson in the spectrum of dimuons in the reaction p + nucleus $\to \mu^+ \mu^- + X$.

CH. BERGER et al., *Phys. Lett.* 76B (1978) 243;
C. W. DARDEN et al., *Phys. Lett.* 76B (1978) 246.
Observation of the Υ-meson in colliding $e^+ e^-$ beams.

M. KOBAYASHI, K. MASKAWA, *Progr. Theor. Phys.* 49 (1973) 652.
The general form of the matrix of left-handed charged currents in the six-quark model.

M. K. GAILLARD, D. V. NANOPOULOS, S. RUDAZ, *Nucl. Phys.* B131 (1977) 285;
R. E. SHROCK, L. L. WANG, *Phys. Rev. Lett.* 41 (1978) 1692.

Phenomenological analysis of the experimental limits for the angles in the Kobayashi-Maskawa matrix.

J. Ellis, M. K. Gaillard, D. V. Nanopoulos, *Nucl. Phys.* B109 (1976) 213.
An analysis of the effects of *CP*-violation in the framework of the six-quark model.

D. Cuffs et al., *Phys. Rev. Lett.* 41 (1978) 363;
R. Vidal et al., *Phys. Lett.* 77B (1978) 344.
A search for long-lived b-hadrons, with negative result.

28.3. Neutrino reactions

28.3.1. *Neutrino-nucleon interactions. Charged currents*

C. L. Cowen, H. W. Cruse, F. B. Harrison, A. D. McGuire, F. Reines, *Science* 124 (1956) 103.
The first measurement of the cross section for the reaction $\bar{\nu}p \to e^+ n$ (with reactor antineutrinos).

G. Danby, J. M. Gaillard, K. Goulianos, L. M. Lederman, M. Mistry, M. Schwartz, J. Steinberger, *Phys. Rev. Lett.* 9 (1962) 36.
The discovery of the muon antineutrino.

J. D. Bjorken, *Phys. Rev.* 179 (1969) 1547.
The prediction of scaling in deep-inelastic lepton-nucleon interactions.

E. D. Bloom et al., *Phys. Rev. Lett.* 23 (1969) 930;
M. Breidenbach et al., *Phys. Rev. Lett.* 23 (1969) 935.
The discovery of scaling in deep-inelastic ep scattering.

The parton model was suggested by R. P. Feynman in 1969 in an unpublished paper. See:
J. D. Bjorken, E. A. Paschos, *Phys. Rev.* 185 (1969) 1975;
R. P. Feynman, *Photon-hadron interactions* (Benjamin, 1972);
R. P. Feynman, in *Neutrino 1974*, ed. C. Baltay (American Inst. of Phys., New York, 1974) p. 299.

Review papers on high-energy neutrino physics:
C. H. Llewellyn Smith, *Phys. Reports* 3 (1972) 261;
B. C. Barish. *Phys. Reports* 39 (1978) *p*. 279,
J. Steinberger, CERN *preprint*, Four Lectures presented at the Summer Institute, Cargèse, 4–23 July, 1977.

28.3.2. Neutrino-nucleon interactions. Neutral currents

F. J. HASERT et al., *Phys. Lett.* B46 (1973) 138;
F. J. HASERT et al., *Nucl. Phys.* B73 (1974) 1.
The discovery of neutral currents in a neutrino experiment in the Gargamelle bubble chamber, at CERN.

A. BENVENUTI et al., *Phys. Rev. Lett.* 32 (1974) 800;
B. AUBERT et al., *Phys. Rev. Lett.* 32 (1974) 1454;
B. C. BARISH et al., *Phys. Rev. Lett.* 34 (1975) 538.
Observation of neutral currents at the Batavia accelerator.

D. CLINE et al., *Phys. Rev. Lett.* 37 (1976) 252, 648;
W. LEE et al., *Phys. Rev. Lett.* 37 (1976) 186;
H. H. WILLIAMS et al., in: *Proceedings Topical Conference on neutrino physics at accelerators, Oxford, 1978.*
M. POHL et al., *Phys. Lett.* 72B (1978) 489.
Observation of neutrino elastic scattering on protons.

V. M. SHEKHTER, *Usp. Fiz. Nauk* 119 (1976) 593 [*Uspekhi* 19 (1976) 645].
A review of neutral currents (experiment and theory).
P. F. ERMOLOV, A. I. MUKHIN, *Usp. Fiz. Nauk* 124 (1978) 385 [*Uspekhi* 21 (1978) 185].
A review of neutrino experiments.

M. HOLDER et al., *Phys. Lett.* 71B (1977) 222; 72B (1977) 254;
P. WANDERER et al., *Phys. Rev.* D17 (1978) 1679.
Experimental investigation of inclusive reactions caused by the neutral current at high energies, carried out at CERN and FNAL, respectively.

H. KLUTTIG et al., *Phys. Lett.*, 77B (1977) 446;
W. KRENZ et al., *Nucl. Phys.*, B135 (1978) 45;
O. ERRIQUES et al., *Phys. Lett.* 73B (1978) 350.
Observation of single pions produced by the neutral neutrino current.

28.3.3. Neutrino-electron interaction

F. REINES et al., *Phys. Rev. Lett.* 37 (1976) 315.
Observation of $\bar{\nu}_e e$ scattering for reactor anti-neutrinos.
F. T. AVIGNONE III, Z. D. GREENWOOD, *Phys. Rev.* D16 (1977) 2383.
Reines' experimental data on $\bar{\nu}_e e$ scattering are re-analyzed.

F. J. HASERT et al., *Phys. Lett.* 46B (1973) 121.

Observation of the first $\nu_\mu e$ scattering event in the Gargamelle bubble chamber.

H. FAISSNER et al., *Phys. Rev. Lett.* 41 (1978) 213;

A. M. CNOPS et al., *Phys. Rev. Lett.* 41 (1978) 357.

Measurement of the $\nu_\mu e$ scattering cross section. The paper cites earlier publications on the subject.

28.3.4. Creation of a muon pair by neutrino in the Coulomb field of a nucleus

M. S. MARINOV, YU. P. NIKITIN, YU. P. OREVKOV, E. P. SHABALIN, *Yad. Fiz.* 3 (1966) 678 [*Sov. J. Nucl. Phys.* 3 (1966) 497]; Erratum: 15 (1972) 1086

Numerical calculations for $E = 50$ GeV.

A. E. ASRATYAN, M. A. KUBANTSEV, *Yad. Fiz.* 25 (1977) 1051 [*Sov. J. Nucl. Phys.* 25 (1977) 558].

It is shown that detection of the process $\nu Z \to \nu \mu \mu Z$ demands an increase in statistics by two orders of magnitude.

28.3.5. Electron-nucleon interactions

YA B. ZELDOVICH, *ZhETF* 36 (1959) 964 [*JETP* 9 (1959) 682].

A theoretical analysis of the P-odd electron-nucleon interaction. Prediction of the effect of weak optical activity of atoms. See also:

S. A. BLUDMAN, *Nuovo Cim.* 9 (1958) 433;

V. N. BAIER, I. B. KHRIPLOVICH, *ZhETF* 39 (1960) 1374 [*JETP* 12 (1960) 959].

M. BOUCHIAT, C. BOUCHIAT, *Phys. Lett.* 48 (1974) 111.

I. B. KHRIPLOVICH, *Pis'ma ZhETF* 20 (1974) 686 [*JETP Lett.* 20 (1974) 315].

Proposal of experiments to detect optical activity of heavy atoms. In particular, the second of the cited papers suggests an experiment to measure the rotation of the plane of polarization of light in thallium and bismuth vapor.

Theoretical reviews on parity non-conservation in atomic physics.

V. A. ALEKSEYEV, B. YA. ZELDOVICH, I. I. SOBELMAN, *Usp. Fiz. Nauk* 118 (1976) 385 [*Uspekhi* 19 (1976) 207];

A. N. MOSKALEV, R. M. RYNDIN, I. B. KHRIPLOVICH, *Usp. Fiz. Nauk* 118 (1976) 409 [*Uspekhi* 19 (1976) 220].

L. M. BARKOV, M. S. ZOLOTOREV, *Pis'ma ZhETF* 27 (1978) 379; 28 (1978) 544 [*JETP Lett*. 27 (1978) 357; 28 (1979) 503].
The angle of rotation of the polarization plane of light in bismuth vapor is measured.

P. E. BAIRD et al., *Phys. Rev. Lett*. 39 (1977) 798.
L. L. LEWIS et al., *Phys. Rev. Lett*. 39 (1977) 795.
Experiments, respectively, in Seattle and Oxford that detected no rotation of the light polarization plane in Bi.

R. CONTI et al., *Phys. Rev. Lett*. 42 (1978) 343.
Observation of optical activity in thallium.

C. Y. PRESCOTT et al., *Phys Lett*. 77B (1978) 347.
Observation of parity non-conservation in inelastic scattering of polarized electrons by nucleons.

28.3.6. *Phenomenological analysis of data on neutral currents*

L. M. SEHGAL, *Phys. Lett*. 71B (1977) 99;
P. Q. HUNG, J. J. SAKURAI, *Phys. Lett*. 72B (1977) 208;
L. F. ABBOTT, R. M. BARNETT, *Phys. Rev. Lett*. 40 (1978) 1303;
D. P. SIDHU, P. LANGACKER, *Phys. Rev. Lett*. 41 (1978) 732;
J. D. BJORKEN, *Phys. Rev.* D18 (1978) 3239;
J. J. SAKURAI, UCLA *preprint* (1978) TEP/27; *Intern. Symp. on high-energy physics with polarized beams and polarized targets*, ANL, October 25–28, 1978, 138;
R. M. BARNETT, SLAC *preprints*, 1978, PUB-2131, PUB-2111; *Proceedings of the Conference Neutrino-78, Purdue Univ., 1978*, ed. E. C. Fowler, p. 699.
Coupling constants of the vector and axial neutral currents are obtained from experimental data, without using the standard model of the electroweak interaction.

28.4. Weak interactions at high energies

28.4.1. *Four-fermion interaction and the unitarity limit*

W. HEISENBERG, *Z. Phys.* 101. (1936) 533.
It is shown that the weak four-fermion interaction increases with particle energy and becomes strong at energies of the order of 10^3 GeV.

L. D. LANDAU, *ZhETF* 10 (1940) 718 (see *Collected Papers of L. D. Landau*, p. 274. Pergamon Press, 1965).
Introduction of the unitarity limit, in particular, in the case of vector bosons.

D. I. BLOKHINTSEV, *Usp. Fiz. Nauk* 62 (1957) 49; *ZhETF* 35 (1958) 53; *Nuovo Cim.* 9 (1958) 925.
A discussion of the role of weak interactions in electromagnetic processes at energies of the order of 10^3 GeV.

B. N. VALUYEV, *ZhETF* 36 (1959) 1578 [*JETP* 9 (1959) 1121];
B. L. IOFFE, *ZhETF* 38 (1960) 1608 [*JETP* 11 (1960) 1158];
M. A. MARKOV, *Proc. Intern. Conf. on high energy phys. at Rochester*, Rochester, 1960 (Interscience) p. 578;
E. P. SHABALIN, *Yad. Fiz.* 6 (1967) 547 [*Sov. J. Nucl. Phys.* 6 (1967) 399].
A discussion of higher-order effects in weak interactions in $\mu \to e\gamma$ and $\mu \to 3e$ processes, and in the equality of constants of β- and μ-decays.

B. L. IOFFE, L. B. OKUN, A. P. RUDIK, *ZhETF* 47 (1964) 1905 [*JETP* 20 (1965) 1281];
A. D. DOLGOV, V. I. ZAKHAROV, L. B. OKUN, *Yad. Fiz.* 14 (1971) 1044; 1247 [*Sov. J. Nucl. Phys.* 14 (1971) 585, 695];
A. D. DOLGOV, L. B. OKUN, V. I. ZAKHAROV, *Nucl. Phys.* B37 (1972) 493, B41 (1972) 197; *Phys. Lett.* 37B (1972) 298.
A theoretical discussion of weak interactions in colliding $e^+ e^-$ beams in the range of the unitarity limit.
See a similar analysis for νe interactions in:
T. APPELQUIST, J. D. BJORKEN, *Phys. Rev.* D4 (1971) 3726.
The dispersion theory of weak processes above the unitarity limit was initiated in:
I. YA. POMERANCHUK, *Yad. Fiz.* 11 (1970) 852. [*Sov. J. Nucl. Phys.* 11 (1970) 477].
Further progress see in:
A. D. DOLGOV, V. N. GRIBOV, L. B. OKUN, V. I. ZAKHAROV, *Nucl. Phys.* B59 (1973) 611.

28.4.2. Gauge symmetry

Non-abelian gauge symmetry was introduced in:
C. N. YANG, R. L. MILLS, *Phys. Rev.* 96 (1954) 191.
Further important papers:

R. Utiyama, *Phys. Rev.* 101 (1956) 1597;
J. J. Sakurai, *Ann. of Phys.* 11 (1960) 1;
S. L. Glashow, M. Gell-Mann, *Ann. of Phys.* 15 (1961) 437;
Y. Ne'eman, *Nucl. Phys.* 26 (1961) 222;
A. Salam, J. C. Ward, *Nuovo Cim.* 11 (1959) 568;19 (1961) 165;
J. Schwinger, *Phys. Rev.* 125 (1962) 397, 1043; 127 (196) 324;
T. W. B. Kibble, *J. Math. Phys.* 2 (1961) 212.

A modern presentation of gauge field theory, as well as a brief discussion of the history of the problem can be found in the following monographs and review papers:
A. A. Slavnov, L. D. Fadde'ev, *Vvedeniye v kvantovuyu teoriyu kalibrovochnikh polei (Introduction to the quantum theory of gauge fields)* (Nauka, Moscow, 1978) (in Russian);
V. N. Popov, *Kontinual'niye integraly v kvantovoi teorii polya i statisticheskoi fizike (Path integrals in quantum field theory and statistical physics)* (Atomizdat, Moscow, 1976) (in Russian);
S. Coleman, in *Laws of hadronic matter*, Proc. of the 11th Course, Ettore Majorana Intern. School, ed. A. Zichichi (Academic Press, 1975);
J. Bernstein, *Rev. Mod. Phys.* 46 (1974) 7;
E. S. Abers, B. W. Lee, *Phys. Reports* C9 (1973) 1.

The presentation of gauge symmetry in this book is in the spirit of the paper:
A. I. Vainshtein, I. B. Khriplovich, *Yad. Fiz.* 13 (1971) 199 [*Sov. J. Nucl. Phys.* 13 (1971) 111].

28.4.3. Spontaneous breaking of gauge symmetry

Higgs effect:
P. W. Higgs, *Phys. Rev. Lett.* 12 (1964) 132; 13 (1964) 508;
P. W. Higgs, *Phys. Rev.* 145 (1966) 1156;
F. Englert, R. Brout, *Phys. Rev. Lett.* 13 (1964) 321;
G. S. Guralnik, C. R. Hagen, T. W. B. Kibble, *Phys. Rev. Lett.* 13 (1964) 585;
T. W. B. Kibble, *Phys. Rev.* 155 (1967) 1554.

Non-relativistic analogue of the Higgs effect: non-vanishing photon mass in a superconductor:
V. L. Ginzburg, L. D. Landau, *ZhETF* 20 (1950) 1064 [*Collected Papers of L. D. Landau*, p. 546. (Pergamon Press, 1965)].

G. W. 'T HOOFT, *Nucl. Phys.* B79 (1974) 276;
A. M. POLYAKOV, *Pis'ma ZhETF* 20 (1974) 430 [*JETP Lett.* 20 (1974) 194].
The magnetic monopole as a classical solution in a theory with spontaneously broken local SU(2) symmetry.

E. B. BOGOMOLNIY, M. S. MARINOV, *Yad. Fiz.* 23 (1976) 676 [*Sov. J. Nucl. Phys.* 23 (1976) 355].
Calculations of the monopole mass.

28.4.4. Standard model of the electroweak interaction

S. L. GLASHOW, *Nucl. Phys.* 22 (1961) 579.
S. WEINBERG, *Phys. Rev. Lett.* 19 (1967) 1264;
A. SALAM, in *Elementary particle theory*, ed. N. SVARTHOLM (Almquist and Wiksell, Stockholm, 1968) 367.

Other papers which discussed the unification of the weak and electromagnetic interactions:

A. SALAM, J. C. WARD, *Phys. Lett.* 13 (1964) 168; *Nuovo Cim.*, 11 (1959) 568.
J. L. LOPES, *Nucl. Phys.* 8 (1958) 234.
J. SCHWINGER, *Ann. of Phys.* 2 (1957) 407.

The history of the problem:
M. VELTMAN, in Proc. of the 6th Int. Symp. on electron and photon interactions at high energies, Bonn, August 1973, ed. H. ROLLNIK and W. PFEIL (North-Holland, Amsterdam, 1974) p. 429.

G. W.'T HOOFT, *Nucl. Phys.* B33 (1971) 173; B35 (1971) 167.
Proof of renormalizability of the standard model.

J. D. BJORKEN, C. H. LLEWELLYN SMITH, *Phys. Rev.* D7 (1973) 887.
Classification of various generalizations of the standard model.

S. L. GLASHOW, S. WEINBERG, *Phys. Rev.* D15 (1977) 1958.
Introduction of the concept of natural conservation of quantum numbers for the generalized standard model.

A. SIRLIN, *Rev. Mod. Phys.* 50 (1978) 573.
Calculation of corrections of the order of αG to amplitudes of weak decays in the standard model.

28.4.5. Production and decay of the W- and Z-bosons

Early calculations of the production of W-bosons:
L. B. OKUN, *Yad. Fiz.* 3 (1966) 590 [*Sov. J. Nucl. Phys.* 3 (1968) 426];
Y. YAMAGUCHI, *Nuovo Cim.* 43 (1966) 193.
Calculations of the production of W- and Z-bosons in the framework of the parton model:
R. Palmer, E. Paschos, N. Samios, L. L. Wang, *Phys. Rev.* D14 (1976) 118;
R. F. PEIERLS, T. L. TRUEMAN, L. L. WANG, *Phys. Rev.* D16 (1977) 1397;
L. B. OKUN, M. B. VOLOSHIN, *Nucl. Phys.* B120 (1977) 459;
C. QUIGG, *Rev. Mod. Phys.* 49 (1977) 297;
J. D. BJORKEN, in *Proc. of the Int. Symp. on lepton and photon interactions*, Hamburg, 1977.

28.4.6. Future facilities designed to search for the W- and Z-bosons

A. ASTBURY et al., CERN *report*, 1978, CERN/SPSC/78-06, SPSC/P92.
A proposed detector for a $p\bar{p}$ collider with center-of-mass energy of 540 GeV.

Collected papers: The acceleration-storage complex UNK based on the IHEP accelerator, *Proc. of the 21st Session of the IHEP coordination council*, May 1976 (Serpukhov, 1977) (in Russian).

Proc. of the LEP Summer Study, CERN 79-01, vol. 1, 2, February 14, 1979.
A detailed analysis of physical and technical problems connected with the project of the electron-positron colliding beam facility 2×80 GeV.

ISABELLE, A proton-proton colliding beam facility, BNL report 50648 (1977).

DUMAND Project:
V. S. BEREZINSKY, G. T. ZATSEPIN, *Usp. Fiz. Nauk* 122 (1977) 3 [*Uspekhi* 20 (1977) 361].
A review of the DUMAND project.
V. S. BEREZINSKY, A. Z. GAZIZOV, *Pis'ma ZhETF* 25 (1977) 276 [*JETP Lett.* 25 (1977) 254].
Theoretical treatment of the reaction $\bar{\nu}e^- \to W^-$ in the DUMAND facility.

28.4.7. Properties of Higgs bosons

Early phenomenological discussions of H-bosons as physical particles:
E. B. BOGOMOL'NIY, *Yad. Fiz.* 18 (1973) 574; 20 (1974) 984 [*Sov. J. Nucl. Phys.* 18 (1973) 297; 20 (1975) 522];
L. RESNIK, M. K. SUNDARESAN, P. J. S. WATSON, *Phys. Rev.* D8 (1973) 172;
J. ELLIS, M. K. GAILLARD, D. V. NANOPOULOS, *Nucl. Phys.* B106 (1976) 292.

The lower bound on the H-boson mass:
A. D. LINDE, *Pis'ma ZhETF* 23 (1976) 73 [*JETP Lett.* 23 (1976) 64];
A. D. LINDE, *Phys. Lett.* 70B (1977) 306;
S. WEINBERG, *Phys. Rev. Lett.* 36 (1976) 294;
P. H. FRAMPTON, *Phys. Rev. Lett.* 37 (1976) 1378.

The theory of unstable vacuum:
M. B. VOLOSHIN, I. YU. KOBZAREV, L. B. OKUN, *Yad. Fiz.* 20 (1974) 1229 [*Sov. J. Nucl. Phys.* 20 (1975) 644];
P. H. FRAMPTON, *Phys. Rev.* D15 (1977) 2922;
S. COLEMAN, *Phys. Rev.* D15 (1977) 2929; D16 (1977) 1248 (E);
C. CALLAN, S. COLEMAN, *Phys. Rev.* D16 (1977) 1762.

The role of H-bosons at high energies:
E. B. BOGOMOL'NIY, V. I. ZAKHAROV, L. B. OKUN, in: *Elementarniye chastitsi* (Pervaya shkola fiziki ITEP) [*Elementary particles*, The First ITEP Physics School] (Atomizdat, Moscow, 1973) No. 1, p. 49 (in Russian);
D. A. DICUS, V. S. MATHUR, *Phys. Rev.* D7 (1973) 311;
B. W. LEE, C. QUIGG, H. THACKER, *Phys. Rev. Lett.* 38 (1977) 883.

M. VELTMAN, *Acta Phys. Polonica* B8 (1977) 475; *Phys. Lett.*, 70B (1977) 253.
It is shown that the strong interaction in the Higgs sector gives negligibly small corrections (of the order of m_f^2/m_W^2, where m_f is the fermion mass) to the effective four-fermion Lagrangian at low energies.

S. WEINBERG, *Phys. Rev.* D19 (1979) 1277.
The relation $m_W^2 = m_Z^2 \cos^2 \theta_W$ in the case of strong interactions in the Higgs sector.

Interaction of H-bosons with gluons:
F. WILCZEK, *Phys. Rev. Lett.* 39 (1977) 1304.
This paper also discusses the decays $\Upsilon \to H + \gamma$.

M. A. SHIFMAN, A. I. VAINSHTEIN, V. I. ZAKHAROV, *Phys. Lett.* 78B (1978) 443.
This paper also gives a solution of the problem of the Higgs charge of the nucleon.

H. M. GEORGI, S. L. GLASHOW, M. E. MACHACEK, D. V. NANOPOULOS, *Phys. Rev. Lett.* 40 (1978) 692.
Parton calculations of H-boson production by two gluons in pp collisions.

Interaction of H-bosons with Z- and W-bosons:
B. L. IOFFE, V. A. KHOZE, Problems of elementary particle and atomic nucleus physics, *Particles and Nucleus* 9 (1978) 178 (in Russian);
J. ELLIS, M. K. GAILLARD, D. V. NANOPOULOS *Nucl. Phys.* B106 (1976) 292;
S. L. GLASHOW, D. V. NANOPOULOS, A. YILDIZ, *Phys. Rev.* 18 (1978) 1724.
Calculations for the process $e^+ e^- \to Z + H$. The last paper also calculates the cross section of the associated production of $W + H$ and $Z + H$ in pp and $\bar{p}p$ collisions.

Schemes with two or more doublets of H-bosons.
A. I. VAINSHTEIN, M. B. VOLOSHIN, V. I. ZAKHAROV, M. A. SHIFMAN, *Yad. Fiz.* 30 (1979) 1368. [*Sov. J. Nucl. Phys.* 30 (1979) 711]
This paper treats $H \to 2\gamma$ decays.

F. WILCZEK, *Phys. Rev. Lett.* 39 (1977) 1304; E. GOLOWICH, T. C. YANG, *Phys. Lett.* 80B (1979) 245.
A discussion of the properties of light H-bosons, including charged ones.

Reviews on H-bosons:
M. K. GAILLARD, *Comments Nucl. Part. Phys.* 8 (1978) 31;
S. WEINBERG, *Physics Today*, 1977, April, p. 42;
A. I. VAINSHTEIN, V. I. ZAKHAROV, M. A. SHIFMAN, *Usp. Fiz. Nauk* 131 (1980) 537 [*Uspekhi* 23 (1980)].

Composite Higgs bosons and the model of technicolor:
L. SUSSKIND, *reprint* SLAC PUB 2142 (June 1978);
S. DIMOPOULOS, L. SUSSKIND, *preprint* ITP-626 (Jan. 1979); *Nucl. Phys.* B155 (1979) 237.
See also:
S. WEINBERG, *Phys. Rev.* D13 (1976) 247;
A. A. MIGDAL, A. M. POLYAKOV, *ZhETF* 51 (1966) 135 [*JETP* 24 (1967) 91];
J. SCHWINGER, *Phys. Rev.* 125 (1962) 397.

28.5. Models of grand and super unification

28.5.1. Model based on the group $SU(4) \times SU(2) \times SU(2)$

J. C. PATI, A. SALAM, *Phys. Rev.* D8 (1973) 1240.

28.5.2. Model based on the group $SU(5)$

H. GEORGI, S. L. GLASHOW, *Phys. Rev. Lett.* 32 (1974) 438;
H. GEORGI, H. R. QUINN, S. WEINBERG, *Phys. Rev. Lett.* 33 (1974) 451;
A. J. BURAS, J. ELLIS, M. K. GAILLARD, D. V. NANOPOULOS, *Nucl. Phys.* B135 (1978) 66.

D. V. NANOPOULOS, CERN preprint (1978) TH 2534.
A brief review of the SU(5) model.

C. JARLSKOG, F. J. YNDURAIN, *Nucl. Phys.* B149 (1979) 29.
T. J. GOLDMAN, D. A. ROSS, *Phys. Lett.* 84B (1979) 208.
A detailed estimate of the lifetime and decay channels of the proton.

E. GILDENER, *Phys. Rev.* D14 (1976) 1667;
K. T. MAHANTHAPPA, D. G. UNGER, *Phys. Lett.* 78B (1978) 604;
S. WEINBERG, *Phys. Lett.* 82B (1979) 387.
The problem of hierarchy of vacuum-expectation values in grand synthesis models.

28.5.3. Model based on the group $SO(10)$

H. FRITZSCH, P. MINKOWSKI, *Ann. of Phys.* 93 (1975) 193;
M. S. CHANOWITZ, J. ELLIS, M. K. GAILLARD, *Nucl. Phys.* B128 (1977) 506;
H. GEORGI, D. V. NANOPOULOS, *Nucl. Phys.* B155 (1979) 152.

28.5.4. Models based on exceptional groups

F. GÜRSEY, in: *Kyoto Symp. on mathematical problems in theoretical physics*, ed. H. ARAKI (Springer, 1975).
M. GÜNAYDIN, M. GÜRSEY, *Phys. Rev.* D9 (1974) 3387.
On the role of exceptional groups in the physics of elementary particles.

F. GÜRSEY, P. SIKIVIE, *Phys. Rev. Lett.* 36 (1976) 775.
P. RAMOND, *Nucl. Phys.* B126 (1977) 509.
An analysis of representations of the group E_7.

Y. ACHIMAN, B. STECH, *Phys. Lett.* 77B (1978) 389.
A model based on the group E_6.

V. I. OGIEVETSKY, V. YU. TSEITLIN, *Yad. Fiz.* 28 (1978) 1616 [*Sov. J. Nucl. Phys.* 28 (1978) 832].
Models based on exceptional groups, with the stable proton.

D. I. DYAKONOV, *Yad. Fiz.* 26 (1977) 845 [*Sov. J. Nucl. Phys.* 26 (1977) 443].
A formula for the Weinberg angle in a unified theory with an arbitrary symmetry group, and with the group E_7 in particular.

28.5.5. Models of simple unification

S. WEINBERG, *Phys. Rev.* D5 (1972) 1962;
J. D. BJORKEN, K. LANE, *Proc. Int. Conf. Neutrino '77*, Moscow (Nauka, Moscow, 1978) vol. 2, p. 412;
H. FRITZSCH, preprint CERN (1977) TH 2309, Invited talk at the XII Recontre de Moriond.

28.5.6. Reviews of grand unification models

M. GELL-MANN, P. RAMOND, R. SLANSKY, *Rev. Mod. Phys.* 50 (1978) 721.
A review with a detailed analysis of representations of exceptional groups and of a number of versions of unified gauge theories.

H. HARARI, *Phys. Reports* 42 (1978) 235.
A review of various schemes of the grand synthesis.

28.5.7. Superunification

YU. A. GOL'FAND, E. P. LIKHTMAN, *Pis'ma ZhETF* 13 (1971) 452 [*JETP Lett.* 13 (1971) 323];
J. GERVAIS, B. SAKITA, *Nucl. Phys.* B34 (1971) 632;
D. V. VOLKOV, V. P. AKULOV, *Pis'ma ZhETF* 16 (1972) 621 [*JETP Lett.* 16 (1972) 438];

J. Wess, B. Zumino, *Nucl. Phys.* B70 (1974) 39. First papers on supersymmetry.

D. Z. Freeman, P. van Nieuwenhuizen, S. Ferrara, *Phys. Rev.* D13 (1976) 3214;
S. Deser, B. Zumino, *Phys. Lett.* 62B (1976) 335.
First papers on supergravity.

S. Ferrara, P. van Nieuwenhuizen, *Phys. Rev. Lett.* 37 (1976) 1669.
First paper on extended supergravity.

A. Salam, J. Strathdee, *Fortschr. Phys.* 26 (1978) 57;
P. Fayet, S. Ferrara, *Phys. Reports* C32 (1977) 249;
A. A. Slavnov, *Usp. Fiz. Nauk* 124 (1978) 487 [*Uspekhi* 21 (1978) 240];
V. I. Ogiyevetsky, L. Mezinchesku, *Usp. Fiz. Nauk* 117 (1975) 637 [*Uspekhi* 18 (1975) 960];
D. Z. Freedman, P. van Nieuwenhuizen, *Scientific American* 238 (1978) no. 2, 126. Review papers.

28.6. Particles and the universe

28.6.1. The general theory of relativity and the theory of the hot universe

A. Einstein, *Ann. der Phys.* 49 (1916) 769.
The general theory of relativity.

A. A. Friedmann, *Z. Phys.* 11 (1922) 377.
A. A. Friedmann, *ZhRFKhO* 56 (1924) 59 (Reprinted in *Usp. Fiz. Nauk* 80 (1963) 439 [*Uspekhi* 6 (1964) 475].
The theory of the non-stationary universe.

E. P. Hubble, *Proc. Nat. Acad. Sci. USA* 15 (1929) 168. Linear relation between radial velocities of galaxies and distances to these galaxies. According to the initial Hubble estimate, the proportionality coefficient $H = 513 \pm 60$ km/s $\cdot 10^6$ parsec.

G. Gamov, *Phys. Rev.* 70 (1946) 572.
The hypothesis of the hot universe and an estimate of the expected temperature of the cosmic microwave radiation background.

A. A. Penzias, R. W. Wilson, *Astrophys. J.* 142 (1965) 419.
The discovery of the cosmic microwave radiation background.

Monographs and reviews on cosmology:

S. WEINBERG, *The first three minutes*, A modern view of the origin of the universe (Basic Books, Inc, New York, 1977);
YA. B. ZELDOVICH, *Usp. Fiz. Nauk*, 89 (1966) 647 [*Uspekhi* 9 (1967) 602].
Popular presentation of the theory of the hot universe.

YA. B. ZELDOVICH, D. I. NOVIKOV, *Relativistic astrophysics* VII, Structure and evolution of the universe (Chicago Univ. Press, 1980) translated from the Russian.

S. WEINBERG, *Gravitation and cosmology* (Wiley, 1972).

Confrontation of cosmological theories with observational data, Symposium No. 63 IAU (Copernicus Symposium II), held in Cracow, Poland, 10–12 September 1973, ed. by M. Longair (Reidel, Dordrecht, Holland, 1974).

28.6.2. The hot universe and leptons

S. S. GERSHTEIN, YA. B. ZELDOVICH, *Pis'ma ZhETF* 4 (1966) 174 [*JETP Lett.* 4 (1966) 120];
R. COWSIK, J. MCCLELLAND, *Phys. Rev. Lett.* 29 (1972) 669;
A. S. SZALAY, G. MARX, *Astronomy and Astrophysics* 49 (1976) 437.
Cosmological bounds on the mass of the neutrino.

M. I. VYSOTSKY, A. D. DOLGOV, YA. B. ZELDOVICH, *Pis'ma ZhETF* 26 (1977) 200 [*JETP Lett.* 26 (1977) 188];
A. D. DOLGOV, M. I. VYSOTSKY, YA. B. ZELDOVICH, in: *Neutrino '77. Proc. of the Int. Conf. on neutrino physics and neutrino astrophysics* (Nauka, Moscow) vol. 1, p. 42.
B. W. LEE, S. WEINBERG, *Phys. Rev. Lett.* 39 (1977) 165;
P. HUT, *Phys. Lett.* 69B (1977) 85;
K. SATO, M. KOBAYASHI, *Progr. Theor. Phys.* 58 (1977) 1775;
J. E. GUNN, B. W. LEE, I. LERCHE, D. N. SCHRAMM, G. STEIGMAN, *Astrophys. J.* 223 (1977) 1015.
Cosmological bounds on masses of heavy neutral leptons (both stable and unstable). See also:
D. A. DICUS, E. N. KOLB, V. L. TEPLITZ, *Phys. Rev. Lett.* 39 (1977) 168; *Astrophys. J.*, 221 (1978) 327;
D. A. DICUS, E. N. KOLB, V. L. TEPLITZ, R. V. WAGONER, *Phys. Rev.* D17 (1978) 1529.

V. F. SHVARTSMAN, *Pis'ma ZhETF*, 9 (1969) 315 [*JETP Lett.* 9 (1969) 184].
G. STEIGMAN, D. N. SCHRAMM, J. E. GUNN, *Phys. Lett.* 66B (1977) 202.
Cosmological restriction on the number of different types of neutrinos derived from the data on the primordial helium abundance, on the basis of the theory of the primordial nucleosynthesis.

Leptons and the hot universe (review paper):
D. N. SCHRAMM, preprint Fermi Institute, 78-25 (1978).

Gravitational interaction of right-handed neutrinos:
M. GELL-MANN, in: *La théorie quantique de champs*, Deuxième Conseil de Physique, ed. R. STOOPS (Bruxelles, 1962) p. 135;
I. YU. KOBZAREV, L. B. OKUN, *ZhETF* 43 (1962) 1904 [*JETP* 16 (1963) 1343]. See also collected essays dedicated to I. E. Tamm: *Problemi teoreticheskoi fiziki* (*Problems of theoretical physics*) (Nauka, Moscow, 1972) p. 219 (in Russian).

Neutrinos and astrophysics:
M. A. RUDERMAN, *Reports on Progress in Physics*, 1965, p. 411.

28.6.3. The hot universe and quarks

YA. B. ZELDOVICH, L. B. OKUN, S. B. PIKELNER, *Usp. Fiz. Nauk* 87 (1965) 115 [*Uspekhi* 8 (1966) 702].
Estimates of the concentration of relic quarks.

A. DE RUJULA, R. C. GILES, R. L. JAFFE, *Phys. Rev.* D17 (1978) 285.
A hypothesis of the existence of free fractionally charged quarks with large mass and large radius.

A review of the search for free fractionally charged quarks:
L. G. LANDSBERG, *Usp. Fiz. Nauk* 109 (1973) 695 [*Uspekhi* 16 (1973) 251].

D. D. OGORODNIKOV, I. M. SAMOILOV, A. M. SOLNTSEV, *ZhETF* 72 (1977) 1633 [*JETP* 45 (1977) 857].
Mass-spectroscopic search: less than one quark per 10^{27} nucleons.

G. S. LA RUE, W. M. FAIRBANK, A. F. HEBARD, *Phys. Rev. Lett.* 38 (1977) 1011 (Erratum: p. 1019);
G. S. LA RUE, W. M. FAIRBANK, J. D. PHILLIPS, *Phys. Rev. Lett.* 42 (1979) 142.
Reports of the detection of fractionally charged particles in small niobium spheres.

G. M. KUKAVADZE, L. YA. MEMELOVA, L. YA. SUVOROV, ZhETF 49 (1965) 389 [JETP 22 (1966) 272];
T. ALVÄGER, R. A. NAUMAN, Phys. Lett. 24B (1967) 647;
R. A. MILLER, L. W. ALWAREZ, W. R. HOLLEY, E. J. STEPHENSON, Science 196 (1977) 521;
R. F. SMITH, J. R. J. BENNET, Nucl. Phys. B149 (1979) 525.
Mass-spectrometric search for anomalously heavy hydrogen (integer-charge quarks).

28.6.4. Cosmology and spontaneous symmetry breaking

T. D. LEE, Phys. Rev. D8 (1973) 1226.
T. D. LEE, G. C. WICK, Phys. Rev. D9 (1974) 2291.
A model of spontaneous breaking of CP invariance.

YA. B. ZELDOVICH, I. YU. KOBZAREV, L. B. OKUN, ZhETF 67 (1974) 3 [JETP 40 (1975) 1]; Phys. Lett. 50B (1974) 340.
Spontaneous CP breaking and vacuum domains.

M. B. VOLOSHIN, I. YU. KOBZAREV, L. B. OKUN, Yad. Fiz. 20 (1974) 1229 [Sov. J. Nucl. Phys. 20 (1975) 644].
The theory of the unstable vacuum (further references in section 28.4.7 "Properties of the Higgs bosons").

D. A. KIRZHNITS, Pis'ma ZhETF 15 (1972) 745 [JETP Lett. 15 (1972)];
D. A. KIRZHNITS, A. D. LINDE, Phys. Lett. 42B (1972) 471; A. D. LINDE, Reports on Progress in Physics 42 (1979) 389.
Restoration of spontaneously broken symmetry in the hot universe.

I. YU. KOBZAREV, in: Elementarniye chastitsi, I. Shkola ITEF (Elementary particles, The First ITEP Physics School) (Atomizdat, Moscow, 1975) no. 3, p. 44.
Review lecture.

28.6.5. Astrophysics and photons

I. YU. KOBZAREV, L. B. OKUN, Usp. Fiz. Nauk, 95 (1968) 131 [Uspekhi 11 (1968) 338];
A. S. GOLDHABER, M. M. NIETO, Rev. Mod. Phys. 43 (1971) 277.
Reviews discussing the photon mass.

L. Davies, Jr., A. S. Goldhaber, M. M. Nieto, *Phys. Rev. Lett.* 35 (1975) 1402.
Restriction on the photon mass on the basis of the data on Jupiter's magnetic field: $1/m_\gamma \gtrsim 5 \cdot 10^{10}$ cm.

G. V. Chibisov, *Usp. Fiz. Nauk* 119 (1976) 551 [*Uspekhi* 119 (1976) 624].
Restriction on the photon mass on the basis of the data on intergalactic magnetic fields: $1/m_\gamma \gtrsim 10^{22}$ cm.

T. D. Lee, C. N. Yang, *Phys. Rev.* 98 (1955) 1501;
L. B. Okun, *Yad. Fiz.* 10 (1969) 358 [*Sov. J. Nucl. Phys.* 10 (1970) 206].
Geophysical and astrophysical restrictions on the possible properties of hypothetical baryonic, leptonic, and muonic photons, obtained, in particular, from the experiments:

R. V. Eötvös, D. Pekar, E. Fekete, *Ann. der Phys.* 68 (1922) 11;
R. H. Dicke, P. G. Roll, G. Krotkov, *Ann. of Phys.* 26 (1964) 442;
V. B. Braginsky, V. I. Panov, *ZhETF* 61 (1971) 873 [*JETP* 34 (1972) 463].

28.6.6. Possible non-conservation of the baryonic charge, and baryonic asymmetry of the universe

A. D. Sakharov, *Pis'ma ZhETF* 5 (1967) 32 [*JETP Lett.* 5 (1967) 24];
V. A. Kuzmin, *Pis'ma ZhETF* 12 (1970) 335 [*JETP Lett.* 12 (1970) 228].
The hypothesis ascribing the observed baryonic asymmetry of the universe to *CP* non-invariant processes which violate conservation of the baryonic charge at the early stages of the expanding universe.

J. C. Pati, A. Salam, *Phys. Rev. Lett.* 31 (1973) 661.
Non-conservation of the baryonic charge in the model with integer-charge quarks.

F. Reines, M. F. Crouch, *Phys. Rev. Lett.* 32 (1974) 493.
J. Learned, F. Reines, A. Soni, *Phys. Rev. Lett.* 43 (1979) 907.
The experimental lower bound for the proton lifetime.

L. B. Okun, Ya. B. Zeldovich, *Comments Nucl. Part. Phys.* 6 (1976) 69.
This paper points out that non-conservation of the baryonic charge and *CP* non-invariance do not result in an excess of baryons over antibaryons, if statistical equilibrium takes place.

A. Yu. Ignatiev, N. V. Krasnikov, V. A. Kuzmin, A. N. Tavkhelidze, *Phys. Lett.* 76B (1978) 436; see also *Neutrino '77, Proc. of the Int. Conf.*

on neutrino physics and astrophysics (Nauka, Moscow, 1977) vol. 2, p. 293;

H. YOSHIMURA, *Phys. Rev. Lett.* 41 (1978) 281 (Errata: 42 (1979) 746);
S. DIMOPOULOS, L. SUSSKIND, *Phys. Lett.* 81B (1979) 416;
D. TOUSSAINT, S. B. TREIMAN, F. WILCZEK, A. ZEE, *Phys. Rev.* D19 (1979) 1036;
D. TOUSSAINT, F. WILCZEK, *Phys. Lett.* 81B (1979) 294;
S. WEINBERG, *Phys. Rev. Lett.* 42 (1979) 850;
A. D. SAKHAROV, *ZhETF* 76 (1979) 1172 [*JETP* 52 (1980) 349];
A. D. DOLGOV, *Pis'ma ZhETF* 29 (1979) 254 [*JETP Lett.* 29 (1979) 288];
J. ELLIS, M. K. GAILLARD, D. V. NANOPOULOS, *Phys. Rev. Lett.* 80B (1979) 360 (Erratum: 82B (1979) 464).
A discussion of various mechanisms which enhance or suppress baryonic asymmetry in the hot universe. An attempt to obtain the observed ratio $n_p/n_\gamma \simeq 10^{-9}$ on the basis of a number of model lagrangians, including models of grand unification.

D. CLINE, C. RUBBIA, preprint (1978) Harvard University and University of Wisconsin.
A proposal of an experimental facility which would be capable of recording the decay $p \to \pi^0 e^+$ if its probability were greater than 10^{-34} year^{-1}.

28.6.7. *Possible non-conservation of the leptonic charge*

D. BRYMAN, C. PICCIOTTO, *Rev. Mod. Phys.* 50 (1978) 11.
A review of the experimental data and theoretical results on the double β-decay.

B. M. PONTECORVO, *ZhETF* 33 (1958) 549 [*JETP* 6 (1958) 429];
V. GRIBOV, B. PONTECORVO, *Phys. Lett.* 28B (1969) 493;
S. M. BILEN'KY, B. M. PONTECORVO, *Usp. Fiz. Nauk* 23 (1977) 181 [*Uspekhi* 20 (1977) 776].
The theory of neutrino oscillations.

28.6.8. *Concentration of relic monopoles*

T. W. B. KIBBLE, *J. Phys.* A9 (1976) 1387.
YA. B. ZELDOVICH, M. YU. KHLOPOV, *Phys. Lett.* 79B (1978) 237.

28.6.9. On the conservation of the electric charge

A. A. POMANSKY, in: *Proc. of the Int. neutrino conference*, Aachen, 1976, ed. H. FAISSNER, H. REITHLER, P. ZERWAS (Vieweg, Braunschweig, 1977) p. 671;
L. B. OKUN, YA. B. ZELDOVICH, *Phys. Lett.* 78B (1978) 597;
M. B. VOLOSHIN, L. B. OKUN, *Pis'ma ZhETF* 28 (1978) 156 [*JETP Lett.* 28 (1978) 145].

28.6.10. Time independence of the fundamental constants

A. I. SHLYAKHTER, *Nature* 264 (1976) 340.
An analysis of the resonance neutron absorption in the natural nuclear pile at Oklo.
The Oklo pile is reviewed in:
YU. V. PETROV, *Usp. Fiz. Nauk* 123 (1977) 473 [*Uspekhi* 20 (1977) 937].

28.7. Supplementary bibliography (autumn 1980)

The list that follows was added in proof. It mostly comprises journal publications of 1980. Unfortunately, it does not include the rapporteur and invited papers presented at the Neutrino-80 Conference (Erice) and the XX International Conference on High Energy Physics (Madison) since the Proceedings of these conferences were not yet available by the time this bibliography was completed.

28.7.1. Electroweak processes. Review papers

S. WEINBERG, *Rev. Mod. Phys.* 52 (1980) 515;
A. SALAM, *Rev. Mod. Phys.* 52 (1980) 525;
S. L. GLASHOW, *Rev. Mod. Phys.* 52 (1980) 539;
Nobel Lectures on physics, 1979.
B. DESPLANQUES, *Nucl. Phys.* A335 (1980) 147. Parity-violating nuclear forces.

P. Depommier, *Nucl. Phys.* A335 (1980) 97. Rear decays of the pion and muon.

J. Egger, *Nucl. Phys.* A335 (1980) 87. Verification of conservation of leptonic numbers.

Yu. G. Zdesenko, *Particles and Nucleus* 11 (1980) Pt. 6, 1369. Double β-decay and conservation of leptonic charge.

M. L. Perl, *Annual Rev. Nucl. Particle Sci.* 30 (1980), *preprint* SLAC-PUB-2446. December 1979.
τ-lepton.

28.7.2. Decays of strange particles

J. Finjord, M. K. Gaillard, *Phys. Rev.* D22 (1980) 778. Calculations of non-leptonic and leptonic decays of hyperons.

F. J. Gilman, M. B. Wise, *Phys. Rev.* D21 (1980) 3150;
M. I. Vysotsky, *Lett. Nuovo Cim.* 26 (1979) 1979. Analysis of the $K \to \pi e^+ e^-$ decay.

R. E. Shrock, M. B. Voloshin, *Phys. Lett.* 87B (1979) 375. Analysis of the $K_L \to \mu\bar{\mu}$ decay in the six-quark model.

J. Donoghue, E. Golowich, W. Ponce, B. Holstein, *Phys. Rev.* D21 (1980) 186. Analysis of non-leptonic decays and the $\Delta T = 1/2$ rule.

28.7.3. Violation of CP invariance

M. I. Vysotsky, *Yad. Fiz.* 31 (1980) 1535 [*Sov. J. Nucl. Phys.* 31 (1980) 797];

B. Guberina, R. Peccei, *Nucl. Phys.* B163 (1980) 289.

F. G. Gilman, M. B. Wise, *Phys. Lett.* 93B (1980) 129. The $K \to \tilde{K}^0$ transition in the six-quark model with gluonic corrections taken into account.

E. Ma, W. A. Simmons, S. F. Taun, *Phys. Rev.* D20 (1979) 2888. Breaking of CP in systems $D^0 - \tilde{D}^0$ and $B^0 - \tilde{B}^0$ in the six-quark model.

E. P. Shabalin, *Yad. Fiz.* 32 (1980) 443 [*Sov. J. Nucl. Phys.* 32 (1980) 228] Estimates of the neutron dipole moment in the six-quark model, with Higgs bosons and strong interaction taken into account.

V. Baluni, *Phys. Rev.* D19 (1979) 2227;
R. Crewter, P. Di Vecchia, G. Veneziano, E. Witten, *Phys. Lett.* 88B (1979) 123;
M. A. Shifman, A. I. Vainshtein, V. I. Zakharov, *Nucl. Phys.* B166 (1980) 493. P- and T-non-invariant effects due to so-called θ-term in the quantum chromodynamics lagrangian.

B. V. Martemyanov, *Yad. Fiz.* 30 (1979) 1364 [*Sov. J. Nucl. Phys.* 30 (1979) 708] Parametrization of CP violation for n quark doublets.

E. Eichten', K. Lane, J. Preskill, *Phys. Rev. Lett.* 45 (1980) 225; CP violation in models without elementary scalar fields.

M. Bander, D. Silverman, A. Soni, *Phys. Rev. Lett.* 43 (1979) 242;
A. R. Zhitnitsky, *Yad. Fiz.* 32 (1980) 434. Analysis of possible CP-odd effects in decays of heavy particles.

J. D. Christenson et al., *Phys. Rev. Lett.* 43 (1979) 1209; 1212. Measurements of η_{00} and η_{+-}.

28.7.4. Decays of c- and b-hadrons

W. Bacino et al., *Phys. Rev. Lett.* 45 (1980) 329. Experimental data on the difference in relative widths of semi-leptonic decays of D^+ and D^0 mesons.

N. Ushida et al., *Phys. Rev. Lett.* 45 (1980) 1049. Measurement of the D^0 lifetime.

N. Ushida et al., *Phys. Rev. Lett.* 45 (1980) 1053. Measurement of the D^+, F^+ and Λ_c^+ charmed particle lifetimes.

M. Bander, D. Silverman, A. Soni, *Phys. Rev. Lett.* 44 (1980) 7;
H. Fritzsch, P. Minkowski, *Phys. Lett.* 90B (1980) 455;
V. Barger, J. P. Leveille, P. M. Stevenson, *Phys. Rev.* D22 (1980) 693. Gluonic enhancement of non-leptonic decays of D^0 mesons.

L. Dulyan, A. Khodzamirian, *preprint* of the Yerevan Physics Inst. 410 (17)–80. The difference in the lifetimes of D^0 and D^+ mesons is interpreted in the model of quark graphs corresponding to the c-quark decay.

S. P. Rosen, *Phys. Rev.* D22 (1980) 776. Corollaries of the $\Delta T = 1$ rule for the decays of charmed mesons.

Y. Tosa, S. Okubo, *Phys. Rev.* D22 (1980) 168. Analysis of inclusive decays of charmed mesons.

S. P. Rosen, *Phys. Rev. Lett.* 44 (1980) 1. Sextet enhancement and differences in lifetimes of charmed mesons.

H. Lipkin, *Phys. Rev. Lett.* 44 (1980) 710. Interaction in final states in decays of charmed mesons.

G. S. Abrams et al., *Phys. Rev. Lett.* 43 (1979) 481. Observation of the $D^0 \to \pi^+ \pi^-$ and $D^0 \to K^+ K^-$ decays. For a discussion of these decays see:

V. Barger, S. Pacvasa, *Phys. Rev. Lett.* 43 (1979) 812;

L. L. Wang, F. Wilczek, *Phys. Rev. Lett.* 43 (1979) 816.

H. Georgi, S. Glashow, *Nucl. Phys.* B167 (1980) 17. An analysis of decays of b-hadrons in models not including the t-quark.

S. K. Bose, E. Paschos, *Nucl. Phys.* B169 (1980) 384. The $K^0 \to \tilde{K}^0$, $D^0 \to \tilde{D}^0$ and $B^0 \to \tilde{B}^0$ transitions in the eight-quark model.

F. Bucella, L. Oliver, *Nucl. Phys.* B162 (1980) 237. Phenomenological restrictions on charm-changing neutral currents.

28.7.5. Neutral currents

A. A. Varfolomeyev, *Yad. Fiz.* 31 (1980) 1268 [*Sov. J. Nucl. Phys.* 31 (1980) 655] Coherent interaction of the neutrino with a dense medium.

Riazzudin, *Phys. Rev. Lett.* 45 (1980) 976. Weak neutral currents without electroweak unification.

E. Pasierb, A. S. Gurr, J. Lathrop, F. Reines, H. W. Sobel, *Phys. Rev. Lett.* 43 (1979) 96. Discovery of neutral currents in the interaction of reactor neutrinos $\bar{\nu}_e$ with deuterons.

L. M. Barkov, M. S. Zolotorev, *Phys. Lett.* 85B (1979) 308. New measurements of rotation of the light polarization plane in atomic bismuth vapor are in agreement with earlier results of the same authors and with the standard electroweak theory.

Yu. V. Bogdanov, I. I. Sobelman, I. I. Struk, *Pis'ma ZhETF* 31 (1980) 234; 556 [*JETP Lett.* 31 (1980) 214; 522] The expected rotation of the light polarization plane is not observed.

28.7.6. Electroweak radiative corrections

A. SIRLIN, *Phys. Rev.* D22 (1980) 971. Radiative corrections in the SU(2) × U(1) theory.

M. GREEN, M. VELTMAN, *Nucl. Phys.* B169 (1980) 137. Radiative corrections to μ-decay and νe-scattering.

M. GRECO, G. PANCHERI-SRIVASTAVA, Y. SRIVASTAVA, *Nucl. Phys.* B171 (1980) 118. Radiative corrections to the $e^+ e^- \to Z^0 \to \mu^+ \mu^-$ process.

M. VELTMAN, *Phys. Lett.* 91B (1980) 95;
F. ANTONELLI, M. CONSOLI, G. CORBO, *Phys. Lett.* 91B (1980) 90. Radiative corrections to $m_{W,Z}(\Delta m \simeq +3 \text{ GeV})$.

28.7.7. Virtual Higgs bosons

T. APPLEQUIST, C. BERNARD, *Phys. Rev.* D22 (1980) 200;
A. C. LONGHITANO, *Phys. Rev.* D22 (1980) 1166. Detailed analysis of the effect of virtual strongly interacting heavy Higgs bosons on weak interactions at low energies.

28.7.8. Dynamical symmetry breaking

S. RABY, S. DIMOPOULOS, L. SUSSKIND, *Nucl. Phys.* B169 (1980) 373. A scheme of tumbling symmetries.

G.'T HOOFT, *Lectures* given at the Cargèse Summer Institute, 1979. The equality of anomalies for compound massless fermions and subquarks is formulated as a necessary condition for a self-consistent theory of metacolor.

E. EICHTEN, K. LANE, *Phys. Lett.* 90B (1980) 125.
P. DI VECCHIA, G. VENEZIANO, *preprint* TH 2868-CERN, May 1980.
S. DIMOPOULOS, S. RABY, G. KANE, *Nucl. Phys.* B182 (1981) 77.
S. DIMOPOULOS, *Nucl. Phys.* B168 (1980) 69. Discussion of the properties of technihadrons.

28.7.9. Properties of free neutrinos

J. N. BAHCALL, *Rev. Mod. Phys.* 50 (1978) 881. Review: solar neutrinos.

J. N. BAHCALL et al., *Phys. Rev. Lett.* 45 (1980) 945. The number (2.2 ± 0.4)SNU experimentally observed by Davis is compared with the theoretical result: (7.5 ± 0.5)SNU (1 SNU = 10^{-36} ν-captures in ^{37}Cl per one target atom per second).

F. REINES, H. SOBEL, E. PASIERB, *Phys. Rev. Lett* 45 (1980) 1307. Data on neutral and charged currents in the $\tilde{\nu}_e d$ interaction are interpreted as an indication of ν-oscillations.

F. BOHEM et al., *Phys. Lett.* 97B (1980) 310. Experimental search for oscillations of reactor neutrinos. No oscillations were recorded.

V. S. KOZIK, V. A. LYUBIMOV, E. G. NOVIKOV, V. Z. NOZIK, E. F. TRETYAKOV, *Yad. Fiz.* 32 (1980) 301 [*Sov. J. Nucl. Phys.* 32 (1980) 154];

V. A. LYUBIMOV, E. G. NOVIKOV, V. Z. NOZIK, E. F. TRETYAKOV, V. S. KOZIK, *Phys. Lett.* 94B (1980) 266. An estimate of the $\tilde{\nu}_e$ mass from the tritium decay spectrum.

A. DE RUJULA, M. LUSIGNOLI, L. MAIANI, S. PETCOV, R. PETRONZIO, *Nucl. Phys.* B168 (1980) 54;

V. BARGER, K. WHISNANT, D. CLINE, R. PHILLIPS, *Phys. Lett.* 93B (1980) 194.
Phenomenological analysis of data on possible neutrino oscillations.

S. M. BILENKY, B. PONTECORVO, *Phys. Lett.* 95B (1980) 233;

S. M. BILENKY, J. HOŠEK, S. T. PETCOV, *Phys. Lett.* 94B (1980) 495;

I. YU. KOBZAREV, B. V. MARTEM'YANOV, L. B. OKUN, M. G. SHCHEPKIN, *Yad. Fiz.* 32 (1980) 1286 [*Sov. J. Nucl. Phys.* 32 (1980) 828];

V. BARGER, P. LANGACKER, J. LEVEILLE, S. PAKVASA, *Phys. Rev. Lett.* 45 (1980) 692. Neutrinos with mixed Dirac and Majorana masses.

B. V. MARTEM'YANOV, M. YU. KHLOPOV, M. G. SHCHEPKIN, *Pis'ma ZhETF* 32 (1980) 484 [*JETP Lett.* 32 (1980) 464]. Effect of ν-oscillations on the $\nu_\mu e$ and $\nu_e e$ scattering.

G. T. ZATSEPIN, A. YU. SMIRNOV, *Yad. Fiz.* 28 (1978) 1569 [*Sov. J. Nucl. Phys.* 28 (1978) 807];

A. DE RUJULA, S. L. GLASHOW, *Phys. Rev. Lett.* 45 (1980) 942. Discussion of the photon decays of the neutrino.

28.7.10. Neutrino masses in grand unification models

R. Barbieri, J. Ellis, M. K. Gaillard, *Phys. Lett.* 90B (1980) 249. Neutrino masses and oscillations is SU(5).

M. Gell-Mann, P. Ramond, R. Slansky, in: *Supergravity*, ed. P. van Nieuwenhuizen, D. Z. Freeman (North-Holland, Amsterdam, 1979) p. 315; *preprint* TH 2855-CERN, January 1980;
R. Barbieri, D. V. Nanopoulos, G. Morchio, F. Strocchi, *Phys. Lett.* 90B (1980) 91.

E. Witten, *Phys. Lett.* 91B (1980) 81. Neutrino masses in SO(10)

R. N. Mohapatra, G. Senjanović, *Phys. Rev. Lett.* 44 (1980) 912. Neutrino mass in the SU(2) × SU(2) × U(1) model.

28.7.11. Grand unification models and proton decay

T. Goldman, D. Ross, *Nucl. Phys.* B171 (1980) 273;
J. Ellis, M. K. Gaillard, D. V. Nanopoulos, S. Rudaz, *Nucl. Phys.* B176 (1980) 61. Uncertainties in the theoretical precitions of the proton lifetime.

V. S. Berezinsky, A. Yu. Smirnov, *Phys. Lett.* 97B (1980) 371. A possibility of a practically stable proton in the SU(5) model.

A. Yu. Smirnov, *Pis'ma ZhETF* 31 (1980) 781 [*JETP Lett.* 31 (180) 737] Superheavy fermions and proton lifetime.

M. Mahacek, *Nucl. Phys.* B159 (1979) 37. Channels of proton decay.

G. Segrè, H. A. Weldon, *Phys. Rev. Lett.* 44 (1980) 1737. U(5) model with stable proton.

G. M. Asatryan, S. G. Matinyan, *Yad. Fiz.* 31 (1980) 1381 [*Sov. J. Nucl. Phys.* 31 (1980) 711]. Flavor mixing and proton decay.

R. E. Marshak, R. N. Mohapatra, *Phys. Rev. Lett.* 44 (1980) 1316.
M. V. Kazarnovsky, V. A. Kuz'min, K. G. Chetyrkin, M. E. Shaposhnikov, *Pis'ma ZhETF* 32 (1980) 88 [*JETP Lett.* 32 (1980) 82]. Neutron-antineutron oscillations. See also:
V. A. Kuzmin, *Pis'ma ZhETF* 12 (1970) 335 [*JETP Lett.* 12 (1970) 228].

F. Wilczeck, A. Zee, *Phys. Rev. Lett.* 43 (1979) 1571;

S. WEINBERG, *Phys. Rev. Lett.* 43 (1979) 1566. Phenomenological operator analysis of proton decay.

For discussions of various problems associated with the search for proton decay see the quarterly:
Progress in Cosmic Ray Deep Mine Experiments and Baryon Decay Detection experiments. Quarterly News Letter. Wisconsin University, No. 1, February 1980.

28.7.12. Various aspects of grand unification

S. G. MATINYAN, *Usp. Fiz. Nauk* 130 (1980) 3 [*Uspekhi* 23 (1980) 1] Review of the SU(5) model.

D. V. NANOPOULOS, *preprint* TH 2896-CERN, 1980. Review of the SU(5), SO(10) and E_6 models.

YU. F. PIROGOV, *Yad. Fiz.* 31 (1980) 547 [*Sov. J. Nucl. Phys.* 31 (1980) 283] $SU(8)_L \times SU(8)_R$ model.

P. FRAMPTON, S. NANDI, *Phys. Rev. Lett.* 43 (1979) 1460. SU(9) model.

R. BARBIERI, D. V. NANOPOULOS, *Phys. Lett.* 91B (1980) 369. E_6 model.

R. LEDNICKÝ, V. TSEITLIN, *Yad. Fiz.* 31 (1980) 1036 [*Sov. J. Nucl. Phys.* 31 (1980) 534] Neutral currents in the E_7 model.

I. BARS, M. GÜNAYDIN, *Phys. Rev. Lett.* 45 (1980) 859. E_8 model.

J. ELLIS, M. K. GAILLARD, A. PETERMAN, C. SACHRAIDA, *Nucl. Phys.* B164 (1980) 253. Hierarchy of gauge hierarchies.

E. GILDENER, *Phys. Lett.* 92B (1980) 111. The problem of hierarchies.

P. H. FRAMPTON, *Phys. Rev. Lett.* 43(1979) 1912. On the impossibility of including technicolor into grand unification.

A. DAVIDSON, K. WALI, P. MANHEIM, *Phys. Rev. Lett.* 45 (1980) 1135. $SO(10) \times SO(10)$ model including four fermion generations and technicolor.

I. LIEDE, M. ROSS, *Nucl. Phys.* B167 (1980) 397. Analysis of experimental data on neutral currents points to a contradiction with the SU(5) model.

S. WEINBERG, *Phys. Lett.* 91B (1980) 51. Threshold effects in the dependence of gauge charges on momentum.

N. G. DESHPANDE, D. ISKANDAR, *Phys. Lett.* 87B (1979) 383; *Nucl. Phys.* B167 (1980) 223. Electroweak models with arbitrary violation of parity in atoms.

28.7.13. Superunification

O. V. TARASOV, A. A. VLADIMIROV, *preprint* E2-80-483, Dubna;
L. V. AVDEEV, O. V. TARASOV, A. A. VLADIMIROV, *Phys. Rev. Lett.* 96B (1980) 94;
M. GRISARU, M. ROČEK, W. SIEGEL, *Phys. Rev. Lett.* 45 (1980) 1063. Vanishing of three loop charge renormalization function in a supersymmetric gauge theory.

R. KALLOSH, *Nucl. Phys.* B165 (1980) 119. Elimination of divergences for the $N = 1$ supergravity in self-dual external fields.

E. S. FRADKIN, M. A. VASILIEV, *Phys. Lett.* 85B (1979) 47;
P. BREITENLOHNER, M. SOHNIUS, *Nucl. Phys.* B165 (1980) 483. Formulation of the $N = 2$ supergravity off the mass shell.

G. BARBIELLINI et al., *preprint* DESY 79/67, October 1979. A discussion of the possibilities of experimental search for super-symmetric particles at LEP.

P. FAYET, *preprint* TH 2864-CERN, May 1980. Possibilities of search for goldstino, photino and gluino with accelerators are discussed.

J. SCHERK, *Phys. Lett.* 88B (1979) 265. Supergravity-based antigravity.

V. I. OGIYEVETSKY, E. S. SOKACHEV, *Yad. Fiz.* 32 (1980) 862; 870. [*Sov. J. Nucl. Phys.* 32 (1980) 443; 447] Torsion and curvature in terms of the axial gravitational superfield.

M. SOHNIUS, K. S. STELLE, P. C. WEST, *Phys. Lett.* 92B (1980) 123. SU(4) supersymmetry off the mass shell.

F. A. BEREZIN, *Yad. Fiz.* 29 (1979) 1670 [*Sov. J. Nucl. Phys.* 29 (1979) 857];

A. S. SCHWARZ, *Nucl. Phys.* B171 (1980) 154. Attempts of quantum field theory with coordinates and fields treated on an equal footing.

E. CREMMER, B. JULIA, *Nucl. Phys.* B159 (1979) 141; Hidden local SU(8) symmetry in the $N = 8$ supergravity.

T. CURTRIGHT, P. FREUND, in: *Supergravity*. Proc. of Super-gravity Workshop at Stony Brook, Sept. 1979, ed. P. VAN NIEUWENHUIZEN and D. Z. FREEDMAN (North-Holland, Amsterdam) 1979, p. 197. Phenomenology of SU(8) symmetry.

J. ELLIS, M. K. GAILLARD, B. ZUMINO, *Phys. Lett.* 94B (1980) 343. SU(5) from broken $N = 8$ supergravity.

28.7.14. Cosmology and elementary particles. Review papers

A. D. DOLGOV, YA. B. ZELDOVICH, *Usp. Fiz. Nauk* 130 (1980) 559; *Rev. Mod. Phys.* 53 (1981) 1.

G. STEIGMAN, *Ann. Rev. Nucl. Part. Sci.* 29 (1979) 313.

J. ELLIS, M. K. GAILLARD, D. V. NANOPOULOS, *preprint* TH 2858-CERN, LAPP-TH-19, 1980.

28.7.15. Neutrino mass and hidden mass in the universe

YA. B. ZELDOVICH, P. A. SUNYAYEV, *Pis'ma v Astronom. Zhurn.* 6 (1980) 451;

A. G. DOROSHKEVICH, YA. B. ZELDOVICH, R. A. SUNYAYEV, M. YU. KHLOPOV, *Pis'ma v Astronom. Zhurn.* 6 (1980) 457; 465.

J. BOND, G. ESTATIO, J. SILK, *preprint* Univ. of Calif., 1980.

H. SATO, F. TAKAHARA, *preprint* RISP-400, 1980.

G. S. BISNOVATIY-KOGAN, I. D. NOVIKOV, *Astronom. Zhurnal* 57 (1980) 899. The role of massive neutrinos is discussed in the formation of galaxy clusters and of their hidden mass. $M \approx m_p/m_\nu$ where M is the mass of so-called pancakes, m_p is the Planck mass, and m_ν is the neutrino mass. An observational estimate of hidden mass in coronas of galaxies see in

J. EINASTO, A. KAASIK, E. SAAR, *Nature* 250 (1974) 309.

T. YANAGIDA, M. YOSHIMURA, *Phys. Rev. Lett.* 45 (1980) 71. Majorana masses of the neutrinos and cosmology.

YA. B. ZELDOVICH, A. A. KLYPIN, M. YU. KHLOPOV, V. M. CHECHETKIN, *Yad. Fiz.* 31 (1980) 1286. [*Sov. J. Nucl. Phys.* 31 (1980) 664] Astrophysical restrictions on the masses of heavy neutral leptons.

P. FRAMPTON, S. L. GLASHOW, *Phys. Rev. Lett.* 44 (1980) 1481. Restrictions on the possible properties of hypotetical long-lived heavy particles.

28.7.16. Calculation of n_B/n_γ

E. Kolb, S. Wolfram, *Nucl. Phys.* B172 (1980) 224.

V. A. Kuz'min, M. E. Shaposhnikov, *Phys. Lett.* 92B (1980) 115.

T. Yanagida, M. Yoshimura, *Nucl. Phys.* B168 (1980) 534. n_B/n_γ in grand unification models.

28.7.17. Magnetic monopoles and cosmology

J. P. Preskill, *Phys. Rev. Lett.* 43 (1979) 1365.

P. Langacker, S-Y. Pi, *Phys. Rev. Lett.* 45 (1980) 1. Monopoles and cosmology.

Y. M. Cho, *Phys. Rev. Lett.* 44 (1980) 1115. Colored monopoles.

M. Daniel, G. Lazarides, Q. Shafi, *Nucl. Phys.* 170 (1980) 156. SU(5) monopoles and confinement.

D. M. Scott, *Nucl. Phys.* 171 (1980) 95; 109. Monopoles in the grand unification theories.

A. D. Linde, *Phys. Lett.* 96B (1980) 293. Confinement of monopoles at high temperatures.

28.7.18. Search for new particles and interactions

M. Marinelli, G. Morpurgo, *Phys. Lett.* 94B (1980) 427; 433. Negative results of search for fractionally charged quarks: $n_q/n_B \lesssim 10^{-21}$.

S. Chaudhuri, D. D. Coon, G. E. Derkits, Jr., *Phys. Rev. Lett.* 45 (1980) 1374. A new method of search for fractionally charged particles is suggested, with sensitivity $\simeq 10^{-21} - 10^{-24}$ particles per nucleon.

M. Basile et al., *Lett. Nuovo Cim.* 29 (1980) 251. Quark search experiment in high energy neutrino interactions.

G. Barbiellini et al., *preprint* DESY 80/42, May 1980. Review of possible experiments on quark and monopole search in colliding LEP beams.

L. B. Okun, *Nucl. Phys.* B173 (1980) 1.

L. B. Okun, *Pis'ma ZhETF* 31 (1980) 156 [*JETP Lett.* 31 (1980) 144]. Hypothesis of the existence of θ-gluons with macroscopic confinement radius.

L. B. OKUN, *ZhETF* 79 (1980) 694 [*JETP* 52 (1980) 351] Discussion of hypothetical long-range forces.

28.7.19. Search for deviations from Newton's law

R. SPERO, J. K. HOSKINS, R. NEWMAN, J. PELLAM, J. SCHULTZ, *Phys. Rev. Lett.* 44 (1980) 1645. Dependence $1/r^2$ for the gravity force is verified in the range from 2 to 5 cm.

V. I. PANOV, V. N. FRONTOV, *ZhETF* 77 (1979) 1701 [*JETP* 50 (1979) 852]. Independence of the gravitational interaction constant on distance is verified in the range from 0.4 to 10 m to the accuracy of about one per cent.

S. I. BLINNINKOV, *Astrophysics and Space Science* 59 (1978) 13. The relation between the mass and radius of white dwarfs excludes variation of G_N greater than 10% in the range from 10m to 1 km.

28.7.20. Verification of electric charge conservation

I. R. BARABANOV et al., *Pis'ma ZhETF* 32 (1980) 384 [*JETP Lett.* 32 (1980) 359]. Upper bound on ^{71}Ga lifetime with respect to decay $^{71}\text{Ga}_{31} \to {}^{71}\text{Ge}_{32} + \gamma$ is established: $T \geq 2.3 \cdot 10^{23}$ years.

B. E. NORMAN, A. G. SEAMSTER, *Phys. Rev. Lett.* 43 (1979) 1226. $T(^{87}\text{Rb}_{37} \to {}^{87}\text{Sr}_{38} + \gamma) \geq 1.9 \cdot 10^{18}$ years.

28.7.21. Future accelerators

Proceedings of the Workshop on possibilities and limitations of accelerators and detectors, April 1979, FNAL, Batavia, Ill. USA.

Open Presentation of the Results of the second ICFA Workshop on Possibilities and Limitations of Accelerators and Detectors held October 4–9, 1979 at Les Diablerets, Switzerland.

R. WILSON, *Scientific American* 242 (1980) No. 1, 26; The next generation of particle accelerators.

D. CLINE, C. RUBBIA, *Phys. Today* 33 (1980) No. 8, 44. Antiproton-proton colliders and intermediate bosons.

CHAPTER 29

Appendix (Some useful formulas)

This appendix is a compilation of some notations, definitions and formulas which are frequently mentioned in the text of the book. The appendix is divided into the following sections:

29.1. Pseudoeuclidean metric. 308
29.2. Groups 309
    29.2.1. Some definitions 309
    29.2.2. The groups SU(n) 310
    29.2.3. The group SU(2) 310
    29.2.4. Fierz identities for σ-matrices 312
    29.2.5. Clebsch-Gordan coefficients, spherical harmonics, and d functions 313
    29.2.6. The group SU(3) 314
    29.2.7. Fierz identities for λ matrices. 315
    29.2.8. SU(3) multiplets. 315
29.3. Properties of Dirac matrices. 318
    29.3.1. γ- matrices 318
    29.3.2. Trace of γ- matrices 319
    29.3.3. Dirac bispinor 320
    29.3.4. Four-fermion invariants and Fierz identities for Dirac matrices 321
    29.3.5. Derivation of Fierz identities for Dirac matrices. 323
29.4. Rules for the calculation of probabilities. 325
    29.4.1. S- and T-matrices 325
    29.4.2. Probability and cross sections 326
    29.4.3. Particles with non-zero spin. 327

29.1. Pseudoeuclidean metric

Contravariant vector a^μ; examples:

$$x^\mu \equiv x^0, x^1, x^2, x^3 \equiv t, x, y, z \equiv t, \boldsymbol{x},$$
$$p^\mu \equiv p^0, p^1, p^2, p^3 \equiv E, p_x, p_y, p_z \equiv E, \boldsymbol{p}.$$

Covariant vector a_μ; examples:

$$x_\mu \equiv x_0, x_1, x_2, x_3 \equiv t, -\mathbf{x},$$
$$p_\mu \equiv p_0, p_1, p_2, p_3 \equiv E, -\mathbf{p}.$$

Metric tensor:

$$a_\mu = g_{\mu\nu} a^\nu, \qquad a^\mu = g^{\mu\nu} a_\nu,$$

where $g_{\mu\nu}$ is a metric tensor with only diagonal components not equal to zero:

$$g^{\mu\nu} = g_{\mu\nu}, \qquad g_{00} = -g_{11} = -g_{22} = -g_{33} = 1.$$

Scalar product:

$$a^\mu b_\mu = g^{\mu\nu} a_\mu b_\nu = g_{\mu\nu} a^\mu b^\nu$$
$$= a^0 b_0 + a^i b_i = a^0 b_0 - a^i b^i = a^0 b^0 - \mathbf{a} \cdot \mathbf{b},$$
$$\mu, \nu = 0, 1, 2, 3; \qquad i, k = 1, 2, 3.$$

Example:

$$p^\mu x_\mu = Et - \mathbf{p} \cdot \mathbf{x}.$$

The momentum operator in the coordinate representation has the form:

$$p^\mu = i\frac{\partial}{\partial x_\mu} = i\frac{\partial}{\partial t}, -i\frac{\partial}{\partial \mathbf{x}} = i\frac{\partial}{\partial t}, \frac{1}{i}\frac{\partial}{\partial \mathbf{x}}.$$

29.2. Groups

29.2.1. Some definitions.

A group $\mathcal{G}$ is a set of elements in which the operation of associative multiplication and a unit element are defined and for each element the set contains the corresponding inverse element. A group is called an abelian group if all elements commute. The representation G of group $\mathcal{G}$ is defined as a group of linear transformations (of matrices) in a linear space (referred to as the basis of the representation, or a multiplet), whose elements are in one-to-one correspondence with the elements of group $\mathcal{G}$. The groups whose elements are analytical functions of a finite number of parameters are called Lie groups. The number of independent parameters is called the dimension of the group. The generators of a given representation are operators I_i which realize transformations infinitesimally close to the identity transformation:

$G = 1 + id\omega_i I_i$. The maximum number of mutually commuting generators is called the rank of the group. The number of linearly independent vectors in the basis (the number of components in the multiplet) is called the dimension of the representation (the dimension of a representation is equal to the order of matrices realizing this representation). If a specific choice of basis breaks the representation into a sum of independent subgroups, the representation is called reducible; it is irreducible if this separation cannot be achieved by any choice of basis. Representations are called fundamental if all other representations of the group can be constructed from them by multiplication. The dimension of a regular (adjoint) representation equals the order of the group.

29.2.2. The groups SU(n)

SU(n) is a group of complex matrices U satisfying the conditions of unitarity ($U^+ U = 1$) and of unimodularity (det $U = 1$), whose fundamental multiplet is an n-component spinor and whose fundamental representation is given by matrices of order n.

29.2.3. The group SU(2)

The fundamental representation of the group SU(2) is given by the matrices

$$U = \exp(i\omega_i \tfrac{1}{2}\sigma_i), \quad i = 1, 2, 3,$$

where σ_i are Pauli matrices, and ω_i are three real parameters. The symbol σ_i is used to describe particle spins; in the description of isospin the same matrices are usually denoted by τ_i. Pauli matrices satisfy the commutation relations

$$\left[\tfrac{1}{2}\sigma_i, \tfrac{1}{2}\sigma_j\right] = i\varepsilon_{ijk}\tfrac{1}{2}\sigma_k,$$

where ε_{ijk} is a completely antisymmetric alternating tensor. The components of this tensor are structure constants of the group SU(2). The Pauli matrices are usually taken in the form

$$\sigma_1 = \begin{pmatrix} 0 & 1 \\ 1 & 0 \end{pmatrix}, \quad \sigma_2 = \begin{pmatrix} 0 & -i \\ i & 0 \end{pmatrix}, \quad \sigma_3 = \begin{pmatrix} 1 & 0 \\ 0 & -1 \end{pmatrix}.$$

An arbitrary representation of the group SU(2) has three generators satisfying the condition

$$\left[I_i, I_j\right] = i\varepsilon_{ijk}I_k.$$

For a regular (three-dimensional) representation,

$$I_1 = \begin{pmatrix} 0 & 0 & 0 \\ 0 & 0 & -i \\ 0 & i & 0 \end{pmatrix},$$

$$I_2 = \begin{pmatrix} 0 & 0 & i \\ 0 & 0 & 0 \\ -i & 0 & 0 \end{pmatrix},$$

$$I_3 = \begin{pmatrix} 0 & -i & 0 \\ i & 0 & 0 \\ 0 & 0 & 0 \end{pmatrix}.$$

Pauli matrices satisfy the following conditions:

$$\sigma_i \sigma_k = \delta_{ik} I + i\varepsilon_{ikl}\sigma_l, \quad \text{where } I = \begin{pmatrix} 1 & 0 \\ 0 & 1 \end{pmatrix},$$

$$\delta_{ik} = \begin{cases} 1 & \text{if } i = k, \\ 0 & \text{if } i \neq k, \end{cases}$$

$$\delta_{ik}\delta_{ik} = 3;$$

$$\varepsilon_{ikl} = \begin{cases} 1 & \text{if } ikl = 123, 312, 231, \\ -1 & \text{if } ikl = 132, 213, 321, \\ 0 & \text{if } i = k, \text{ or } k = l, \text{ or } l = i; \end{cases}$$

$$\varepsilon_{ikl}\varepsilon_{i'k'l'} = \begin{vmatrix} \delta_{ii'} & \delta_{ik'} & \delta_{il'} \\ \delta_{ki'} & \delta_{kk'} & \delta_{kl'} \\ \delta_{li'} & \delta_{lk'} & \delta_{ll'} \end{vmatrix},$$

$$\varepsilon_{ikl}\varepsilon_{i'k'l} = \begin{vmatrix} \delta_{ii'} & \delta_{ik'} \\ \delta_{ki'} & \delta_{kk'} \end{vmatrix} = \delta_{ii'}\delta_{kk'} - \delta_{ik'}\delta_{ki'},$$

$$\varepsilon_{ikl}\varepsilon_{i'kl} = 2\delta_{ii'},$$

$$\varepsilon_{ikl}\varepsilon_{ikl} = 6.$$

Trace of σ-matrices:

$$\operatorname{Tr} I = 2,$$

$$\operatorname{Tr} \sigma_i = 0,$$

$$\operatorname{Tr} \sigma_i \sigma_k = 2\delta_{ik},$$

$$\operatorname{Tr} \sigma_i \sigma_k \sigma_l = 2i\varepsilon_{ikl},$$

$$\operatorname{Tr} \sigma_i \sigma_k \sigma_l \sigma_m = 2[\delta_{ik}\delta_{lm} + \delta_{im}\delta_{kl} - \delta_{il}\delta_{km}].$$

29.2.4. Fierz identities for σ-matrices:

$$\delta_b^a \delta_d^c = \tfrac{1}{2}\delta_d^a \delta_b^c + \tfrac{1}{2}\boldsymbol{\sigma}_d^a \cdot \boldsymbol{\sigma}_b^c,$$

$$\boldsymbol{\sigma}_b^a \boldsymbol{\sigma}_d^c = \tfrac{3}{2}\delta_d^a \delta_b^c - \tfrac{1}{2}\boldsymbol{\sigma}_d^a \boldsymbol{\sigma}_b^c,$$

where

$$\boldsymbol{\sigma} \cdot \boldsymbol{\sigma} = \sigma_i \sigma_i, \qquad i = 1, 2, 3.$$

These relations are easily verified by multiplying both equalities by $\delta_b^a \delta_d^c$ and $\delta_d^a \delta_b^c$. It follows from these equalities that

$$3\delta_b^a \delta_d^c + \boldsymbol{\sigma}_b^a \cdot \boldsymbol{\sigma}_d^c = +(3\delta_d^a \delta_b^c + \boldsymbol{\sigma}_d^a \cdot \boldsymbol{\sigma}_b^c),$$

$$\delta_b^a \delta_d^c - \boldsymbol{\sigma}_b^a \cdot \boldsymbol{\sigma}_d^c = -(\delta_d^a \delta_b^c - \boldsymbol{\sigma}_d^a \cdot \boldsymbol{\sigma}_b^c).$$

The first of these expressions, acting on the product of spinors $\psi^b \varphi^d$, yields a state with spin 1, and the second yields a state with spin 0. This is in agreement with the form of the operator which gives zero when applied to a state with spin S:

$$S^2 - S(S+1) = \left(\tfrac{1}{2}(\boldsymbol{\sigma}_1 + \boldsymbol{\sigma}_2)\right)^2 - S(S+1)$$

$$= \tfrac{3}{2} + \tfrac{1}{2}\boldsymbol{\sigma}_1 \cdot \boldsymbol{\sigma}_2 - S(S+1)$$

$$= \begin{cases} \tfrac{3}{2} + \tfrac{1}{2}\boldsymbol{\sigma}_1 \cdot \boldsymbol{\sigma}_2 & \text{for } S = 0, \\ -\tfrac{1}{2} + \tfrac{1}{2}\boldsymbol{\sigma}_1 \cdot \boldsymbol{\sigma}_2 & \text{for } S = 1. \end{cases}$$

29.2.5. Clebsch-Gordan coefficients, spherical harmonics, and d functions

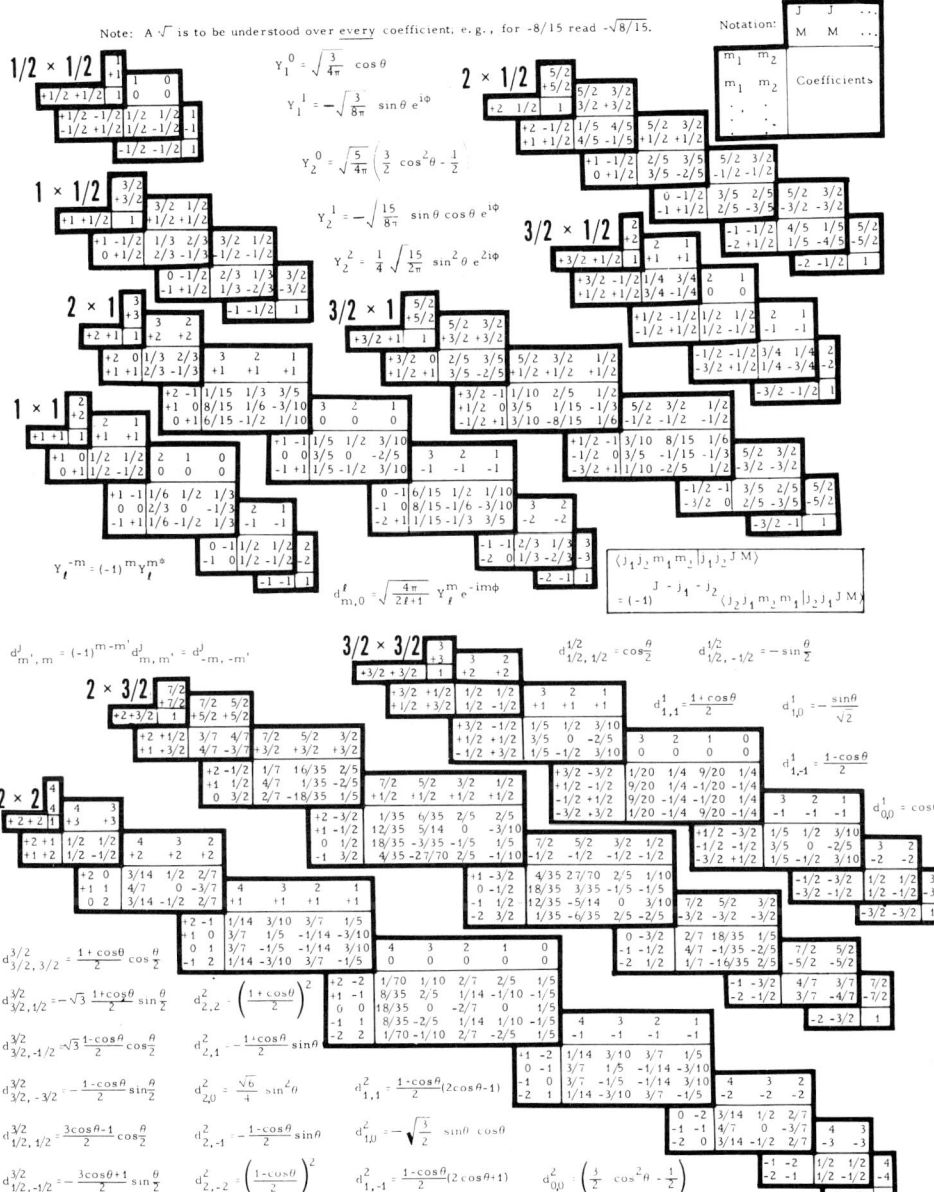

Reprinted from Rev. Mod. Phys. 52, No. 2 (1980) page 34.

29.2.6. The group SU(3)

The fundamental representation of the group SU(3) is given by the matrices

$$U = \exp(\tfrac{1}{2}\lambda_i \omega_i), \qquad i = 1, 2, \ldots, 8,$$

where λ_i are the Gell-Mann matrices, and ω_i are eight real parameters. Usually the matrices λ_i are chosen in the form:

$$\lambda_1 = \begin{pmatrix} 0 & 1 & 0 \\ 1 & 0 & 0 \\ 0 & 0 & 0 \end{pmatrix}, \quad \lambda_2 = \begin{pmatrix} 0 & -i & 0 \\ i & 0 & 0 \\ 0 & 0 & 0 \end{pmatrix}, \quad \lambda_3 = \begin{pmatrix} 1 & 0 & 0 \\ 0 & -1 & 0 \\ 0 & 0 & 0 \end{pmatrix},$$

$$\lambda_4 = \begin{pmatrix} 0 & 0 & 1 \\ 0 & 0 & 0 \\ 1 & 0 & 0 \end{pmatrix}, \quad \lambda_5 = \begin{pmatrix} 0 & 0 & -i \\ 0 & 0 & 0 \\ i & 0 & 0 \end{pmatrix}, \quad \lambda_6 = \begin{pmatrix} 0 & 0 & 0 \\ 0 & 0 & 1 \\ 0 & 1 & 0 \end{pmatrix},$$

$$\lambda_7 = \begin{pmatrix} 0 & 0 & 0 \\ 0 & 0 & -i \\ 0 & i & 0 \end{pmatrix}, \quad \lambda_8 = \frac{1}{\sqrt{3}} \begin{pmatrix} 1 & 0 & 0 \\ 0 & 1 & 0 \\ 0 & 0 & -2 \end{pmatrix}.$$

The matrices λ satisfy the following relations:

$$\operatorname{Tr} \lambda_i \lambda_j = 2\delta_{ij},$$

$$[\lambda_i \lambda_j] = 2if_{ijk}\lambda_k,$$

$$[\lambda_i \lambda_j]_+ = \tfrac{4}{3}\delta_{ij} + 2d_{ijk}\lambda_k, \qquad \text{where } i, j, k = 1, 2, \ldots, 8.$$

Here f_{ijk} are structure constants of the group SU(3), d_{ijk} are symmetrical and f_{ijk} are antisymmetrical with respect to permutations of any pair of indices. Direct calculations easily give 54 non-zero constants f_{ijk} and 58 non-zero constants d_{ijk}:

ijk	f_{ijk}	ijk	d_{ijk}	ijk	d_{ijk}
123	1	118	$1/\sqrt{3}$	355	$\tfrac{1}{2}$
147	$\tfrac{1}{2}$	146	$\tfrac{1}{2}$	366	$-\tfrac{1}{2}$
156	$-\tfrac{1}{2}$	157	$\tfrac{1}{2}$	377	$-\tfrac{1}{2}$
246	$\tfrac{1}{2}$	228	$1/\sqrt{3}$	448	$-\tfrac{1}{2}\sqrt{3}$
257	$\tfrac{1}{2}$	247	$-\tfrac{1}{2}$	558	$-\tfrac{1}{2}\sqrt{3}$
345	$\tfrac{1}{2}$	256	$\tfrac{1}{2}$	668	$-\tfrac{1}{2}\sqrt{3}$
367	$-\tfrac{1}{2}$	338	$1/\sqrt{3}$	778	$-\tfrac{1}{2}\sqrt{3}$
458	$\tfrac{1}{2}\sqrt{3}$	344	$\tfrac{1}{2}$	888	$-1/\sqrt{3}$
678	$\tfrac{1}{2}\sqrt{3}$				

($54 = 9 \times 6$ where 6 is the number of permutations of indices $i \neq j \neq k$, and $58 = 4 \times 6 + 11 \times 3 + 1$). Note that $d_{ijk} = 0$ if the number of indices 2, 5, 7 is odd. On the other hand, $f_{ijk} = 0$ if the number of these indices is even. These indices, 2, 5, 7, are special because the corresponding matrices λ are antisymmetric.

29.2.7. Fierz identities for λ matrices

Using the completeness of the nine matrices δ^α_β, λ^α_β, we can write:

$$\delta^\alpha_\beta \delta^\gamma_\delta = A \delta^\alpha_\delta \delta^\gamma_\beta + B \lambda^\alpha_\delta \lambda^\gamma_\beta,$$

$$\lambda^\alpha_\beta \lambda^\gamma_\delta = C \delta^\alpha_\delta \delta^\gamma_\beta + D \lambda^\alpha_\delta \lambda^\gamma_\beta,$$

where A, B, C and D are coefficients to be determined and where

$$\lambda \cdot \lambda = \lambda_i \lambda_i, \quad i = 1, 2, \ldots, 8.$$

Multiplication of these two equalities by $\delta^\delta_\alpha \delta^\beta_\gamma$ yields

$$3 = 9A, \quad 16 = 9C,$$

and multiplication by $\delta^\beta_\alpha \delta^\delta_\gamma$ yields

$$9 = 3A + 16B, \quad 0 = 3C + 16,$$

whence

$$\delta^\alpha_\beta \delta^\gamma_\delta = \tfrac{1}{3} \delta^\alpha_\delta \delta^\gamma_\beta + \tfrac{1}{2} \lambda^\alpha_\delta \cdot \lambda^\gamma_\beta,$$

$$\lambda^\alpha_\beta \cdot \lambda^\gamma_\delta = \tfrac{16}{9} \delta^\alpha_\delta \delta^\gamma_\beta - \tfrac{1}{3} \lambda^\alpha_\delta \cdot \lambda^\gamma_\beta.$$

Now it is not difficult to show that

$$8 \delta^\alpha_\beta \delta^\gamma_\delta + 3 \lambda^\alpha_\beta \cdot \lambda^\gamma_\beta = + \left(8 \delta^\alpha_\delta \delta^\gamma_\beta + 3 \lambda^\alpha_\delta \cdot \lambda^\gamma_\beta \right),$$

$$4 \delta^\alpha_\beta \delta^\gamma_\delta - 3 \lambda^\alpha_\beta \cdot \lambda^\gamma_\delta = - \left(4 \delta^\alpha_\delta \delta^\gamma_\beta - 3 \lambda^\alpha_\delta \cdot \lambda^\gamma_\beta \right).$$

Applied to the product of two triplet spinors, the first of these expressions selects the state 6, and the second one selects the state $\bar{3}$ (recall that $3 \times 3 = 6 + \bar{3}$).

29.2.8. SU(3) multiplets

A contravariant three-component spinor t^α is transformed by the matrices $U = \exp(\tfrac{1}{2} i \omega_i \lambda_i)$; it is denoted by 3. A covariant spinor t_α is transformed by complex conjugate matrices $U^* = \exp(-\tfrac{1}{2} i \omega_i \lambda_i^*)$; it will be denoted by $\bar{3}$. Representations of higher dimensions can be constructed out of 3 and $\bar{3}$ by making use of the invariant tensors δ_β^α, $\varepsilon_{\alpha\beta\gamma}$, and $\varepsilon^{\alpha\beta\gamma}$:

$3 \times \bar{3} = 8 + 1$:
  singlet, $\quad 1 \sim t^\alpha t_\beta \delta_\alpha^\beta$;
  octet, $\quad 8 \sim T_\beta^\alpha = t^\alpha t_\beta - \tfrac{1}{3}\delta_\beta^\alpha(t^\gamma t_\gamma)$.

$3 \times 3 = 6 + \bar{3}$:
  antitriplet, $\quad \bar{3} \sim T_\gamma = t^\alpha t^\beta \varepsilon_{\alpha\beta\gamma}$;
  sextet, $\quad 6 \sim T^{\alpha\beta} = t^\alpha t^\beta + t^\beta t^\alpha$.

$3 \times 6 = 8 + 10$:
  $\quad 8 \sim T_\delta^\gamma = t^\alpha t^{\beta\gamma} \varepsilon_{\alpha\beta\delta}$;
  decuplet, $\quad 10 \sim T^{\alpha\beta\gamma}$.

$\bar{3} \times 6 = 3 + 15$:
  $\quad 3 \sim T^\gamma = t_\alpha T^{\alpha\gamma}$;
  $\quad 15 \sim T_\alpha^{\beta\gamma}$.

$8 \times 8 = 1 + 8 + 8 + 10 + \overline{10} + 27$:
  $\quad \overline{10} \sim T_{\alpha\beta\gamma}$;
  $\quad 27 \sim T_{\alpha\beta}^{\gamma\delta}$.

An arbitrary tensor can be written in the form

$$T_p^q = T_{\alpha_1\alpha_2\ldots\alpha_p}^{\beta_1\beta_2\ldots\beta_q},$$

where symmetrization is carried out separately over all upper and lower indices, and the trace for any pair $\alpha_i\beta_k$ is zero. The total number of components of the multiplet T_p^q is found easily:

$$N = \tfrac{1}{2}(p+1)(q+1)(p+q+2).$$

Examples of physical SU(3) multiplets:

$q^\alpha = \begin{pmatrix} u \\ d \\ s \end{pmatrix}\quad$ quark triplet,

$\bar{q}_\alpha = (\bar{u}, \bar{d}, \bar{s})\quad$ antiquark (anti)triplet,

$$P_\beta^\alpha = \begin{pmatrix} \sqrt{\tfrac{1}{6}}\eta^0 + \sqrt{\tfrac{1}{2}}\pi^0 & \pi^+ & K^+ \\ \pi^- & \sqrt{\tfrac{1}{6}}\eta^0 - \sqrt{\tfrac{1}{2}}\pi^0 & K^0 \\ K^- & \tilde{K}^0 & -\dfrac{2\eta^0}{\sqrt{6}} \end{pmatrix}\quad\text{octet of pseudo-scalar mesons,}$$

$$B_\beta^\alpha = \begin{pmatrix} \sqrt{\tfrac{1}{6}}\Lambda^0 + \sqrt{\tfrac{1}{2}}\Sigma^0 & \Sigma^+ & p \\ \Sigma^- & \sqrt{\tfrac{1}{6}}\Lambda^0 - \sqrt{\tfrac{1}{2}}\Sigma^0 & n \\ \Xi^- & \Xi^0 & -\sqrt{\tfrac{1}{6}}2\Lambda^0 \end{pmatrix}\quad\text{octet of baryons.}$$

When the isotopic subgroup SU(2) of group SU(3) is singled out, it is convenient to plot the particles of the multiplet on the so-called $T_3 Y$ diagrams. Examples are given in figs. 29.1, 2, 3.

By combining d and s (or s and u) quarks, instead of u and d, into an SU(2) doublet we single out the U (or V) spin subgroup* of SU(3) (see fig. 29.4). Figs. 29.1–4 demonstrate that particles within one U-multiplet have identical charges. The composition of U-multiplets is obvious in these figures, with the exception of the central particles on the $T_3 Y$ diagram for the octet. The point is that the Σ^0 and Λ^0 states possess a definite T-spin but no definite U-spin. It is their linear superpositions

$$\Sigma_U^0 = -\tfrac{1}{2}\Sigma^0 + \tfrac{1}{2}\sqrt{3}\,\Lambda^0,\ \Lambda_U^0 = -\tfrac{1}{2}\sqrt{3}\,\Sigma^0 - \tfrac{1}{2}\Lambda^0,$$

that possess definite U-spin: unity for the first and zero for the second.

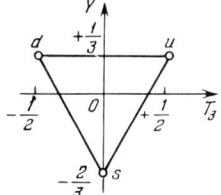

Fig. 29.1

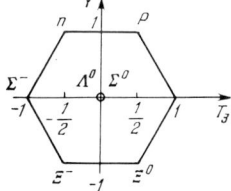

Fig. 29.2

---

*Sometimes the minus sign is assigned to some of the particles of the SU(3) multiplet in order to make positive the matrix elements of the ladder operators of a given SU(2) subgroup (see J. J. de Swart, *Rev. Mod. Phys.* 35 (1963) 916).

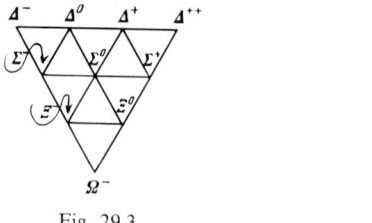
Fig. 29.3

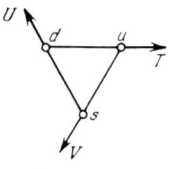
Fig. 29.4

29.3. Properties of Dirac matrices

29.3.1. γ-Matrices

The matrices γ^μ satisfy the condition

$$\gamma^\mu \gamma^\nu + \gamma^\nu \gamma^\mu = 2g^{\mu\nu}, \qquad \mu, \nu = 0, 1, 2, 3.$$

We use the representation

$$\gamma^0 = \begin{pmatrix} 1 & 0 \\ 0 & -1 \end{pmatrix}, \qquad \gamma^i = \begin{pmatrix} 0 & \sigma_i \\ -\sigma_i & 0 \end{pmatrix}, \qquad i = 1, 2, 3.$$

The last three matrices can be written in vector form:

$$\gamma = \begin{pmatrix} 0 & \boldsymbol{\sigma} \\ -\boldsymbol{\sigma} & 0 \end{pmatrix};$$

$$\gamma_5 = \gamma^5 = i\gamma_0\gamma_1\gamma_2\gamma_3 = -i\gamma^0\gamma^1\gamma^2\gamma^3 = -\begin{pmatrix} 0 & 1 \\ 1 & 0 \end{pmatrix}.$$

(The only difference with the notation of the well-known book by Bjorken and Drell is the sign in the definition of γ_5.)

$$\gamma^\mu \gamma^5 = -\gamma^5 \gamma^\mu, \quad (\gamma^5)^2 = 1.$$

The "scalar product" of γ^μ matrices by a 4-vector A^μ:

$$\hat{A} \equiv A^\mu \gamma_\mu \equiv A_\mu \gamma^\mu \equiv A^0 \gamma^0 - \boldsymbol{A} \cdot \boldsymbol{\gamma}.$$

By multiplying $\gamma^\mu \gamma^\nu + \gamma^\nu \gamma^\mu = 2g^{\mu\nu}$ by $g_{\mu\nu}$, $A_\mu B_\nu$, and finally by A_ν, we

Dirac matrices

obtain, respectively:

$$\gamma^\mu \gamma_\mu = 4,$$
$$\hat{A}\hat{B} + \hat{B}\hat{A} = 2AB,$$
$$\gamma^\mu \hat{A} + \hat{A}\gamma^\mu = 2A^\mu.$$

If this last equality is multiplied from the right by $R\gamma_\mu$, where R is an arbitrary matrix expression, we obtain

$$\gamma^\mu \hat{A} R \gamma_\mu + \hat{A}\gamma^\mu R \gamma_\mu = 2R\hat{A},$$

whence

for $R = 1$: $\quad \gamma^\mu \hat{A} \gamma_\mu = -2\hat{A},$

for $R = \hat{B}$: $\quad \gamma^\mu \hat{A}\hat{B} \gamma_\mu = 2\hat{A}\hat{B} + 2\hat{B}\hat{A} = 4AB,$

for $R = \hat{B}\hat{C}$: $\quad \gamma^\mu \hat{A}\hat{B}\hat{C} \gamma_\mu = -2\hat{C}\hat{B}\hat{A}.$

29.3.2. Trace of γ-matrices

The trace, Tr, is the sum of diagonal elements of a matrix.

$$\text{Tr}\,\gamma^\mu = 0, \quad \text{Tr}\,\gamma^5 = 0, \quad \text{Tr}\,I = 4.$$

By definition,

$$\text{Tr}\,\gamma^{\alpha_1}\gamma^{\alpha_2}\cdots\gamma^{\alpha_{n-1}}\gamma^{\alpha_n} = \text{Tr}\,\gamma^{\alpha_2}\cdots\gamma^{\alpha_{n-1}}\gamma^{\alpha_n}\gamma^{\alpha_1},$$

whence, using that $(\gamma^5)^2 = 1$ and $\gamma^5 \gamma^\mu = -\gamma^\mu \gamma^5$, it is not difficult to derive that the trace of a product of an odd number of matrices is zero:

$$\text{Tr}\,\gamma^{\alpha_1}\gamma^{\alpha_2}\cdots\gamma^{\alpha_{n-1}}\gamma^{\alpha_n} = \text{Tr}\,\gamma^5\gamma^5\gamma^{\alpha_1}\gamma^{\alpha_2}\cdots\gamma^{\alpha_{n-1}}\gamma^{\alpha_n}$$
$$= \text{Tr}\,\gamma^5\gamma^{\alpha_1}\gamma^{\alpha_2}\cdots\gamma^{\alpha_{n-1}}\gamma^{\alpha_n}\gamma^5$$
$$= -\text{Tr}\,\gamma^{\alpha_1}\gamma^{\alpha_2}\cdots\gamma^{\alpha_{n-1}}\gamma^{\alpha_n} = 0.$$

The following reduction formula for the case of even n can be obtained by using the relation $\gamma^\alpha \gamma^\beta + \gamma^\beta \gamma^\alpha = 2g^{\alpha\beta}$ and by a method similar to that used above:

$$\text{Tr}\,\gamma^{\alpha_1}\gamma^{\alpha_2}\cdots\gamma^{\alpha_{n-1}}\gamma^{\alpha_n} = g^{\alpha_1\alpha_2}\,\text{Tr}\,\gamma^{\alpha_3}\gamma^{\alpha_4}\cdots\gamma^{\alpha_{n-1}}\gamma^{\alpha_n}$$
$$- g^{\alpha_1\alpha_3}\,\text{Tr}\,\gamma^{\alpha_2}\gamma^{\alpha_4}\cdots\gamma^{\alpha_{n-1}}\gamma^{\alpha_n} + \cdots$$
$$+ (-1)^n g^{\alpha_1\alpha_n}\,\text{Tr}\,\gamma^{\alpha_2}\gamma^{\alpha_3}\cdots\gamma^{\alpha_{n-1}}.$$

In the case $n = 2$ and $n = 4$ we obtain

$$\text{Tr } \gamma^\alpha \gamma^\beta = 4g^{\alpha\beta},$$

$$\text{Tr } \gamma^\alpha \gamma^\beta \gamma^\gamma \gamma^\delta = 4(g^{\alpha\beta}g^{\gamma\delta} + g^{\alpha\delta}g^{\beta\gamma} - g^{\alpha\gamma}g^{\beta\delta}).$$

As follows from the definition $\gamma^5 = -i\gamma^0\gamma^1\gamma^2\gamma^3$

$$\text{Tr } \gamma^5 \gamma^\alpha \gamma^\beta \gamma^\gamma \gamma^\delta = 4i\varepsilon^{\alpha\beta\gamma\delta},$$

where $\varepsilon^{\alpha\beta\gamma\delta}$ is a completely antisymmetric tensor of rank four ($\varepsilon^{0123} = 1$).

$$\varepsilon^{\alpha\beta\gamma\delta}\varepsilon_{\mu\nu\rho\sigma} = - \begin{Vmatrix} \delta^\alpha_\mu & \delta^\alpha_\nu & \delta^\alpha_\rho & \delta^\alpha_\sigma \\ \delta^\beta_\mu & \delta^\beta_\nu & \delta^\beta_\rho & \delta^\beta_\sigma \\ \delta^\gamma_\mu & \delta^\gamma_\nu & \delta^\gamma_\rho & \delta^\gamma_\sigma \\ \delta^\delta_\mu & \delta^\delta_\nu & \delta^\delta_\rho & \delta^\delta_\sigma \end{Vmatrix}.$$

Multiplication of this expression by δ^σ_δ yields

$$\varepsilon^{\alpha\beta\gamma\delta}\varepsilon_{\mu\nu\rho\delta} = - \begin{Vmatrix} \delta^\alpha_\mu & \delta^\alpha_\nu & \delta^\alpha_\rho \\ \delta^\beta_\mu & \delta^\beta_\nu & \delta^\beta_\rho \\ \delta^\gamma_\mu & \delta^\gamma_\nu & \delta^\gamma_\rho \end{Vmatrix}.$$

(We have made use of the relations

$$\delta^\alpha_\sigma \delta^\sigma_\delta = \delta^\alpha_\delta, \qquad \delta^\delta_\sigma \delta^\sigma_\delta = 4.)$$

Likewise, we can derive

$$\varepsilon^{\alpha\beta\gamma\delta}\varepsilon_{\mu\nu\gamma\delta} = -2 \begin{Vmatrix} \delta^\alpha_\mu & \delta^\alpha_\nu \\ \delta^\beta_\mu & \delta^\beta_\nu \end{Vmatrix},$$

$$\varepsilon^{\alpha\beta\gamma\delta}\varepsilon_{\mu\beta\gamma\delta} = -6\delta^\alpha_\mu,$$

$$\varepsilon^{\alpha\beta\gamma\delta}\varepsilon_{\alpha\beta\gamma\delta} = -24.$$

29.3.3. Dirac bispinor

The Dirac equation for a free particle with mass m and 4-momentum p has the form

$$(\hat{p} - m)u = 0,$$

where u is a four-component spinor (bispinor)

$$u = \begin{pmatrix} u_1 \\ u_2 \\ u_3 \\ u_4 \end{pmatrix}.$$

Let us define the conjugate bispinor as

$$\bar{u} = u^+ \gamma^0 = (u_1^*, u_2^*, -u_3^*, -u_4^*).$$

It satisfies the equation

$$\bar{u}(\hat{p} - m) = 0.$$

29.3.4. Four-fermion invariants and Fierz identities for Dirac matrices

In all, 16 bilinear terms can be constructed using bispinors $\bar{a}$ and b which, in the general case, correspond to two different particles. The linear combinations of these terms form five different Lorentz-covariant quantities

Covariant quantity		Number of components
$\bar{a}b$	scalar	1
$\bar{a}\gamma^\alpha b$	vector	4
$\sqrt{\tfrac{1}{2}}\,\bar{a}\sigma^{\alpha\beta}b$	tensor	6
$\bar{a}\gamma^5\gamma^\alpha b$	axial vector	4
$\bar{a}\gamma^5 b$	pseudoscalar	1

Here $\sigma^{\alpha\beta} = \tfrac{1}{2}(\gamma^\alpha\gamma^\beta - \gamma^\beta\gamma^\alpha)$. The factor $\sqrt{\tfrac{1}{2}}$ is introduced into the definition of the tensor in order to normalize each of the six components of the tensor to unity (actually, to -1):

$$\sqrt{\tfrac{1}{2}}\,\sigma_{\alpha\beta}\sqrt{\tfrac{1}{2}}\,\sigma^{\alpha\beta} = \tfrac{1}{4}(\gamma_\alpha\gamma_\beta - \gamma_\beta\gamma_\alpha)\gamma^\alpha\gamma^\beta$$

$$= \tfrac{1}{4}\left(\gamma_\alpha\gamma_\beta\gamma^\alpha\gamma^\beta - \gamma_\beta\gamma_\alpha\gamma^\alpha\gamma^\beta\right) = -6.$$

The remaining covariants are normalized in the same manner:

$$1 \cdot 1 = 1, \qquad \gamma_\alpha\gamma^\alpha = 4, \qquad \gamma_\alpha\gamma_5\gamma^\alpha\gamma^5 = -4, \qquad \gamma_5\gamma^5 = 1.$$

It is possible, of course, to normalize all five variants to $+1$, by introducing i into the definitions of the tensor and the axial vector; traditionally, however, this is not done.

The Lorentz scalar can be constructed out of four bispinors $\bar{a}, b, \bar{c}, d$ in five different ways:

$$(\bar{a}b)(\bar{c}d) \qquad \text{S-variant,}$$
$$(\bar{a}\gamma_\alpha b)(\bar{c}\gamma^\alpha d) \qquad \text{V-variant,}$$
$$\tfrac{1}{2}(\bar{a}\sigma_{\alpha\beta} b)(\bar{c}\sigma^{\alpha\beta} d) \qquad \text{T-variant,}$$
$$(\bar{a}\gamma_\alpha\gamma_5 b)(\bar{c}\gamma^\alpha\gamma^5 d) \qquad \text{A-variant,}$$
$$(\bar{a}\gamma_5 b)(\bar{c}\gamma^5 d) \qquad \text{P-variant.}$$

Sixteen matrices $(1, \gamma_\alpha, \sigma_{\alpha\beta}, \gamma_5\gamma_\alpha, \gamma_5)$ form a complete system so that any one of the above variants can be expressed as a linear superposition of variants with a changed sequence of spinors:

$$(\bar{a}O_i b)(\bar{c}O^i d) = \sum_k c_{ik}(\bar{a}O_k d)(\bar{c}O^k b),$$

where

$$O_S = 1, \qquad O_V = \gamma_\alpha, \qquad O_T = \sigma_{\alpha\beta}/\sqrt{2}, \qquad O_A = \gamma_5\gamma_\alpha, \qquad O_P = \gamma_5.$$

It can be shown (see the next section) that the coefficients c_{ik} have the values given in the table below (Fierz table):

	S	V	T	A	P
S	$\tfrac{1}{4}$	$\tfrac{1}{4}$	$-\tfrac{1}{4}$	$-\tfrac{1}{4}$	$\tfrac{1}{4}$
V	1	$-\tfrac{1}{2}$	0	$-\tfrac{1}{2}$	-1
T	$-\tfrac{3}{2}$	0	$-\tfrac{1}{2}$	0	$-\tfrac{3}{2}$
A	-1	$-\tfrac{1}{2}$	0	$-\tfrac{1}{2}$	1
P	$\tfrac{1}{4}$	$-\tfrac{1}{4}$	$-\tfrac{1}{4}$	$\tfrac{1}{4}$	$\tfrac{1}{4}$

Signs in the table correspond to the case of commuting spinors (c-numbers); in the case of q-numbers all coefficients change signs. Note that the table is symmetric with respect to inversion relative to the central element c_{TT}. Note also that the table must be read from left to right, but not downward. The reason is that our five invariant amplitudes are not orthogonal to one another: the Fierz matrix is not a matrix of orthogonal rotation. It is easy to verify, by looking at the Fierz matrix, that the permutation $b \leftrightarrow d$ transforms two combinations of variants into themselves with the plus sign, and three combinations with the minus sign (under the assumption of commuting bispinors):

$$3(S + P) - T, \qquad 2(S - P) + V - A \qquad \text{symmetrical,}$$

$$V + A, \quad S + P + T, \qquad 2(S - P) - (V - A) \qquad \text{antisymmetrical.}$$

In the non-relativistic limit, symmetric combinations transform into $3 + \boldsymbol{\sigma} \cdot \boldsymbol{\sigma}$, and anisymmetric ones into $1 - \boldsymbol{\sigma} \cdot \boldsymbol{\sigma}$ (indeed, in this limit $S \to V \to 1$, $A \to T \to -\boldsymbol{\sigma} \cdot \boldsymbol{\sigma}$, $P \to 0$).

We recall that the expression $\tfrac{1}{4}(3 + \boldsymbol{\sigma}_1 \cdot \boldsymbol{\sigma}_2)$ is a projection operator of a state with total spin S equal to unity, and the expression $\tfrac{1}{4}(1 - \boldsymbol{\sigma}_1 \cdot \boldsymbol{\sigma}_2)$ is a projection operator of a state with $S = 0$. This is readily shown if we raise to the second power the equality

$$S = \tfrac{1}{2}\boldsymbol{\sigma}_1 + \tfrac{1}{2}\boldsymbol{\sigma}_2;$$

$$S(S+1) = \tfrac{1}{4}\boldsymbol{\sigma}_1^2 + \tfrac{1}{4}\boldsymbol{\sigma}_2^2 + \tfrac{1}{2}\boldsymbol{\sigma}_1 \cdot \boldsymbol{\sigma}_2 = \tfrac{1}{2}(3 + \boldsymbol{\sigma}_1 \cdot \boldsymbol{\sigma}_2).$$

We find, therefore, that the non-relativistic and relativistic symmetry properties are in complete agreement. Indeed, the state with $S = 0$ is antisymmetric under permutation of the spinors forming it, while that with $S = 1$ is symmetric.

So far we have been discussing five Lorentz scalars. The same Fierz matrix relates five Lorentz pseudoscalars. This is easily verified if we substitute, for instance, d by $\gamma_5 d$ in the above four-fermion expressions.

Two relations are encountered especially often in calculations of weak processes:

$$\bar{a}\gamma_\alpha(1+\gamma_5)b \cdot \bar{c}\gamma^\alpha(1+\gamma_5)d = -\bar{a}\gamma_\alpha(1+\gamma_5)d \cdot \bar{c}\gamma^\alpha(1+\gamma_5)b,$$

$$\bar{a}\gamma_\alpha(1+\gamma_5)b \cdot \bar{c}\gamma^\alpha(1-\gamma_5)d = +2\bar{a}(1-\gamma_5)d \cdot \bar{c}(1+\gamma_5)b.$$

These formulas are readily obtained from the Fierz coefficient of the V-variant. The first of them is obtained by means of the substitutions

$$\bar{a} \to \bar{a}(1-\gamma_5), \quad b \to (1+\gamma_5)b,$$
$$\bar{c} \to \bar{c}(1-\gamma_5), \quad d \to (1+\gamma_5)d.$$

The second one is obtained by means of the substitutions

$$\bar{a} \to \bar{a}(1-\gamma_5), \quad b \to (1+\gamma_5)b,$$
$$\bar{c} \to \bar{c}(1+\gamma_5), \quad d \to (1-\gamma_5)d.$$

29.3.5. Derivation of Fierz identities for Dirac matrices

In this section, we give an explicit derivation of the Fierz matrix discussed in sect. 29.3.4. First we consider some auxiliary relations.

An arbitrary 4×4 γ-matrix can be presented as a linear combination of

16 γ_A matrices:

$$\gamma = \tfrac{1}{4}\sum_A C_A \gamma_A, \qquad A = 1,\ldots,16.$$

$A = 1$	: 1 unit matrix, S-variant;
$A = 2, 3, 4, 5$	: $\gamma^0, \gamma^1, \gamma^2, \gamma^3$, V-variant;
$A = 6, 7, 8, 9, 10, 11$	: $\sigma^{10}, \sigma^{20}, \sigma^{30}, \sigma^{12}, \sigma^{23}, \sigma^{31}$, T-variant;
$A = 12, 13, 14, 15$	: $\gamma^5\gamma^0, \gamma^5\gamma^1, \gamma^5\gamma^2, \gamma^5\gamma^3$, A-variant;
$A = 16$	: γ^5, P-variant.

Hence,

$$C_A = \Delta_A \operatorname{Tr} \gamma\gamma_A,$$

where

$$\Delta_A = \tfrac{1}{4} \operatorname{Tr} \gamma_A \gamma_A.$$

It is not difficult to demonstrate that

$$\Delta_A = +1 \quad \text{for } A = 1, 2, 6, 7, 8, 13, 14, 15, 16,$$
$$\Delta_A = -1 \quad \text{for } A = 3, 4, 5, 9, 10, 11, 12.$$

This gives:

$$\gamma = \tfrac{1}{4} \sum_A \Delta_A \operatorname{Tr}(\gamma\gamma_A) \gamma_A,$$

or, if the matrix indices are written explicitly,

$$\gamma^i_k = \tfrac{1}{4} \sum_A \Delta_A (\gamma^l_m \gamma^m_{Al}) \gamma^i_{Ak}.$$

The last equality holds if

$$\tfrac{1}{4} \sum_A \Delta_A \gamma^m_{Al} \gamma^i_{Ak} = \delta^m_k \delta^i_l.$$

Substituting $m \to m'$ and $l \to l'$ in the above expression, and multiplying by a tensor $F^{m'}_m G^{l'}_l$ in which F and G are matrix expressions, we arrive at the basic relation yielding the coefficients of the Fierz matrix:

$$F^m_k G^i_l = \tfrac{1}{4} \sum_A \Delta_A (F\gamma_A G)^m_l (\gamma_A)^i_k.$$

We shall obtain exactly what we need if we multiply this expression by the spinors $\bar{a}^k, b_m, \bar{c}^l, d_i$; the spinors b_m and d_i change partners when we go from the left-hand side of the equation to the right-hand side.

Let us now derive the Fierz matrix row by row.

Scalar variant. In this case $F = G = 1$. Negative values of $\Delta_3, \Delta_4, \Delta_5$ give the correct sign in the scalar product $\gamma^\alpha \gamma_\alpha = \gamma_0 \gamma_0 - \boldsymbol{\gamma} \cdot \boldsymbol{\gamma}$. The negative sign of c_{SA} appears because $\Delta_{12} = -1$. The negative sign of c_{ST} is easily confirmed if we take into account that the term $\sigma^{10}\sigma^{10}$ enters the sum $\sigma^{\alpha\beta}\sigma_{\alpha\beta}$ with a minus sign (due to the pseudo-euclidean metric).

Pseudoscalar variant. $F = G = \gamma_5$. Since γ^α commutes with γ^5, $c_{PV} = -c_{SV}$, $c_{PA} = -c_{SA}$.

Vector variant. $F = \gamma^\alpha$, $G = \gamma_\alpha$. The following relations determine the coefficients of the second row of the Fierz matrix:

$$\gamma^\alpha \gamma_A \gamma_\alpha = 4\gamma_A \quad \text{for} \quad \gamma_A = 1,$$
$$= -2\gamma_A \quad \text{for} \quad \gamma_A = \gamma^\mu,$$
$$= 0 \quad \text{for} \quad \gamma_A = \sigma^{\mu\nu},$$
$$= 2\gamma_A \quad \text{for} \quad \gamma_A = \gamma^5 \gamma^\mu,$$
$$= -4\gamma_A \quad \text{for} \quad \gamma_A = \gamma^5.$$

Axial variant. $F = \gamma^5 \gamma^\alpha$, $G = \gamma_5 \gamma_\alpha$. Calculations are similar to those of the preceding case.

Tensor variant. $F = \sqrt{\tfrac{1}{2}}\,\sigma^{\alpha\beta}$, $G = \sqrt{\tfrac{1}{2}}\,\sigma_{\alpha\beta}$. The following relations are used to calculate the Fierz coefficients:

$$\sigma^{\alpha\beta} \gamma_A \sigma_{\alpha\beta} = -12\gamma_A \quad \text{for} \quad \gamma_A = 1,$$
$$= 0 \quad \text{for} \quad \gamma_A = \gamma^\mu,$$
$$= -4\gamma_A \quad \text{for} \quad \gamma_A = \sigma^{\mu\nu},$$
$$= 0 \quad \text{for} \quad \gamma_A = \gamma^5 \gamma^\mu,$$
$$= -12\gamma_A \quad \text{for} \quad \gamma_A = \gamma^5.$$

The Fierz relations for longitudinal spinors are obtained by taking $F = \gamma^\alpha(1 + \gamma_5)$, $G = \gamma_\alpha(1 + \gamma_5)$, or $F = \gamma^\alpha(1 + \gamma_5)$, $G = \gamma_\alpha(1 - \gamma_5)$.

29.4. Rules for the calculation of probabilities

29.4.1. S- and T-matrices

Consider a set of physical states which are transformed into one another due to interactions. We shall characterize the transition from state i to state f by a quantity S_{fi}. The set of all values of S_{fi} forms the scattering matrix, also referred to as S-matrix. With all interactions switched off, the S-matrix turns into a unit matrix I: each state is transformed into itself. This means that physical processes occur if the T-matrix is non-zero, with T defined by the relation

$$S = I + iT.$$

Let us introduce the process amplitude M_{fi} as the quantity defined by the relation

$$T_{fi} = (2\pi)^4 \delta^4(p_f - p_i) M_{fi},$$

where p_i and p_f are 4-momenta of the initial and final states, and the δ-function expresses in an explicit form the energy-momentum conservation law:

$$\delta^4(p_f - p_i) = \delta(p_f^x - p_i^x)\delta(p_f^y - p_i^y)\delta(p_f^z - p_i^z)\delta(E_f - E_i).$$

For the sake of brevity the subscripts f and i in M_{fi} are hereafter dropped.

29.4.2. Probability and cross sections

The square of the absolute value of T_{fi} determines the transition probability from the initial state i to the final state f:

$$\overline{w_{fi}} = |T_{fi}|^2 = (2\pi)^4 \delta^4(p_f - p_i)(2\pi)^4 \delta(0) |M|^2.$$

In order to calculate $\overline{w_{fi}}$, introduce a four-dimensional normalization volume VT, which of course will be eliminated from the final result. As follows from the definition of δ^4, for $V \to \infty$ and $T \to \infty$

$$(2\pi)^4 \delta^4(0) = VT.$$

In order to calculate the transition probability into a group of states instead of a single final state f, we must multiply $\overline{w_{fi}}$ by the element of the phase-space volume $d\overline{\Phi}$ which is written as

$$d\overline{\Phi} = \prod_{l=1}^{n} \frac{d\mathbf{k}_l V}{(2\pi)^3},$$

where n is the number of particles in the final state, and $\mathbf{k}_l$ is the 3-momentum of the lth particle.

Now we must take care of the correct normalization of the expression for the transition probability. Wave functions of particles will be normalized in such a manner that each unit volume contains $2E$ particles, where E is the energy of a particle. It is clear that this normalization corresponds, in the case of scalar particles, to the wave function $\varphi = \exp(-ikx)$. Indeed, in this case the particle density is

$$i\left(\varphi^* \frac{\partial \varphi}{\partial t} - \varphi \frac{\partial \varphi^*}{\partial t}\right) = 2E.$$

The normalized probability is obtained by dividing $\overline{w_{fi}}$ by N:

$$N = \prod_{l=1}^{n} (2E_l V) \prod_{i=1}^{k} (2E_i V),$$

where k is the number of particles in the initial state. A decay corresponds to $k = 1$, and a collision of two particles to $k = 2$.

As a result, the following expression is obtained for the normalized probability transition per unit time:

$$dw_{fi} = \frac{\overline{w}_{fi}}{T} \frac{d\overline{\Phi}}{N} = \frac{V|M|^2}{\prod_{i=1}^{k}(2E_i V)} d\Phi,$$

where

$$d\Phi = (2\pi)^4 \delta^4(p_f - p_i) \prod_{l=1}^{n} \frac{dk_l}{(2\pi)^3 2E_l}.$$

We obtain for particle decay rate ($k = 1$):

$$d\Gamma = \frac{1}{2E_a} |M|^2 d\Phi,$$

where E_a is the energy of the decaying particle. A collision of two particles ($k = 2$) is usually characterized by the cross section defined by

$$dw_{fi} = d\sigma \cdot j,$$

where j is the flux density. The flux density in the laboratory reference frame where particle a is at rest and particle b impinges on it at a velocity v_b, is

$$j = \frac{v_b}{V}.$$

As a result, the cross section is

$$d\sigma = \frac{dw_{fi}}{j} = \frac{1}{2m_a 2E_b v_b} |M|^2 d\Phi.$$

The quantity $I = m_a E_b v_b = m_a |p_b|$ can be written in an invariant form:

$$I = \sqrt{(p_a p_b)^2 - p_a^2 p_b^2}$$

and finally we obtain:

$$d\sigma = \frac{1}{4\sqrt{(p_a p_b)^2 - m_a^2 m_b^2}} |M|^2 d\Phi.$$

29.4.3. Particles with non-zero spin

So far we have been discussing the case of zero-spin particles. The formulas derived above are readily generalized to the case of particles with arbitrary spin. The most typical situation is the calculation of Γ and σ when the polarization states of the initial particles are not fixed, and those of the final particles are not measured. In this case

$$d\Gamma = \frac{1}{2J_a + 1} \frac{1}{2E_a} \overline{|M|^2} \, d\Phi,$$

where J_a is the spin of the decaying particle, and

$$d\sigma = \frac{1}{(2J_a + 1)(2J_b + 1)} \frac{1}{4\sqrt{(p_a p_b)^2 - m_a^2 m_b^2}} \overline{|M|^2} \, d\Phi,$$

where J_a and J_b are spins of colliding particles. The bar over $|M|^2$ denotes summation over spin states of both the initial and final particles. The factors $1/(2J_a + 1)$ and $1/(2J_a + 1)(2J_b + 1)$ take into account that we actually need to carry out averaging over polarization states of the initial particles and not the summation.

Summation over polarization states of particles with spin $\frac{1}{2}$ is easily realized by means of the relativistic-invariant density matrix:

$$\sum_s u_i(s) \bar{u}^k(s) = (\hat{p} + m)_i^k,$$

where p is the 4-momentum of the particle, and m is its mass. If the spin state, s, of the particle is fixed, then

$$u(s)\bar{u}(s) = \tfrac{1}{2}(\hat{p} + m)(1 - \gamma_5 \hat{s})$$

(and $v(s)\bar{v}(s) = \tfrac{1}{2}(\hat{p} - m)(1 - \gamma_5 \hat{s})$ for an antiparticle with 4-momentum p), where

$$s^\mu = \begin{cases} s^0 = \mathbf{p} \cdot \boldsymbol{\xi}/m \\ \mathbf{s} = \boldsymbol{\xi} + (\mathbf{p} \cdot \boldsymbol{\xi})\mathbf{p}/m(m + E) \end{cases}$$

and $\boldsymbol{\xi}$ is a unit vector in the direction of polarization of the particle in its rest frame. It can easily be shown that

$$s^2 = -1, \qquad sp = 0.$$

The relativistic-invariant density matrix, summed over spin states of a massive spin-1 particle, has the form:

$$\sum_s \varphi_\mu(s) \varphi_\nu^*(s) = -\left(g_{\mu\nu} - \frac{p_\mu p_\nu}{m^2}\right),$$

while for the photon

$$\sum_s e_\mu(s) e_\nu^*(s) = -g_{\mu\nu}.$$

CHAPTER 30

Tables of experimental data*

CONTENTS

30.1. Physical and numerical constants and parameters 330
    30.1.1. Physical and numerical constants 330
    30.1.2. Electroweak interaction parameters 331
30.2. Tables of particle properties 332
    30.2.1. Stable particle table 332
    30.2.2. Addendum to Stable particle table 339
    30.2.3. Notes to Stable particle table 341
    30.2.4. Additional notes to Stable particle table 342
30.3. Weak decays and $\Delta I = \frac{1}{2}$ rule 343
    30.3.1. Test of $\Delta I = \frac{1}{2}$ rule for K-decays...343
    30.3.2. Tests of $\Delta I = \frac{1}{2}$ rule for hyperon decays.....................................346
30.4. CP and CPT invariances 350
    30.4.1. Violation of CP invariance in K^0 decays 350
    30.4.2. Electric dipole moment of the neutron 350
    30.4.3. Search for T-odd muon polarization in $K^0_{\mu 3}$ decay 350
    30.4.4. Test of CPT invariance 351
30.5. Conservation of leptonic numbers 352
    30.5.1. Neutrino oscillations 352
    30.5.2. $\mu \leftrightarrow e$ and $\tau \to \mu(e)$ processes 353
30.6. Selection rules for the weak current 353
    30.6.1. Absence of semileptonic decays with $\Delta S \neq \Delta Q$ 353
    30.6.2. Suppression of hadronic decays with $\Delta S > 1$ or $\Delta C = -\Delta S$ 354
    30.6.3. Absence of decays with $\Delta Q = 0$ for $\Delta S \neq 0$ 354
    30.6.4. Absence of processes with $\Delta Q = 0$ for $\Delta S \neq 0$ or $\Delta C \neq 0$ 355

Sections 30.1.1, 30.2.1, 2, 3, and 30.3 are reproduced without modifications from the review [1] (referred to below as PDG); some important new results are added in section 30.2.4 (see also section 30.6.3). Sections 30.4.1 and 30.4.4 are based on the Data Card Listings of PDG and the table of ref. [2], respectively. The upper bounds of the physical values listed in tables of this chapter are given at 90% confidence level. The data cited from PDG are given without references throughout the chapter.

References
[1] Particle Data Group, Rev. Mod. Phys. 52 (1980) No. 2, Part II
[2] H. Poth, Phys. Lett. 77B (1978) 321

*This chapter was complied by I. S. Tsukerman. The author expresses his thanks to the Particle Data Group (N. Barash-Schmidt, C. Bricman, R. L. Crawford, C. Dionisi, C. P. Horne, R. L. Kelly, M. J. Losty, M. Mazzucato, L. Montanet, A. Rittenberg, M. Roos, T. Shimada, T. G. Trippe, C. G. Wohl, and G. P. Yost) and to Reviews of Modern Physics for kind permission to reprint from Review of Particle Properties the material contained in sections 30.1.1, 30.2.1, 30.2.2, 30.2.3, and 30.3.

30.1. Physical and numerical constants and parameters

30.1.1. Physical and numerical constants*

PHYSICAL CONSTANTS

		Uncert. (ppm)
N_A	$= 6.022\,045(31) \times 10^{23}$ mole^{-1}	5.1
V_m	$= 22413.83(70)$ cm^3 mole^{-1} = molar volume of ideal gas at STP	31
c	$= 2.997\,924\,58(1.2) \times 10^{10}$ cm sec^{-1}	0.004
e	$= 4.803\,242(14) \times 10^{-10}$ esu $= 1.602\,189\,2(46) \times 10^{-19}$ coulomb	2.9; 2.9
1 MeV	$= 1.602\,189\,2(46) \times 10^{-6}$ erg	2.9
$\hbar = h/2\pi$	$= 6.582\,173(17) \times 10^{-22}$ MeV sec $= 1.054\,588\,7(57) \times 10^{-27}$ erg sec	2.6; 5.4
$\hbar c$	$= 1.973\,285\,8(51) \times 10^{-11}$ MeV cm $= 197.32858(51)$ MeV fermi	2.6; 2.6
$(\hbar c)^2$	$= 0.389\,385\,7(20)$ GeV^2 mb	5.2
α	$= e^2/\hbar c = 1/137.03604(11)$	0.82
$k_{\text{Boltzmann}}$	$= 1.380\,662(44) \times 10^{-16}$ erg °K^{-1}	32
	$= 8.61735(28) \times 10^{-11}$ MeV °K^{-1} = 1 eV/11604.50(36) °K	32; 31
$\sigma_{\text{Stef. Boltz.}}$	$= 5.67032(71) \times 10^{-5}$ erg sec^{-1} cm^{-2} °K^{-4}	125
	$= 3.53911(44) \times 10^{7}$ eV sec^{-1} cm^{-2} °K^{-4}	125
m_e	$= 0.511\,003\,4(14)$ MeV $= 9.109\,534(47) \times 10^{-28}$ g	2.8; 5.1
m_p	$= 938.2796(27)$ MeV $= 1836.15152(70)m_e = 6.722\,795(61)m_{\pi\pm}$	2.8; 0.38; 9.0
	$= 1.007\,276\,470(11)$ amu	0.011
1 amu	$= 1/12\,m_{C^{12}} = 931.5016(26)$ MeV	2.8
m_d	$= 1875.6280(53)$ MeV	2.8
r_e	$= e^2/m_e c^2 = 2.817\,938\,0(70)$ fermi (1 fermi $= 10^{-13}$ cm)	2.5
$\lambda_e \equiv \hbar/m_e c$	$= r_e \alpha^{-1} = 3.861\,590\,5(64) \times 10^{-11}$ cm	1.6
$a_{\infty \text{Bohr}}$	$= \hbar^2/m_e e^2 = r_e \alpha^{-2} = 0.529\,177\,06(44)$ Å (1 Å $= 10^{-8}$ cm)	0.82
σ_{Thomson}	$= (8/3)\pi r_e^2 = 0.665\,244\,8(33)$ barn (1 barn $= 10^{-24}$ cm^2)	4.9
μ_{Bohr}	$= e\hbar/2m_e c = 0.578\,837\,85(95) \times 10^{-14}$ MeV gauss^{-1}	1.6
μ_N	$= e\hbar/2m_p c = 3.152\,451\,5(53) \times 10^{-18}$ MeV gauss^{-1}	1.7
μ_p/μ_{Bohr}	$= 0.001\,521\,032\,209(16)$	0.011
$\tfrac{1}{2}\omega^e_{\text{cyclotron}}$	$= e/2m_e c = 8.794\,024(25) \times 10^{6}$ rad sec^{-1} gauss^{-1}	2.8
$\tfrac{1}{2}\omega^p_{\text{cyclotron}}$	$= e/2m_p c = 4.789\,378(14) \times 10^{3}$ rad sec^{-1} gauss^{-1}	2.8

Hydrogen-like atom (nonrelativistic, μ = reduced mass):

$$\left(\frac{v}{c}\right)_{\text{rms}} = \frac{z\alpha}{n}\,; \quad E_n = \frac{\mu}{2}v^2 = \frac{\mu}{2}\left(\frac{cz\alpha}{n}\right)^2\,; \quad a_n = \frac{n^2 \hbar}{\mu z c \alpha}$$

$R_\infty = m_e e^4/2\hbar^2 = m_e c^2 \alpha^2/2 = 13.605\,804(36)$ eV (Rydberg)		2.6
$\quad\quad = m_e c \alpha^2/2h = 109\,737.3177(83)$ cm^{-1}		0.075

$pc = 0.3\,H\rho$ (MeV, kilogauss, cm)

1 year (sidereal) $= 365.256$ days $= 3.1558 \times 10^{7}$ sec ($\approx \pi \times 10^{7}$ sec)
density of dry air $= 1.204$ mg cm^{-3} (at 20°C, 760 mm)
acceleration by gravity $= 980.62$ cm sec^{-2} (sea level, 45°)
gravitational constant $= 6.6720(41) \times 10^{-8}$ cm^3 g^{-1} sec^{-2} 615
1 calorie (thermochemical) $= 4.184$ joules
1 atmosphere $= 1.01325$ bar (1 bar $= 10^{6}$ dynes cm^{-2})
1 eV per particle $= 11604.50(36)$ °K (from $E = kT$) 31

*Reprinted from PDG

NUMERICAL CONSTANTS

$\pi = 3.141\,592\,7$	1 rad = 57.295 779 5 deg	$\sqrt{\pi} = 1.772\,453\,85$
$e = 2.718\,281\,8$	$1/e = 0.367\,879\,4$	$\sqrt{2} = 1.414\,213\,6$
$\ln 2 = 0.693\,147\,2$	$\ln 10 = 2.302\,585\,1$	$\sqrt{3} = 1.732\,050\,8$
$\log_{10} 2 = 0.301\,030\,0$	$\log_{10} e = 0.434\,294\,5$	$\sqrt{10} = 3.162\,277\,7$

Notes to Physical Constants

Revised April 1980 by Barry N. Taylor. Originally prepared by Stanley J. Brodsky, based mainly on the "1973 Least-Squares Adjustment of the Fundamental Constants," by E. R. Cohen and B. N. Taylor, J. Phys. Chem. Ref. Data 2, 663 (1973). The figures in parentheses correspond to the one-standard-deviation uncertainty in the last digits of the main number. The equivalent uncertainty in parts per million (ppm) is given in the last column. Note that the uncertainties of the output values of a least-squares adjustment are in general correlated, and the general law of error propagation must be used in calculating additional quantities.

The set of constants resulting from the 1973 adjustment of Cohen and Taylor has been recommended for international use by CODATA (Committee on Data for Science and Technology), and is the most up-to-date, generally accepted set currently available. However, since the publication of the 1973 adjustment, a number of new experiments have been completed, yielding improved values for some of the consants: $N_A = 6.022\,097\,8(63) \times 10^{23}$ mole^{-1} (1.04 ppm); $\alpha^{-1} = 137.035\,963(15)$ (0.11 ppm) [obtained using the Josephson effect]; and $R_\infty = 109\,737.314\,76(32)$ cm^{-1} (0.003 ppm). But it must be realized that, since the output values of a least-squares adjustment are related in a complex way and a change in the measured value of one constant usually leads to corresponding changes in the adjusted values of others, one must be cautious in carrying out calculations using both the output values from the 1973 adjustment and the results of more recent experiments. A new adjustment is planned for completion by early 1982.

30.1.2. Electroweak interaction parameters

$$G = G_\mu = (1.16632 \pm 0.00004) \cdot 10^{-5} \text{ GeV}^{-2}$$

(without higher order electroweak corrections (see review [1]));

$$|\sin \theta_C| = |\sin \theta_1 \cos \theta_3| = 0.219 \pm 0.011,$$
$$|\cos \theta_1| = 0.9737 \pm 0.0025;$$
$$\sin^2 \theta_W = 0.230 \pm 0.009, \qquad \rho = 1.010 \pm 0.098 \ [2],$$
$$\sin^2 \theta_W = 0.238 \pm 0.011, \qquad \rho = 1 \text{ (fixed) } [3].$$

Phenomenological neutral current couplings [2]:

g_L^u	g_L^d	g_R^u	g_R^d	g_V^e	g_A^e
0.351	−0.415	−0.179	−0.010	0.043	−0.545
±.037	±.055	±.032	±.046	±.066	±.045

References

[1] M. M. Nagels et al., Nucl. Phys. B147 (1979) 189
[2] K. Winter, Proc. Int. Symp. on Lepton and Photon Interactions at High Energies, Batavia, 1979, p. 258
[3] I. Liede, M. Roos, Nucl. Phys. B167 (1980) 397

30.2. Tables of particle properties
30.2.1. Stable Particle Table

Reprinted from PDG. For additional parameters, see section 30.2.2.
Quantities in italics have changed by more than one (old) standard deviation since April 1978.

Particle	$I^G(J^P)C_n$ [a]	Mass (MeV) Mass^2 (GeV)^2	Mean life (sec) $c\tau$ (cm)	Partial decay mode Mode	Fraction [b]	p or p_{max} [c] (MeV/c)
γ	$0,1(1^-)-$	$0(<6\times10^{-22})$	—	stable		
ν_e	$J=\frac{1}{2}$	$0(<0.00006)$	stable ($>3\times10^8\, m_{\nu_e}$(MeV))			
e	$J=\frac{1}{2}$	0.5110034 $\pm.0000014$	stable ($>5\times10^{21}$y)			
ν_μ	$J=\frac{1}{2}$	$0(<0.57)$	stable ($>2.6\times10^4\, m_{\nu_\mu}$(MeV))			
μ	$J=\frac{1}{2}$	105.65946 $\pm.00024$ $m^2=0.01116392$ $m_\mu-m_{\pi^\pm}=-33.9074$ $\pm.0012$	2.197120×10^{-6} $\pm.000077$ $c\tau=6.5868\times10^4$	$\mu^- \xrightarrow{d} e^-\bar{\nu}\nu$ $e^-\bar{\nu}\nu\gamma$ $e^-\gamma\gamma$ $e^-e^+e^-$ $e^-\gamma$ $e^-\nu_e\bar{\nu}_\mu$	$(98.6\pm0.4)\%$ $^e(\ 1.4\ )\%$ $(<4\ )\times10^{-6}$ $(<1.9\ )\times10^{-9}$ $(<1.9\ )\times10^{-10}$ $(<25\ )\%$	53 53 53 53 53 53
τ	$J=\frac{1}{2}$ [f]	1784 ±4 $m^2=3.18$	$<2.3\times10^{-12}$ $c\tau<0.07$	$\tau^- \xrightarrow{d} \mu^-\bar{\nu}\nu$ $e^-\bar{\nu}\nu$ hadron$^-$ neutrals $\dagger[\pi^-\nu]$ $\dagger[\rho^-\nu]$ $\dagger[K^-$ neutrals$]$	$(17.9\pm1.5)\%$ $(17.0\pm1.1)\%$ $(33\pm10)\%$ $(8.2\pm2.6)\%]$ $(22\pm4)\%]$ $(\text{small})]$	889 892 887 723

Tables of particle properties 333

			3(hadron$^\pm$) neutrals	(35 ±11)%	715
			†[$\pi^-\rho^0\nu$	(4.2 ± 1.3)%]	864
			†[$\pi^-\pi^-\pi^+\nu$ (incl. $\pi\rho\nu$)	(7 ± 5)%]	864
			†[$\pi^-\pi^-\pi^+(\geq 0\pi^0)\nu$	(18 ± 7)%]	
			($\geq$ 3chgd.) neutrals	(32 ± 5)%	
			†[e$^-$ chgd. parts.		
			+μ^- chgd. parts.	(<4)%]	
$\pi^\pm$	$1^-(0^-)$	139.5669	$\pi^+ \longrightarrow$		
		±.0012	$\mu^+\nu$	100 %	30
			e$^+\nu$	(1.267 ± 0.023) × 10^{-4}	70
		$m^2 = 0.0194789$	$\mu^+\nu\gamma$	^{e}(1.24 ± 0.25) × 10^{-4}	30
		$c\tau = 780.4$	e$^+\nu\pi^0$	(1.02 ± 0.07) × 10^{-8}	5
		$(\tau^+ - \tau^-)/\bar\tau =$	e$^+\nu\gamma$	^{e}(5.6 ± 0.7) × 10^{-8}	70
		(0.05 ± 0.07)%	e$^+\nu e^+ e^-$	(<5) × 10^{-9}	70
		(test of CPT)			
π^0	$1^-(0^-)+$	134.9626	$\gamma\gamma$	(98.85 ± 0.05)%	67
		±.0039	γe$^+$e$^-$	(1.15 ± 0.05)%	67
		$m^2 = 0.0182149$	$\gamma\gamma\gamma$	(<1.5) × 10^{-6}	67
		$m_{\pi^\pm} - m_{\pi^0} = 4.6043$	e$^+$e$^-$e$^+$e$^-$	^{g}(3.32) × 10^{-5}	67
		±.0037	$\gamma\gamma\gamma\gamma$	(<4) × 10^{-5}	67
			e$^+$e$^-$	(2.2$^{+2.4}_{-1.1}$) × 10^{-7}	67
η	$0^+(0^-)+$	548.8	$\gamma\gamma$	(38.0 ± 1.0)% S = 1.2*	274
		±0.6	$\pi^0\gamma\gamma$	^{h}(3.1 ± 1.1)% S = 1.2*	258
		S = 1.4*	$3\pi^0$	(29.9 ± 1.1)% S = 1.1*	180
		$m^2 = 0.3012$	$\pi^+\pi^-\pi^0$	(23.6 ± 0.6)% S = 1.1*	175
		$\Gamma = (0.85 \pm 0.12)$ keV	$\pi^+\pi^-\gamma$	(4.89 ± 0.13)% S = 1.1*	236
		Neutral decays	e$^+$e$^-\gamma$	(0.50 ± 0.12)%	274
		(71.0 ± 0.7)%	π^0e$^+$e$^-$	(<4) × 10^{-5}	258
		S = 1.1*	$\pi^+\pi^-$	(<0.15)%	236
			e$^+$e$^-\pi^+\pi^-$	(0.1 ± 0.1)%	236
		Charged decays	$\pi^+\pi^-\pi^0\gamma$	(<6)%	175
		(29.0 ± 0.7)%	$\pi^+\pi^-\gamma\gamma$	(<0.2)%	236
		S = 1.1*	$\mu^+\mu^-$	(2.2 ± 0.8) × 10^{-5}	253
			$\mu^+\mu^-\gamma$	(1.5 ± 0.8) × 10^{-4}	253
			$\mu^+\mu^-\pi^0$	(<5) × 10^{-4}	211
			e$^+$e$^-$	(<3) × 10^{-4}	274

334   30. Tables of experimental data

Particle	$I^G(J^P)C_n{}^a$	Mass (MeV) Mass^2 (GeV)^2	Mean life (sec) $c\tau$ (cm)	Partial decay mode		p or $p_{max}{}^c$ (MeV/c)
				Mode	Fraction^b	
$K^\pm$	$\tfrac{1}{2}(0^-)$	493.669 ± 0.015 $m^2 = 0.24371$	1.2371×10^{-8} $\pm .0026$ S = 1.9* $c\tau = 370.9$ $(\tau^+ - \tau^-)/\bar\tau =$ $(.11 \pm .09)\%$ (test of CPT) S = 1.2*	$K^+ \xrightarrow{d}$		
				$\mu^+\nu$	$(63.50 \pm 0.16)\%$	236
				$\pi^+\pi^0$	$(21.16 \pm 0.15)\%$	205
				$\pi^+\pi^+\pi^-$	$(5.59 \pm 0.03)\%$ S = 1.1*	125
				$\pi^+\pi^0\pi^0$	$(1.73 \pm 0.05)\%$ S = 1.3*	133
				$\mu^+\nu\pi^0$	$(3.20 \pm 0.09)\%$ S = 1.7*	215
				$e^+\nu\pi^0$	$(4.82 \pm 0.05)\%$ S = 1.1*	228
		$m_{K^\pm} - m_{K^0} = -4.01$ ± 0.13 S = 1.1*		$\mu^+\nu\gamma$	$^e(\ 5.8 \pm 3.5\ ) \times 10^{-3}$	236
				$e^+\nu\pi^0\pi^0$	$(\ 1.8^{+2.4}_{-0.6}\ ) \times 10^{-5}$	207
				$e^+\nu\pi^+\pi^-$	$(\ 3.90 \pm 0.15\ ) \times 10^{-5}$	203
				$e^+\bar\nu\pi^+\pi^-$	$(<5\ ) \times 10^{-7}$	203
				$\mu^+\nu\pi^+\pi^-$	$(\ 0.9 \pm 0.4\ ) \times 10^{-5}$	151
				$\mu^-\bar\nu\pi^+\pi^+$	$(<3.0\ ) \times 10^{-6}$	151
				$e^+\nu$	$(\ 1.54 \pm 0.09) \times 10^{-5}$	247
				$e^+\nu\gamma\,(SD+)^i$	$(\ 1.52 \pm 0.23\ ) \times 10^{-5}$	247
				$e^+\nu\gamma\,(SD-)^i$	$(<1.0\ ) \times 10^{-4}$	247
				$\pi^+\pi^0\gamma$	$^{j,e}(\ 2.75 \pm 0.16) \times 10^{-4}$	205
				$\pi^+\pi^+\pi^-\gamma$	$(\ 1.0 \pm 0.4\ ) \times 10^{-4}$	125
				$\mu^+\nu\pi^0\gamma$	$^e(<6\ ) \times 10^{-5}$	215
				$e^+\nu\pi^0\gamma$	$^e(\ 3.7 \pm 1.4\ ) \times 10^{-4}$	228
				$e^+e^-\pi^+$	$(\ 2.6 \pm 0.5\ ) \times 10^{-7}$	227
				$e^+e^+\pi^-$	$(<1\ ) \times 10^{-8}$	227
				$\mu^+\mu^-\pi^+$	$(<2.4\ ) \times 10^{-6}$	172
				$\pi^+\gamma\gamma$	$(<3.5\ ) \times 10^{-5}$	227
				$\pi^+\gamma\gamma\gamma$	$^e(<3.0\ ) \times 10^{-4}$	227
				$\pi^+\nu\bar\nu$	$(<0.6\ ) \times 10^{-6}$	227
				$\pi^+\gamma$	$(<4\ ) \times 10^{-6}$	227
				$e^+\mu^\pm\pi^\mp$	$(<7\ ) \times 10^{-9}$	214
				$e^+\mu^-\pi^+$	$(<5\ ) \times 10^{-9}$	214
				$e^+\nu\nu\bar\nu$	$(<6\ ) \times 10^{-5}$	247
				$\mu^+\nu\nu\bar\nu$	$(<6\ ) \times 10^{-6}$	236
				$\mu^+\nu e^+ e^-$	$(\ 11 \pm 3\ ) \times 10^{-7}$	236
				$\mu^-\nu e^+ e^+$	$(<2.0\ ) \times 10^{-8}$	236
				$e^+\nu e^+ e^-$	$(\ 2^{+2}_{-1}\ ) \times 10^{-7}$	247

Tables of particle properties

K^0 $\overline{K}^0$	$\frac{1}{2}(0^-)$	497.67 ± 0.13 S = 1.1* $m^2 = 0.24768$	50% K_{Short}, 50% K_{Long}		

K_S^0	$\frac{1}{2}(0^-)$	0.8923×10^{-10} $\pm .0022$ $c\tau = 2.675$	$\pi^+\pi^-$ $\pi^0\pi^0$ $\mu^+\mu^-$ e^+e^- $\pi^+\pi^-\gamma$ $\gamma\gamma$	(68.61 ±0.24)% (31.39 ±0.24)% S = 1.1* (<3.2) × 10^{-7} (<3.4) × 10^{-4} ^{e}(1.85 ±0.10) × 10^{-3} (<0.4) × 10^{-3}	206 209 225 249 206 249

K_L^0	$\frac{1}{2}(0^-)$	5.183×10^{-8} ± 0.40 $c\tau = 1554$ $m_{K_L} - m_{K_S} = 0.5349 \times 10^{10} \hbar\,sec^{-1}$ ± 0.0022	$\pi^0\pi^0\pi^0$ $\pi^+\pi^-\pi^0$ $\pi^\pm\mu^\mp\nu$ $\pi^\pm e^\mp\nu$ (incl. $\pi e\nu\gamma$) $\pi e\nu\gamma$ $\pi^+\pi^-$ $\pi^0\pi^0$ $\pi^+\pi^-\gamma$ $\pi^0\gamma\gamma$ $\gamma\gamma$ $e\mu$ $\mu^+\mu^-$ $\mu^+\mu^-\gamma$ $\mu^+\mu^-\pi^0$ e^+e^- $e^+e^-\gamma$ $\pi^+\pi^-e^+e^-$ $\pi^0\pi^\pm e^\mp\nu$	(21.5 ±0.7)% S = 1.3* (12.39 ±0.18)% S = 1.2* (27.0 ±0.5)% S = 1.1* (38.8 ±0.5)% S = 1.1* ^{e}(1.3 ±0.8)% ^{k}(0.203 ± 0.005)% ^{k}(0.094 ± 0.018)% S = 1.5* ^{e}(6.0 ±2.0) × 10^{-5} (<2.4) × 10^{-4} (4.9 ±0.5) × 10^{-4} (<2.0) × 10^{-9} (9.1 ±1.9) × 10^{-9} (<7.8) × 10^{-6} (<5.7) × 10^{-5} (<2.0) × 10^{-9} (<2.8) × 10^{-5} (<8.8) × 10^{-6} (<2.2) × 10^{-3}	139 133 216 229 229 206 209 206 231 249 238 225 225 177 249 249 206 207

336 — 30. Tables of experimental data

Particle	$I^G(J^P)C_n^a$	Mass (MeV) Mass^2 (GeV)^2	Mean life (sec) $c\tau$ (cm)	Partial decay mode Mode	Fraction^b	p or p_{max}^c (MeV/c)
$D^\pm$	$\frac{1}{2}(0^-)^f$	1868.3^f ±0.9	$(2.5^{+3.5}_{-1.5}) \times 10^{-13}$	$D^\pm \xrightarrow{d}$		
				$K^\mp$ anything	(10 ±7)%	845
				$^\dagger[K^-\pi^+\pi^+]$ (incl. $K^*\pi$)	(3.9 ± 1.0)%]	456
		$m^2 = 3.491$	$c\tau = 0.007$	$^\dagger[\bar{K}^*(892)^0\pi^+]$	seen]	743
		$m_{D^\pm} - m_{D^0} = 5.0$		$^\dagger[K^-K^+\pi^+]$	(<0.6)%	
		±0.8		$\bar{K}^0$ anything	(39 ±29)%	862
				$^\dagger[\bar{K}^0\pi^+]$	(1.5 ± 0.6)%	862
				$e^\pm$ anything	^{ml} 8.2 ± 1.2)%	908
				$\pi^+\pi^+\pi^-$	(<0.31)%	
				K^+ anything	(6 ±6)%	
				$K^+\pi^+\pi^-$	(<0.20)%	845
$D^0 \atop \bar{D}^0$	$\frac{1}{2}(0^-)^f$	1863.1^f ±0.9	$(3.5^{+3.5}_{-1.7}) \times 10^{-13}$	$D^0 \xrightarrow{d}$		
				$K^\mp$ anything	(35 ±10)%	860
		$m^2 = 3.471$	$c\tau = 0.01$	$^\dagger[K^-\pi^+]$	(1.8 ± 0.5)%]	843
				$^\dagger[K^-\pi^+\pi^0]$	(12 ±6)%]	812
		$\frac{\Gamma(D^0 \to \bar{D}^0 \to K^+\pi^-)}{\Gamma(D^0 \to K\pi)} < 0.16$		$^\dagger[K^-\pi^+\pi^+\pi^-]$	(3.5 ± 0.9)%]	
				$\bar{K}^0$ anything + K^0 any	(57 ±26)%	859
				$^\dagger[\bar{K}^0\pi^0 + K^0\pi^0]$	(<6)%	
				$^\dagger[\bar{K}^0\pi^+\pi^- + K^0\pi^+\pi^-]$	(4.4 ± 1.1)%	841
				$e^\pm$ anything	^{m} 8.2 ± 1.2)%	
				$\pi^+\pi^-$	(5.9 ± 3.2) × 10^{-4}	921
				K^+K^-	(2.0 ± 0.8) × 10^{-3}	790
p	$\frac{1}{2}(\frac{1}{2}^+)$	938.2796 ±0.0027 $m^2 = 0.880369$	stable (>10^{30}y)	stable	$\|q_p\| - \|q_e\| < 10^{-21}\|q_e\|^n$	1
n	$\frac{1}{2}(\frac{1}{2}^+)$	939.5731 ±0.0027 $m^2 = 0.882798$ $m_p - m_n = -1.29343$	917 ± 14 $c\tau = 2.75 \times 10^{13}$	$pe^-\bar{\nu}$ $p\nu\bar{\nu}$ (chg. noncons.)	100% (<3) × 10^{-19}	1

Λ	$0(\tfrac{1}{2}^+)$	1115.60	2.632×10^{-10}	$p\pi^-$	$(64.2 \pm 0.5\ \ )\%$	100
		± 0.05	$\pm .020\ S=1.6*$	$n\pi^0$	$(35.8 \pm 0.5\ \ )\%$	104
		$S=1.2*$	$c\tau = 7.89$	$pe^-\bar\nu$	$(8.07 \pm 0.28) \times 10^{-4}$	163
		$m^2 = 1.2446$		$p\mu^-\bar\nu$	$(1.57 \pm 0.35) \times 10^{-4}$	131
		$m_\Lambda - m_{\Sigma^0} = -76.86$		$p\pi^-\gamma$	$(0.85 \pm 0.14) \times 10^{-3}$	100
		± 0.08				
Σ^+	$(1\tfrac{1}{2}^+)$	1189.36	0.800×10^{-10}	$p\pi^0$	$(51.64 \pm 0.30\ )\%$	189
		± 0.06	$\pm .004$	$n\pi^+$	$(48.36 \pm 0.30\ )\%$	185
		$S = 1.8*$	$c\tau = 2.40$	$p\gamma$	$(1.24 \pm 0.18) \times 10^{-3}\ S = 1.4*$	225
		$m^2 = 1.4146$		$n\pi^+\gamma$	$(0.93 \pm 0.10) \times 10^{-3}$	185
				$\Lambda e^+\nu$	$^e(2.02 \pm 0.47) \times 10^{-5}$	71
		$m_{\Sigma^+} - m_{\Sigma^-} = -7.98$	$\dfrac{\Gamma(\Sigma^+ \to l^+ n\nu)}{\Gamma(\Sigma^- \to l^- n\nu)} < .04$	$\begin{cases}n\mu^+\nu \\ n e^+\nu\end{cases}$	$(<3.0\ \ )\times 10^{-5}$	202
		$\pm .08$			$(<0.5\ \ )\times 10^{-5}$	224
		$S = 1.2*$		pe^+e^-	$(<7\ \ )\times 10^{-6}$	225
Σ^0	$1(\tfrac{1}{2}^+)^p$	1192.46	5.8×10^{-20}	$\Lambda\gamma$	100 $\%$	74
		± 0.08	± 1.3	Λe^+e^-	$^g(5.45\ \ )\times 10^{-3}$	74
		$m^2 = 1.4220$	$c\tau = 1.7 \times 10^{-9}$	$\Lambda\gamma\gamma$	$(<3\ \ )\%$	74
Σ^-	$1(\tfrac{1}{2}^+)$	1197.34	1.482×10^{-10}	$n\pi^-$	100 $\%$	193
		± 0.05	$\pm .011\ S = 1.3*$	$ne^-\bar\nu$	$(1.08 \pm 0.04) \times 10^{-3}$	230
		$m^2 = 1.4336$	$c\tau = 4.44$	$n\mu^-\bar\nu$	$(0.45 \pm 0.04) \times 10^{-3}$	210
				$\Lambda e^-\bar\nu$	$(0.61 \pm 0.05) \times 10^{-4}$	79
		$m_{\Sigma^0} - m_{\Sigma^-} = -4.88$		$n\pi^-\gamma$	$^e(4.6 \pm 0.6\ \ ) \times 10^{-4}$	193
		$\pm .06$				

338   30. Tables of experimental data

Particle	$I^G(J^P)C_n$ [a]	Mass (MeV) Mass^2 (GeV)^2	Mean life (sec) $c\tau$ (cm)	Partial decay mode Mode	Fraction[b]	p or p_{max}^c (MeV/c)
Ξ^0	$\frac{1}{2}(\frac{1}{2}^+)$ [q]	1314.9 ±0.6 $m^2 = 1.7290$	2.90×10^{-10} ±.10 $c\tau = 8.69$	$\Lambda\pi^0$	100 %	135
				$\Lambda\gamma$	(0.5 ± 0.5)%	184
				$\Sigma^0\gamma$	(<7)%	117
				$p\pi^-$	(<3.6)$\times 10^{-5}$	299
				$pe^-\bar{\nu}$	(<1.3)$\times 10^{-3}$	323
				$\Sigma^+ e^- \bar{\nu}$	(<1.1)$\times 10^{-3}$	120
				$\Sigma^- e^+ \nu$	(<0.9)$\times 10^{-3}$	112
				$\Sigma^+ \mu^- \bar{\nu}$	(<1.1)$\times 10^{-3}$	64
				$\Sigma^- \mu^+ \nu$	(<0.9)$\times 10^{-3}$	49
	$m_{\Xi^0} - m_{\Xi^-} = -6.4$ ±.6			$p\mu^- \bar{\nu}$	(<1.3)$\times 10^{-3}$	309
Ξ^-	$\frac{1}{2}(\frac{1}{2}^+)$ [q]	1321.32 ±0.13 $m^2 = 1.7459$	1.641×10^{-10} ±.016 $c\tau = 4.92$	$\Lambda\pi^-$	100 %	139
				$\Lambda e^- \bar{\nu}$	(2.8 ± 1.2)$\times 10^{-4}$	190
				$\Sigma^0 e^- \bar{\nu}$	(<5)$\times 10^{-4}$	123
				$\Lambda\mu^- \bar{\nu}$	(3.1 ± 1.2)$\times 10^{-4}$	163
				$\Sigma^0\mu^- \bar{\nu}$	(<8)$\times 10^{-4}$	70
				$n\pi^-$	(<1.1)$\times 10^{-3}$	303
				$n e^- \bar{\nu}$	(<3.2)$\times 10^{-3}$	327
				$n\mu^- \bar{\nu}$	(<1.5)$\times 10^{-3}$	313
				$\Sigma^-\gamma$	(<1.2)%	118
				$p\pi^-\pi^-$	(<4)$\times 10^{-3}$	223
				$p\pi^- e^- \bar{\nu}$	(<4)$\times 10^{-4}$	304
				$p\pi^- \mu^- \bar{\nu}$	(<4)$\times 10^{-4}$	250
				$\Xi^0 e^- \bar{\nu}$	(<2.3)$\times 10^{-3}$	6
Ω^-	$0(\frac{3}{2}^+)$ [q]	1672.22 ±.31 $m^2 = 2.7963$	0.82×10^{-10} ±.03 $c\tau = 2.5$	ΛK^-	(68.6 ± 1.3)%	211
				$\Xi^0 \pi^-$	(23.4 ± 1.3)%	293
				$\Xi^- \pi^0$	(8.0 ± 0.8)%	290
				$\Xi^0 e^- \bar{\nu}$	(~ 1)%	319
				$\Xi(1530)^0 \pi^-$	(~ 2)$\times 10^{-3}$	15
				$\Lambda\pi^-$	(<1.3)$\times 10^{-3}$	449
				$\Xi^-\gamma$	(<3.1)$\times 10^{-3}$	314
Λ_c^+	$0(\frac{1}{2}^+)$ [r]	2273 ±6 $S = 1.6^*$ $m^2 = 5.17$	$\sim 7 \times 10^{-13}$ $c\tau \sim 0.02$	$\Lambda\pi^+\pi^+\pi^-$	(seen)	798
				$pK^-\pi^+$	(2.2 ± 1.0)%	814
				$pK^*(892)^0$	(seen)	567
				$\Delta(1232)^{++} K^-$	(seen)	700

Tables of particle properties

	Magnetic Moment $\frac{e\hbar}{2m_e c}$			**μ Decay parameters**[s]	
e	1.001 159 652 41 $\pm$.000 000 000 20				
μ	1.001 165 924 $\pm$.000 000 009 $\frac{e\hbar}{2m_\mu c}$		$\rho = 0.752 \pm 0.003$ $\xi = 0.972 \pm 0.013$ $\|g_A/g_V\| = 0.86^{+0.33}_{-0.11}$	$\eta = -0.12 \pm 0.21$ $\delta = 0.755 \pm 0.009$ $\phi = 180° \pm 15°$	$h = 1.000 \pm 0.13$
η		**Left-right asymmetry** $(0.12 \pm .17)\%$ $(0.88 \pm .40)\%$		**Sextant asymmetry** $(0.19 \pm 0.16)\%$	**Quadrant asymmetry** $(-0.17 \pm 0.17)\%$ $\beta = 0.047 \pm 0.062$

	Mode	Partial rate (sec^{-1})		**Slope parameters for K $\to 3\pi$**[t]	
$K^\pm$	$\mu\nu$	$(51.33 \pm 0.17) \times 10^6$	$S = 1.2*$	$K^+ \to \pi^+\pi^+\pi^-$	$g = -0.215 \pm .004$ $S = 1.4*$
	$\pi\pi^0$	$(17.10 \pm 0.13) \times 10^6$	$S = 1.1*$	$K^- \to \pi^-\pi^-\pi^+$	$g = -0.217 \pm .007$ $S = 2.5*$
	$\pi\pi^+\pi^-$	$(4.52 \pm 0.02) \times 10^6$	$S = 1.1*$	$K^\pm \to \pi^0\pi^0\pi^\pm$	$g = 0.607 \pm .030$ $S = 1.3*$
	$\pi\pi^0\pi^0$	$(1.40 \pm 0.04) \times 10^6$	$S = 1.3*$	$K_L^0 \to \pi^+\pi^-\pi^0$	$g = 0.670 \pm .014$ $S = 1.6*$
	$\mu\pi^0\nu$	$(2.58 \pm 0.07) \times 10^6$	$S = 1.7*$		
	$e\pi^0\nu$	$(3.90 \pm 0.04) \times 10^6$	$S = 1.1*$		
K_S^0	$\pi^+\pi^-$	$^k(0.7689 \pm .0033) \times 10^{10}$		$K_{l3}^+ \begin{cases} \lambda_+^e = 0.029 \pm .004 \\ \lambda_+^\mu = 0.026 \pm .008 \quad S = 1.5* \\ \lambda_0^\mu = -0.003 \pm .007 \quad S = 1.5* \end{cases}$	$K_{l3}^0 \begin{cases} \lambda_+^e = 0.0301 \pm .0016 \quad S = 1.2* \\ \lambda_+^\mu = 0.034 \pm .006 \quad S = 2.5* \\ \lambda_0^\mu = 0.020 \pm .007 \quad S = 2.5* \end{cases}$
	$\pi^0\pi^0$	$^k(0.3517 \pm .0029) \times 10^{10}$	$S = 1.1*$	See Data Card Listings for ξ, f_s, and f_t.	
K_L^0	$\pi^0\pi^0\pi^0$	$(4.14 \pm 0.15) \times 10^6$	$S = 1.3*$	**CP violation parameters**[u,k]	
	$\pi^+\pi^-\pi^0$	$(2.39 \pm 0.04) \times 10^6$	$S = 1.2*$	$\|\eta_{+-}\| = (2.274 \pm .022) \times 10^{-3}$	$\|\eta_{00}\| = (2.33 \pm .08) \times 10^{-3}$ $S = 1.1*$
	$\pi\mu\nu$	$(5.21 \pm 0.10) \times 10^6$	$S = 1.1*$	$\phi_{+-} = (44.6 \pm 1.2)°$	$\phi_{00} = (54 \pm 5)°$
	$\pi e\nu$	$(7.49 \pm 0.11) \times 10^6$	$S = 1.1*$	$\|\eta_{+-0}\|^2 < 0.12$ $\|\eta_{000}\|^2 < 0.28$	$\delta = (0.330 \pm .012) \times 10^{-2}$
	$\pi^+\pi^-$	$^k(3.91 \pm 0.10) \times 10^4$	$S = 1.7*$		
	$\pi^0\pi^0$	$^k(1.81 \pm 0.35) \times 10^4$	$S = 1.5*$	$\Delta S = -\Delta Q$ Re $x = 0.009 \pm 0.020$ $S = 1.4*$	Im $x = -0.004 \pm .026$ $S = 1.1*$

	Magnetic moment ($e\hbar/2m_pc$)		Decay parameters[v]					
				Measured		Derived		
			α	ϕ(degree)	γ	Δ(degree)	g_A/g_V	g_V/g_A
p	2.7928456 ±.0000011							
n[w]	−1.91304184 ±.00000088	$pe^-\nu$					−1.254 ± 0.007 $\delta = (180.11 \pm 0.17)°$	
Λ[w]	−0.614 ±.005	$p\pi^-$ $n\pi^0$ $pe\nu$	0.642 ± 0.013 0.646 ± 0.044	(−6.5 ± 3.5)°	0.76	$(7.7^{+4.0}_{-4.1})°$	−0.62 ± 0.05 S = 1.2*	
Σ^+	2.33 ±.13	$p\pi^0$ $n\pi^+$ $p\gamma$	−0.979 ± 0.016 +0.068 ± 0.013 $-1.03^{+0.52}_{-0.42}$	(36 ± 34)° (167 ± 20)° S = 1.1*	0.17 −0.97	(187 ± 6)° $(-72^{+132}_{-11})°$		
Σ^-	−1.41 ±.25	$n\pi^-$ $ne^-\nu$ $\Lambda e^-\nu$	−0.068 ± 0.008	(10 ± 15)°	0.98	$(249^{+12}_{-115})°$	±(0.385 ± 0.070) S = 2.3*	0.10 ± 0.22 S = 1.5*
Ξ^0	−1.20 ±.06	$\Lambda\pi^0$	−0.47 ± 0.05 S = 1.3*	(21 ± 12)°	0.84	$(216^{+13}_{-19})°$		
Ξ^-	−1.85 ±.75	$\Lambda\pi^-$	−0.403 ± 0.017	(2 ± 6)° S = 1.1*	0.92	(185 ± 13)°		
Ω^-		ΛK^-	−0.26 ± 0.33 S = 1.5*					

30.2.3. Notes to Stable Particle Table*

→ Indicates an entry in the Stable Particle Data Card Listings not entered in the Stable Particle Table. This is the case for ν_τ, for the charmed-strange meson $F^\pm$, and for listings of searches for heavy leptons other than $\tau^\pm$, intermediate boson searches, quark searches, magnetic monopole searches, charm searches, and other particle searches.

* S = Scale factor = $\sqrt{\chi^2/(N-1)}$, where $N \approx$ number of experiments. S should be ≈ 1. If $S > 1$, we have enlarged the error of the mean, $\delta \bar{x}$; i.e., $\delta \bar{x} \to S\delta \bar{x}$. This convention is still inadequate, since if $S \gg 1$ the experiments are probably inconsistent, and therefore the real uncertainty is probably even greater than $S\delta x$. See text, and ideograms in Stable Particle Data Card Listings.

† Square brackets indicate a subreaction of the previous (unbracketed) decay mode.

a. The baryon number B, strangeness S, and charm C of the hadrons which appear in the tables are as follows:

Mesons (B = 0)	S	C	Baryons (B = 1)	S	C
π, η	0	0	p, n	0	0
K^+, K^0	+1	0	Λ, Σ	−1	0
$K^-, \overline{K}^0$	−1	0	Ξ	−2	0
D^+, D^0	0	+1	Ω^-	−3	0
$D^-, \overline{D}^0$	0	−1	Λ_c^+	0	+1

b. Quoted upper limits correspond to a 90% confidence level.
c. In decays with more than two bodies, p_{max} is the maximum momentum that any particle can have.
d. For simplicity, decay mode charge states are written for the particle shown. For antiparticle modes all particles must be charge conjugated.
e. See Stable Particle Data Card Listings for energy limits used in this measurement.
f. Quantum numbers shown are favored but not yet established. See Data Card Listings.
g. Theoretical value; see also Stable Particle Data Card Listings.
h. See note in Stable Particle Data Card Listings.
i. Structure-dependent part with positive (SD +) and negative (SD −) photon helicity.
j. The direct emission branching fraction is $(1.56 \pm .35) \times 10^{-5}$.
k. The $K_S^0 \to \pi\pi$ and $K_L^0 \to \pi\pi$ rates (and branching fractions) are from independent fits and do not include results of K_L^0-K_S^0 interference experiments. The $|\eta_{+-}|$ and $|\eta_{00}|$ values given in the addendum are these rates combined with the $|\eta_{+-}|$ and $|\eta_{00}|$ results from interference experiments.
l. Error does not include 0.13% uncertainty in the absolute SPEAR energy calibration. Assumes $m_\psi = 3095$ MeV.
m. This is a weighted average of $D^\pm$ (44%) and D^0 (56%) branching fractions.
n. Limit from neutrality-of-matter experiments. Assumes $|q_n| = |q_p| - |q_e|$.
p. J^P not measured for Σ^0. Assumed same as $\Sigma^\pm$ to allow isotriplet association.
q. P for Ξ and J^P for Ω^- not yet measured. Values shown are SU(3) predictions.
r. J^P for Λ_c^+ not yet measured. Values shown are SU(4) predictions.
s. $|g_A/g_V|$ defined by $g_A^2 = |C_A|^2 + |C'_A|^2$, $g_V^2 = |C_V|^2 + |C'_V|^2$, and $\Sigma \langle \bar{e} | \Gamma_i | \mu \rangle \langle \bar{\nu} | \Gamma_i(C_i + C'_i \gamma_5) | \nu \rangle$; ϕ defined by $\cos\phi = -\text{Re}(C_A^* C'_V + C'_A C_V^*)/g_A g_V$ [for more details, see text Section VI A].

*Reprinted from PDG

t. The definition of the slope parameter of the Dalitz plot is as follows [see also text Section VI B.1]:

$$|M|^2 = 1 + g\left(\frac{s_3 - s_0}{m_{\pi^+}^2}\right)$$

u. The definition for the CP violation parameters is as follows [see also text Section VI B.3]:

$$\eta_{+-} = |\eta_{+-}|e^{i\phi_{+-}} = \frac{A(K_L^0 \to \pi^+\pi^-)}{A(K_S^0 \to \pi^+\pi^-)}, \quad \eta_{00} = |\eta_{00}|e^{i\phi_{00}} = \frac{A(K_L^0 \to \pi^0\pi^0)}{A(K_S^0 \to \pi^0\pi^0)},$$

$$\delta = \frac{\Gamma(K_L^0 \to l^+) - \Gamma(K_L^0 \to l^-)}{\Gamma(K_L^0 \to l^+) + \Gamma(K_L^0 \to l^-)}, \quad |\eta_{+-0}|^2 = \frac{\Gamma(K_S^0 \to \pi^+\pi^-\pi^0)^{CP\,viol.}}{\Gamma(K_L^0 \to \pi^+\pi^-\pi^0)},$$

$$|\eta_{000}|^2 = \frac{\Gamma(K_S^0 \to \pi^0\pi^0\pi^0)^{CP\,viol.}}{\Gamma(K_L^0 \to \pi^0\pi^0\pi^0)}.$$

v. The definition of these quantities is as follows [for more details on sign convention, see text Section VI B]:

$$\alpha = \frac{2|s||p|\cos\Delta}{|s|^2 + |p|^2} \quad \bigg| \quad \beta = \sqrt{1-\alpha^2}\sin\phi \quad \bigg| \quad g_A/g_V \text{ defined by } \langle B_f | \gamma_\lambda(g_V - g_A\gamma_5) | B_i\rangle$$

$$\beta = \frac{-2|s||p|\sin\Delta}{|s|^2 + |p|^2} \quad \bigg| \quad \gamma = \sqrt{1-\alpha^2}\cos\phi \quad \bigg| \quad \delta \text{ defined by } g_A/g_V = |g_A/g_V|e^{i\delta}$$

w. For limits on electric dipole moment of n and Λ, see Data Card Listings.

30.2.4. Additional notes to Stable Particle Table

Some important results of the recent mass and lifetime measurements are given below:

ν_e	$m_{\nu_e} < 35$ eV [1]; 14 eV $\leq m_{\nu_e} \leq 46$ eV (99% c.l.) [2]
$\tau^\pm$	$\tau_\tau^\pm < 1.4 \cdot 10^{-12}$ sec (95% c.l.) [3]
$D^\pm$	$\tau_{D^\pm} = 11.2 \pm 5.1$ [4], $2.5^{+2.2}_{-1.1}$ [5], $10.4^{+3.9}_{-2.9}$ [6], $10.3^{+10.5}_{-4.1}$ [7] ($\times 10^{-13}$ sec)
D^0	$\tau_{D^0} = 3.7 \pm 2.8$ [4], $0.53^{+.57}_{-.25}$ [5], < 2.1 (95% c.l.) [6], $1.00^{+.52}_{-.31}$ [7] ($\times 10^{-13}$ sec)
n	$\tau_n = 877 \pm 8$ [8], 937 ± 18 [9] (sec)
Λ_c^+	$m_{\Lambda_c^+} = 2.285 \pm 0.003$ GeV [10], $\tau_{\Lambda_c^+} = (1.14^{+.90}_{-.44}) \cdot 10^{-13}$ sec [7]

References
[1] V. A. Lubimov, Proc. 18th Int. Conf. High Energy Phys., Tbilisi (1976), Vol. 2, p. B118
[2] V. A. Lubimov et al., Phys. Lett. 94B (1980) 266
[3] R. Brandelick et al., Phys. Lett. 92B (1980) 199
[4] R. H. Schindler, SLAC Report No. 219 (1979)
[5] D. Allasia et al., Nucl. Phys. B176 (1980) 13
[6] W. Bacino et al., Phys. Rev. Lett. 45 (1980) 329
[7] N. Ushida et al., Phys. Rev. Lett. 45 (1980) 1049 and 1053
[8] L. N. Bondarenko et al., Pis'ma ZhETF 28 (1978) 329 [JETP Lett. 28 (1978) 303]
[9] J. Byrne et al., Phys. Lett. 92B (1980) 274
[10] P. Musset, Proc. Neutrino '80 Conf., Erice (1980)

30.3. Weak decays and $\Delta I = \frac{1}{2}$ rule*

30.3.1. Test of $\Delta I = \frac{1}{2}$ rule for K-decays

The quantities of interest for making tests of theoretical predictions regarding the $\Delta I = \frac{1}{2}$ rule for K decay are usually partial decay rates for single channels or special sums of channels. It is not possible to compute the errors on sums, differences, and ratios of partial decay rates from the information given in the Table of Stable Particles because of the presence of off-diagonal terms in the error matrix. For this reason we give some of these quantities in table 1. Throughout this appendix, numbers in italics are used to indicate that a quantity has changed by more than one (old) standard deviation since our previous edition, and S gives the scale factor included in the quoted error because of inconsistencies in the data (see footnote at end of Stable Particle Table for definition of S).

Table 1
(000) and (+ −0) refer to the sign of the pions into which the K_L decays.

$\Gamma_{K^+_{l3}} = \Gamma_{K^+_{e3}} + \Gamma_{K^+_{\mu 3}}$	$= (6.484 \pm 0.089)10^6 \text{ sec}^{-1}$	
$\Gamma_{K^+_{\mu 3}}/\Gamma_{K^+_{e3}}$	$= 0.663 \pm 0.018$	$S = 1.7^*$
$\Gamma_{K^+_\tau}/\Gamma_{K^+_{\tau'}}$	$= 3.226 \pm 0.082$	
$\Gamma_{K^0_{l3}} = \Gamma_{K^0_{e3}} + \Gamma_{K^0_{\mu 3}}$	$= (12.70 \pm 0.15)10^6 \text{ sec}^{-1}$	$S = 1.1^*$
$\Gamma_{K^0_{\mu 3}}/\Gamma_{K^0_{e3}}$	$= 0.695 \pm 0.017$	
$\Gamma_{K^0(000)}/\Gamma_{K^0(+-0)}$	$= 1.733 \pm 0.076$	$S = 1.3^*$

1. *Leptonic decay rates*

The $\Gamma_{K_{l3}}$ rates are useful in testing the leptonic $\Delta I = \frac{1}{2}$ rule in the way suggested by Trilling [1]. The predictions are

$$\Gamma_{K^0_{l3}}/2\Gamma_{K^+_{l3}} = 1.012, \text{ a phase-space factor [2]},$$

and

$$\Gamma_{K^0_{\mu 3}}/\Gamma_{K^0_{e3}} = \Gamma_{K^+_{\mu 3}}/\Gamma_{K^+_{e3}}.$$

From Table 1,

$$\Gamma_{K^0_{l3}}/2\Gamma_{K^+_{l3}} = 0.979 \pm 0.018$$

and

$$\frac{\Gamma_{K^0_{\mu 3}}}{\Gamma_{K^0_{e3}}} \left[\frac{\Gamma_{K^+_{\mu 3}}}{\Gamma_{K^+_{e3}}} \right]^{-1} = 1.048 \pm 0.038.$$

*Reprinted from PDG

These results seem to show a less than 2σ disagreement with the predictions, but the errors should be regarded with caution in view of the internal disagreements in the data. (Note the ideograms in the Data Listings for the charged K meson.)

2. Three-pion decays

We follow here the tests done by Mast et al. [3], based on the general analysis of K decays suggested by Zemach [4]. Both decay rates (Γ) and slopes (g, the energy dependence of the Dalitz plot distributions) are used. The $\Delta I = 1/2$ rule predicts that the following test quantities are all equal to zero:

$$\text{Test } 1 = \tfrac{2}{3} \frac{\Gamma_{K^0(000)}}{\phi_1} \left[\frac{\Gamma_{K^0(+-0)}}{\phi_2} \right]^{-1} - 1,$$

$$\text{Test } 2 = \tfrac{1}{4} \frac{\Gamma_{K_\tau^+}}{\phi_3} \left[\frac{\Gamma_{K_{\tau'}^+}}{\phi_4} \right]^{-1} - 1,$$

$$\text{Test } 3 = \tfrac{1}{2} \frac{\Gamma_{K_\tau^+}}{\phi_3} \left[\frac{\Gamma_{K^0(+-0)}}{\phi_2} \right]^{-1} - 1,$$

$$\text{Test } 4 = \tfrac{1}{2} g_{K_{\tau'}^+} + g_{K_\tau^+},$$

$$\text{Test } 5 = g_{K^0(+-0)} + g_{K_\tau^+} - \tfrac{1}{2} g_{K_{\tau'}^+}.$$

The ϕ_i are phase-space factors which have been calculated as described in Mast et al. [3] by use of a relativistic formulation and the masses and slopes from this edition. The factors labeled UDP are the relative areas of the Dalitz plots, assuming a uniform distribution. The NUDP include the observed slopes (see below). The CNUDP have been calculated by including the final-state Coulomb interaction. The values are:

		Method	
	UDP	NUDP	CNUDP
$\phi_1(000) =$	1.490	1.490	1.444
$\phi_2(+-0) =$	1.221	1.303	1.287
$\phi_3(++-) =$	1.000	1.000	1.000
$\phi_4(+00) =$	1.247	1.173	1.137

For convenience, we repeat the slope parameters tabulated in the Stable Particle Table. They are as follows:

$g_{K_\tau^+}$ $= -0.215 \pm 0.004$ $S = 1.4^*$
$g_{K_\tau^-}$ $= -0.217 \pm 0.007$ $S = 2.5^*$
$\bar{g}_{K_\tau^\pm}$ $= -0.215 \pm 0.003$
$g_{K_{\tau'}^\pm}$ $= 0.607 \pm 0.030$ $S = 1.7^*$
$g_{K^0(+-0)}$ $= 0.670 \pm 0.014$ $S = 1.6^*$

A difference in the τ^+ and τ^- slopes would be an indication of CP violation in this decay. Since no difference is observed at this time, we average the two and use this value in Test 4 and Test 5.

We use the CNUDP factors and the rates and slopes reported in this edition to compute the five test quantities which the $\Delta I = \frac{1}{2}$ rule predicts to be zero. The results are:

Test 1 $= 0.030 \pm 0.045$
Test 2 $= -0.083 \pm 0.023$
Test 3 $= 0.216 \pm 0.020$
Test 4 $= 0.088 \pm 0.016$
Test 5 $= 0.152 \pm 0.021$

The three-pion final state can be in isospin states $I = 1, 2, 3$. Tests 1 and 2 test the existence of isospin $I = 3$ in the final state. Since the rate tests (Tests 1, 2, and 3) could differ from zero by as much as 0.1 owing to the mass differences and the occurrence of big slopes [5], no evidence for $I = 3$ is found. Test 4 is related to the $I = 2$ amplitude in the final state and indicates the presence of $I = 2$. Tests 3 and 5 give information on the $\Delta I = \frac{3}{2}$ part of the $I = 1$ amplitude relative to the $\Delta I = \frac{1}{2}$ part. Both tests indicate the presence of $\Delta I = \frac{3}{2}$.

References
[1] G. Trilling, K-Meson Decays, UCRL-16473 (updated from Argonne Conference Proceedings, 1965, p. 115)
[2] N. Brene (CERN), private communication. In our Jan. 1968 edition we had erroneously used 1.04
[3] T. S. Mast, L. K. Gershwin, M. Alston-Garnjost, R. O. Bangerter, A. Barbaro-Galtieri, J. J. Murray, F. T. Solmitz, and R. D. Tripp, Phys. Rev. *183*, 1200 (1969)
[4] C. Zemach, Phys. Rev. *133*, B1201 (1964)
[5] C. Bouchiat and M. Veltman, Topical Conference on weak interactions, CERN 69-7 (1969) p. 225

30.3.2. Test of $\Delta I = \frac{1}{2}$ rule for hyperon decays

O. E. Overseth
University of Michigan

1. Non-leptonic decay amplitudes

In this edition we again use the new convention for the amplitudes A and B adopted in 1973. Some theorists have suggested that dimensionless amplitudes are more useful to them than the ones appearing in the literature. Berge [1] used a convention with A and B in units of $\sec^{-1/2}$. Samios [2] used a convention which gave A and B in units of $(\text{MeV} \cdot \sec)^{-1/2}$. Following is the convention suggested by Jackson [3], which gives dimensionless A and B.

The effective Lagrangian density for non-leptonic hyperon decays $(B_1 \to B_2 + \pi)$ can be written

$$\mathcal{L}_{\text{eff}} = G\mu_c^2 [\bar{\psi}_2(A + B\gamma_5)\psi_1]\phi_\pi,$$

where $G = 10^{-5} m_p^{-2}$ is a coupling constant characteristic of first-order weak decays, μ_c is the charged pion mass, and A and B are *dimensionless* complex numbers giving the relative amplitudes of the parity-violating and parity-conserving decays, respectively. The matrix γ_5 is to be taken in the Pauli form, $\gamma_5 = \begin{pmatrix} 0 & -I \\ -I & 0 \end{pmatrix}$. The invariant amplitude for the decay is

$$\mathcal{M} = G\mu_c^2 [\bar{u}(p)(A + B\gamma_5)u(P)],$$

where P is the 4-momentum of the decaying hyperon of mass M, and p is the 4-momentum of the baryon decay product of mass m. With the normalization convention, $\bar{u}_i u_i = 2m_i$, the Pauli form of the matrix element in the rest frame of the decaying hyperon is

$$\mathcal{M} = G\mu_c^2 \langle \chi_2 | \sqrt{2M(E+m)}A + \sqrt{2M(E-m)}B\boldsymbol{\sigma} \cdot \hat{\boldsymbol{q}} | \chi_1 \rangle,$$

where E is the total energy of the final baryon and $\hat{q}$ is a unit vector in the direction of motion of the final baryon. Comparison with Sec. VI D of the text shows that the amplitudes s and p defined there are proportional to A and B:

$$\frac{p}{s} = \left(\frac{E-m}{E+m}\right)^{1/2} \frac{B}{A} = \left[\frac{(M-m)^2 - \mu^2}{(M+m)^2 - \mu^2}\right]^{1/2} \frac{B}{A}.$$

Here μ is the mass of the pion entering the decay. The parameters α, β, and γ can therefore be expressed in terms of A and B, rather than s and p, if desired.

The decay rate for $B_1 \to B_2 + \pi$ is

$$\Gamma = \frac{G^2\mu_c^4}{8\pi}q\left\{\left[\frac{(M+m)^2-\mu^2}{M^2}\right]|A|^2 + \left[\frac{(M-m)^2-\mu^2}{M^2}\right]|B|^2\right\},$$

where q is the c.m. momentum of the decay products. For reference, the dimensionless constant in this expression has the value $G^2\mu_c^4/8\pi = 1.9488 \times 10^{-15}$.

Table 1 summarizes the amplitudes A and B for the nonleptonic decays of the Λ, Σ, and Ξ hyperons. These amplitudes have been calculated by using the experimental data for mean lives, branching ratios, and the decay asymmetry α given in the Stable Particle Table of this review. Time-reversal invariance is assumed and final-state interactions are neglected, so A and B are taken to be relatively real. The subscript on the hyperon refers to the sign of the decaying pion. The statistical correlation coefficient

$$C_{AB} = \frac{\langle \Delta A \Delta B \rangle}{\sqrt{\langle \Delta A^2 \rangle \langle \Delta B^2 \rangle}}$$

is also given. The absolute signs of A and B have been assigned, using the following convention. Taking $A(\Lambda_-^0)$ as positive, the other S-wave decay amplitudes are chosen to give an approximate fit to the triangular relationships

$$\sqrt{2}A(\Sigma_0^+) + A(\Sigma_+^+) = A(\Sigma_-^-)$$

and

$$\sqrt{3}A(\Sigma_0^+) + A(\Lambda_-^0) = 2A(\Xi_-^-).$$

The signs of the B amplitudes relative to those of the corresponding A amplitudes are determined by the sign of the appropriate α decay parameter.

Table 1

$M \to m + \mu$	A	B	C_{AB}
$\Lambda_-^0 \to p + \pi^-$	1.47 ± 0.01	9.98 ± 0.24	-0.289
$\Lambda_0^0 \to n + \pi^0$	-1.07 ± 0.02	-7.14 ± 0.56	-0.741
$\Sigma_+^+ \to n + \pi^+$	0.06 ± 0.01	19.07 ± 0.07	-0.038
$\Sigma_0^+ \to p + \pi^0$	1.48 ± 0.05	-12.04 ± 0.58	0.982
$\Sigma_-^- \to n + \pi^-$	1.93 ± 0.01	-0.65 ± 0.07	0.003
$\Xi_0^0 \to \Lambda + \pi^0$	1.54 ± 0.03	-6.43 ± 0.66	0.188
$\Xi_-^- \to \Lambda + \pi^-$	2.04 ± 0.01	-6.93 ± 0.31	0.268

2. Tests of the $\Delta I = \frac{1}{2}$ rule

(a) Λ Decay

For Λ decay the $\Delta I = \frac{1}{2}$ rule predicts that $\Gamma_0/\Gamma_- = 0.50$ and $\alpha_0 = \alpha_-$. In order to determine the magnitude of possible $\Delta I = \frac{3}{2}$ amplitudes present we write the linear expressions [4] for the $\Delta I = \frac{3}{2}$ A- and B-wave amplitudes in terms of $\Delta\alpha$, where $\Delta\alpha$ is the measured value of α_0/α_- minus the predicted value, and in terms of $\Delta\Gamma$ similarly defined. Evaluating these we find

$$\Delta\alpha = -1.54 \, A_3/A_1 + 1.61 \, B_3/B_1,$$
$$\Delta\Gamma = 1.84 \, A_3/A_1 + 0.25 \, B_3/B_1.$$

Here the $\Delta I = \frac{3}{2}$ amplitudes are expressed relative to the $\Delta I = 1/2$ amplitudes. The numerical values of the coefficients depend on the ratio B/A. The uncertainties in the coefficients are small compared to the uncertainties in $\Delta\alpha$ and $\Delta\Gamma$. Final-state πN interactions have been included in these relations but have a very small effect. From the Stable Particle Table,

$$\Delta\alpha = 0.006 \pm 0.066, \quad \Delta\Gamma = 0.058 \pm 0.012,$$

and hence

$$A_3/A_1 = 0.027 \pm 0.008,$$
$$B_3/B_1 = 0.030 \pm 0.037.$$

The possible 3% $\Delta I = \frac{3}{2}$ A-wave amplitude is due to the disagreement of decay rates with prediction. At this level the results are sensitive to electromagnetic corrections. However, in Λ decay the phase space correction and the other radiative corrections appear to be about equal in magnitude and have opposite signs [5, 6], and hence cancel each other in the correction to the decay rates.

(b) Ξ Decay

The analysis for Ξ decay is very similar to that for Λ decay. If the $\Delta I = \frac{1}{2}$ rule is valid, $\Gamma_0(\Xi^0)/\Gamma_-(\Xi^-) = 0.50$ and $\alpha_0 = \alpha_-$. For this case the expressions linear in $\Delta I = \frac{3}{2}$ A- and B-wave amplitudes are [4]

$$\Delta\alpha = 1.37 \, A_3/A_1 - 1.37 \, B_3/B_1,$$
$$\Delta\Gamma = -1.44 \, A_3/A_1 - 0.06 \, B_3/B_1.$$

From the Stable Particle Table,

$$\Delta\alpha = 0.18 \pm 0.12, \quad \Delta\Gamma = 0.066 \pm 0.020,$$

and we find

$$A_3/A_1 = -0.038 \pm 0.014,$$
$$B_3/B_1 = -0.17 \pm 0.09.$$

(c) Σ Decay

The traditional test of the $\Delta I = 1/2$ rule in Σ decay is that the amplitudes satisfy the relationship

$$\sqrt{2}\, \Sigma_0^+ + \Sigma_+^+ - \Sigma_-^- = 0.$$

Graphically this is equivalent to closing the Σ triangle when the amplitudes are plotted on A, B axes. Including $\Delta I \geq \tfrac{3}{2}$ amplitudes in Σ decay analysis, the "Σ triangle" relationship becomes

$$\sqrt{2}\, A_0 + A_+ - A_- = -3\sqrt{\tfrac{2}{5}}\, A_3 + \frac{2}{\sqrt{15}} A_5,$$

where A_3 and A_5 are $\Delta I = 3/2$ and $\Delta I = 5/2$ amplitudes, respectively. There is a similar equation for the B amplitudes. From table 1,

$$\sqrt{2}\, A_0 + A_+ - A_- = 0.22 \pm 0.09, \quad \sqrt{2}\, B_0 + B_+ - B_- = 2.7 \pm 1.0.$$

If we neglect the $\Delta I = \tfrac{5}{2}$ amplitudes and assume all amplitudes to be real we can solve for possible $\Delta I = \tfrac{3}{2}$ amplitudes. The result is

$$\frac{A_3}{A_-} = -0.061 \pm 0.024, \quad \frac{B_3}{B_+} = -0.074 \pm 0.027.$$

Thus for hyperon decay, present experimental data limit $\Delta I = 3/2$ amplitudes to less than about 5%.

3. The Lee-Sugawara relation

From table 1 the Lee-Sugawara relation [7, 8], $\sqrt{3}\, \Sigma_0^+ + \Lambda_-^0 - 2\Xi_-^- = 0$, is satisfied to -0.07 ± 0.11 for the A amplitudes, and to 3.0 ± 1.9 for the B amplitudes.

References

[1] J. P. Berge, in *Proc. of the 13th Int. Conf. on high-energy physics, Berkeley*, (1966) (University of California Press, Berkeley, 1967), p. 46
[2] N. P. Samios, International Conference on weak interactions, Argonne, (1965), p. 189
[3] J. D. Jackson, private communication (1973).
[4] See O. E. Overseth and S. Pakvasa, Phys. Rev. *184* (1969) 1663. The expression for Γ_0/Γ_- for Λ decay should read

$$\frac{\Gamma_0}{\Gamma_-} \approx \tfrac{1}{2}\left\{1 + 3\sqrt{2} \times \left[\frac{S_{11}S_{33}\cos(\delta_1 - \delta_3) + P_{11}P_{33}\cos(\delta_{11} - \delta_{31})}{S_{11}^2 + P_{11}^2}\right]\right\}.$$

[5] See A. A. Belavin and I. M. Narodetsky, Yad. Fiz. *8* (1968) 978 [Soviet J. Nucl. Phys. *8* (1969) 568]
[6] G. W. Intemann, private communication (1973)
[7] See B. W. Lee, Phys. Rev. Lett. *12* (1964) 83
[8] See H. Sugawara, Prog. Theor. Phys. *31* (1964) 213

30.4. CP and CPT invariances*

30.4.1. Violation of CP invariance in K^0 decays

Experimental data on η_{+-}, η_{00}, and δ characterizing CP violation in K_L^0 decays see in section 30.2.2.

The model of superweak interaction [1] corresponds to the following values of the parameters:

$$\phi_{+-} = \phi_{00} = \arctan\left[2\Delta m\tau(K_S^0)/\hbar\right] = (43.67 \pm 0.14)°,$$

$$\mathrm{Re}\,\varepsilon = |\eta_{+-}|\left[1 + 2\Delta m\tau(K_S^0)/\hbar\right] = (1.645 \pm 0.016) \cdot 10^{-3}.$$

The experimental value of the last quantity,

$$\mathrm{Re}\,\varepsilon = (1.621 \pm 0.088) \cdot 10^{-3},$$

is found from the relation

$$\mathrm{Re}\,\varepsilon = \delta |1-x|^2 / 2(1-|x|^2)$$

where x is the ratio of amplitudes corresponding to transitions with $\Delta Q = -\Delta S$ and $\Delta Q = +\Delta S$, that is

$$x = A(K^0 \to \pi^+ \ell^- \bar{\nu})/A(K^0 \to \pi^- \ell^+ \nu);$$

the value of x will be given below (see the text following table in section 30.6.1).

Reference
[1] L. Wolfenstein, Phys. Lett. 13 (1964) 562

30.4.2. Electric dipole moment of the neutron

$$d_n = \begin{cases} (5 \pm 15) \cdot 10^{-25} e \cdot \mathrm{cm} \\ (4 \pm 7.5) \cdot 10^{-25} e \cdot \mathrm{cm}\;[1]. \end{cases}$$

Reference
[1] I. S. Altaryov et al., Pis'ma ZhETF 29 (1979) 794 [JETP Lett. 29 (1979) 730]

30.4.3. Search for T-odd muon polarization in $K_{\mu 3}^0$ decay

Measurements of the muon polarization, P_{μ^+}, normal to the plane of $K_L^0 \to \pi^- \mu^+ \nu$ decay, give the result [1]: $P_{\mu^+} = 0.0021 \pm 0.0048$.

*Absence of reference means PDG as the source of the corresponding experimental data.

Non-conservation of CP parity corresponds to non-zero relative phase of the form factors f_+ and f_- of $K_{\mu 3}$ decay, that is to Im $\xi \neq 0$ where $\xi = f_-/f_+$. The electromagnetic interactions of π^- and μ^+ in the final state gives rise to the same effect [2, 3].

The experimental value of P_μ given above determines Im $\xi|_{\text{exp}}$, which must be compared with the value of Im $\xi|_{\text{e.m.}}$ which is due to the interaction in the final state. These quantities are [1]:

$$\text{Im } \xi|_{\text{exp}} = 0.012 \pm 0.026, \quad \text{Im } \xi|_{\text{e.m.}} = 0.008.$$

References
[1] M. P. Schmidt et al., Phys. Rev. Lett. 43 (1979) 556
[2] N. Byers, S. W. MacDowell, C. N. Yang, High energy physics and elementary particles (IAEA, Vienna, 1965) p. 953
[3] L. B. Okun, I. B. Khriplovich, Yad. Fiz., 6 (1967) 821 [Sov. J. Nucl. Phys. 6 (1968) 598] (Here the coefficient in front of Im $\xi|_{\text{e.m.}}$, derived in [2], is corrected.)

30.4.4. Test of CPT invariance [1]

	$(m_+ - m_-)/m$	$(\tau_+ - \tau_-)/\tau$	$(g_+ - g_-)/g$
e^-/e^+			$< 1.2 \cdot 10^{-8}$
μ^-/μ^+	$-(2 \pm 5) \cdot 10^{-6}$	$(7.9 \pm 10) \cdot 10^{-4}$ $-(4 \pm 9) \cdot 10^{-4}$	$(2.6 \pm 1.7) \cdot 10^{-8}$
$p/\bar{p}$	$(7.6 \pm 3.9) \cdot 10^{-5}$	a	$-(0.4 \pm 7.2) \cdot 10^{-3\,b}$
π^+/π^-	$-(2.4 \pm 4.7) \cdot 10^{-5}$	$(5.3 \pm 6.8) \cdot 10^{-4c}$	
K^+/K^-	$-(1.3 \pm 1.1) \cdot 10^{-4}$	$(1.1 \pm 0.9) \cdot 10^{-3}$	

Notations: m, τ, g denote mass, lifetime, and g-factor; indices $\pm$ indicate the sign of the particle charge.

[a] The experimental result for the antiproton lifetime for the decay $\bar{p} \to e^+ \pi^0$ is [2]:

$$\tau(\bar{p} \to e^+ \pi^0) > 1700 \text{ hours}.$$

[b] Actually this result was obtained for the relative difference between magnetic moments of the proton and antiproton, $(\mu_+ - \mu_-)/\mu$.

[c] See PDG.

References
[1] H. Poth, Phys. Lett. 77B (1978) 321
[2] M. Bell et al., Phys. Lett. 86B (1979) 215

30.5. Conservation of leptonic numbers

30.5.1. Neutrino oscillations

$\nu_\mu \leftrightarrow \nu_e$, $\bar{\nu}_\mu \leftrightarrow \bar{\nu}_e$, GGM, CERN PS [1]: $P(\nu_\mu \to \nu_e)/P(\nu_\mu \to \nu_\mu) = -0.03(10) \cdot 10^{-2}$, $P(\nu_e \to \nu_e)/P(\nu_\mu \to \nu_\mu) = 0.95(30)$,	$\nu_e \leftrightarrow \nu_i$, $i \neq e$, BNL solar ν_e detector: $P(\nu_e \to \nu_e) = 1.6(4)/(\sim 4.7)$ [2], $\Delta m^2 = 4 \cdot 10^{-10}$ eV^2 [3]
$P(\bar{\nu}_\mu \to \bar{\nu}_e)/P(\bar{\nu}_\mu \to \bar{\nu}_\mu) = 0.02(7) \cdot 10^{-2}$, $P(\bar{\nu}_e \to \bar{\nu}_e)/P(\bar{\nu}_\mu \to \bar{\nu}_\mu) = 0.89(30)$; $\sqrt{\Delta m^2} \lesssim 1 - 2$ eV for $\sin 2\alpha = 1.0 - 0.2$	$\bar{\nu}_e \leftrightarrow \bar{\nu}_i$, $i \neq e$, Savannah River Plant, $\bar{\nu}_e d \to e^+ nn/\bar{\nu}_e d \to \bar{\nu}_e pn$ [4]: $0.5 \leq \sin^2 2\alpha \leq 0.8$ and $0.7 \leq \Delta m^2$ (eV^2) ≤ 1.0 (68% c.l.)
$\bar{\nu}_\mu \leftrightarrow \bar{\nu}_e$, LAMPF: $P(\bar{\nu}_\mu \to \bar{\nu}_e) < 0.098$ [5] $\Delta m^2 \lesssim 0.9 - 1.8$ eV^2 for $\sin^2 2\alpha = 1.0 - 0.3$ [6]	$\bar{\nu}_e \leftrightarrow \bar{\nu}_i$, $i \neq e$, Fission reactor, ILL [7]: $\bar{\nu}_e p \to e^+ n$, $\Delta m^2 \lesssim 0.2 - 0.14$ for $\sin^2 2\alpha = 0.5 - 1.0$
$\nu_\mu \leftrightarrow \nu_\tau$, HES, FNAL: $P(\nu_\mu \to \nu_\tau) < 0.63 \cdot 10^{-2}$; $\Delta m^2 \lesssim 3 - 25$ eV^2 for $\sin^2 2\alpha = 1.0 - 0.02$ [8]	$\nu_e \leftrightarrow \nu_i$, $i \neq e$, LAMPF: $P(\nu_e \to \nu_e) = 1.09(^{37}_{41})$; $\Delta m^2 \lesssim 2.5 - 4.0$ eV^2 for $\sin^2 2\alpha = 1.0 - 0.6$ [6]
$\nu_e \leftrightarrow \nu_\tau$, BEBC, CERN SPS: $P(\nu_e \to \nu_\tau) < 0.35$ [9]	$\nu_e \leftrightarrow \nu_i$, $i \neq e$, BEBC: $P(\nu_e \to \nu_e) = 1.04(15)$; $\Delta m^2 \lesssim 60 - 150$ eV^2 for $\sin 2\alpha = 1.0 - 0.5$ [10]

Notations: (i) GGM: Gargamelle bubble chamber; LAMPF: Los Alamos Meson Physics Facility; ILL: Institut Laue-Langevin; HES: hybrid emulsion spectrometer; BEBC: Big European Bubble Chamber; (ii) P, α, and Δm^2 denote the transition probability averaged, the mixing angle, and the difference of squared masses of neutrinos, respectively; for instance, in the case of $\nu_\mu \leftrightarrow \nu_e$ mixing:

$$\nu_e = \nu_1 \cos \alpha + \nu_2 \sin \alpha,$$
$$\nu_\mu = -\nu_1 \sin \alpha + \nu_2 \cos \alpha, \quad \Delta m^2 = |m^2(\nu_1) - m^2(\nu_2)|;$$

(iii) The figures in parentheses correspond to the one-standard-deviation uncertainty in the last digits of main number.

Note: Limits on "exotic" oscillations $\nu_\mu, \nu_e \leftrightarrow \bar{\nu}_{\mu L}$ may be inferred from ref. [11] if one additionally assumes that $\bar{\nu}_{\mu L}$ could produce μ^+ when interacting with nucleons.

References

[1] E. Bellotti, et al., Lett. Nuovo Cim. 17 (1976) 553; J. Blietschau et al., Nucl. Phys. B133 (1978) 205
[2] J. K. Rowley, B. T. Cleveland, R. Davis, Jr., and C. Evans, Proc. Neutrino '77 Conf., vol. 1, p. 15
[3] R. Ehrlich, Phys. Rev. D18 (1978) 2323
[4] E. Pasierb et al., Phys. Rev. Lett. 43 (1979) 96; F. Reines et al., Phys. Rev. Lett. 45 (1980) 1307
[5] S. E. Willis et al., Phys. Rev. Lett. 44 (1980) 522; 45 (1980) 1370E
[6] P. Némethy et al., Phys. Rev. D23 (1981) 262
[7] F. Boehm et al., Phys. Lett. 97B (1980) 310
[8] N. Ushida et al., Phys. Rev. Lett. 47 (1981) 1694.
[9] P. Fritze et al., Phys. Lett. 96B (1980) 427
[10] H. Deden et al., Phys. Lett. 98B (1980) 310
[11] M. Holder et al., Phys. Lett. 74B (1978) 277

30.5.2. $\mu \leftrightarrow e$ and $\tau \rightarrow \mu(e)$ processes

$\mu^+ \rightarrow e^+ \gamma$, LAMPF: $B < 1.9 \cdot 10^{-10}$ [1]	$\mu^+ \rightarrow e^+ \gamma\gamma$, LAMPF: $B < 5 \cdot 10^{-8}$ [2]	$\mu^+ \rightarrow e^+ e^- e^+$, JINR: $B < 1.9 \cdot 10^{-9}$ [3]
$\mu^{-32}S \rightarrow e^{-32}S^*$, SIN: $R < 1.5 \cdot 10^{-10}$ (99% c.l.) [4]	$\mu^{-32}S \rightarrow e^{+32}Si^*$, SIN: $R < 1.5 \cdot 10^{-9}$ [5]	$\mu^{-127}I \rightarrow e^{+127}Sb^*$, SIN: $R < 3 \cdot 10^{-10a}$ [6]
$e^-e^- \rightarrow \mu^-\mu^{-b,c}$, SLAC: $\sigma < 0.67 \cdot 10^{-32}$ cm^2 [7]	$\mu^+ \rightarrow e^+ \bar{\nu}_e \nu_\mu^c$, LAMPF: $B < 0.098$ [8]	$\bar{\nu}_\mu e^- \rightarrow \mu^- \bar{\nu}_e^c$, CERN SPS: $R < 0.09$ [9]
$\tau \rightarrow e\gamma$, SLAC: $B < 0.026$ [10]	$\tau \rightarrow ee^+e^-$, SLAC: $B < 0.006$ [11]	$\tau \rightarrow \mu\gamma$, SLAC: $B < 0.013$ [10]

Notations: B, R, and σ denote relative decay rate, ratio of the cross sections of the forbidden process to that of corresponding allowed one, and total cross section, respectively.

[a] For the "elastic" process, without the final nucleus breaking.

[b] If the process under consideration is described by four-fermion interaction with the hamiltonian $H = (G'/\sqrt{2})(\bar{\mu}O_\alpha e)(\bar{\mu}O_\alpha e) + \text{h.c.}$, where $O_\alpha = \gamma_\alpha(1 + \gamma_5)$, then the upper bound of σ given above corresponds to the following upper bound on the coupling constant:

$$G' < 6.1 \cdot 10^{-3} m_p^{-2} \quad (m_p \text{ is the proton mass}).$$

[c] The process is allowed by multiplicative but not additive leptonic number conservation.

References

[1] J. D. Bowman et al., Phys. Rev. Lett. 42 (1979) 556
[2] J. D. Bowman et al. Phys. Rev. Lett. 41 (1978) 442
[3] S. M. Korenchenko et al., ZhETF 70 (1976) 3 [JETP 43 (1976) 1]
[4] A. Badertscher et al., Proc. 19th Int. Conf. High Energy Phys., Tokyo, 1978, p. 1019 (contributed paper 950); quoted from the report by G. Altarelli, ibid., p. 441
[5] A. Badertscher et al., Phys. Lett. 79B (1978) 371
[6] R. Abela et al., Phys. Lett. 95B (1980) 318
[7] W. C. Barber et al., Phys. Rev. Lett. 22 (1969) 902
[8] E. Willis et al., Phys. Rev. Lett. 44 (1980) 522; 45 (1980) 1370E
[9] M. Jonker et al., Phys. Lett. 93B (1980) 203
[10] M. L. Perl, Proc. Int. Symp. on lepton and photon interactions at high energies, Hamburg, 1977, p. 145
[11] G. J. Feldman, Proc. 19th Int. Conf. high energy phys., Tokyo, 1978, p. 777

30.6. Selection rules for the weak current*

30.6.1. Absence of semileptonic decays with $\Delta S \neq \Delta Q$

$K^+ \rightarrow \pi^+\pi^+ e^- \bar{\nu}_e$	$< 5 \cdot 10^{-7}$	$\Xi^0 \rightarrow \Sigma^- e^+ (\mu^+)\nu$	$< 9 \cdot 10^{-4}$
$K^+ \rightarrow \pi^+\pi^+ \mu^- \bar{\nu}_\mu$	$< 3 \cdot 10^{-6}$	$\Xi^0 \rightarrow pe^- (\mu^-)\bar{\nu}$	$< 1.3 \cdot 10^{-3}$
$\Sigma^+ \rightarrow ne^+ \nu_e$	$< 5 \cdot 10^{-6}$	$\Xi^- \rightarrow ne^- \bar{\nu}_e$	$< 3.2 \cdot 10^{-3}$
$\Sigma^+ \rightarrow n\mu^+ \nu_\mu$	$< 3 \cdot 10^{-5}$	$\Xi^- \rightarrow n\mu^- \bar{\nu}_\mu$	$< 1.5 \cdot 10^{-2}$

*This section gives upper limits of the branching ratios. Absence of reference means PDG as the source of the corresponding experimental data.

Mean values of x, the ratio of the amplitude of $K^0 \to \pi^+ \ell^- \bar{\nu}(\overline{K}^0 \to \pi^- \ell^+ \nu)$ decays, corresponding to $\Delta Q = -\Delta S$, to that of $K^0 \to \pi^- \ell^+ \nu(\overline{K}^0 \to \pi^+ \ell^- \bar{\nu})$ decays, corresponding to $\Delta Q = \Delta S$, are equal to (see section 30.2.2):

$$\langle \text{Re}\, x \rangle \equiv \left\langle \text{Re}\, \frac{A(\Delta Q = -\Delta S)}{A(\Delta Q = \Delta S)} \right\rangle = 0.009 \pm 0.020,$$

$$\langle \text{Im}\, x \rangle \equiv \left\langle \text{Im}\, \frac{A(\Delta Q = -\Delta S)}{A(\Delta Q = \Delta S)} \right\rangle = -0.004 \pm 0.026.$$

30.6.2. Suppression of hadronic decays with $\Delta S > 1$ or $\Delta C = -\Delta S$

$\Delta S > 1$	$\Xi^0 \to p\pi^-$	$< 3.6 \cdot 10^{-5}$	$\Xi^- \to p\pi^-\pi^-$	$< 4 \cdot 10^{-4}$
	$\Xi^- \to n\pi^-$	$< 1.1 \cdot 10^{-3}$	$\Omega^- \to \Lambda\pi^-$	$< 1.3 \cdot 10^{-3}$
$\Delta C = -\Delta S$	$D^0 \to K^+\pi^-$	$< 5 \cdot 10^{-3}$	$D^+ \to K^+\pi^+\pi^-$	$< 2 \cdot 10^{-3}$

30.6.3. Absence of decays with $\Delta Q = 0$ for $\Delta S \neq 0$[a]

$K^+ \to \pi^+ \nu \bar{\nu}$	$< 5.6 \cdot 10^{-7}$	$K_L^0 \to \mu^+\mu^-$	$(9.1 \pm 1.9) \cdot 10^{-9}$
$K^+ \to \pi^+ e^+ e^-$	$(2.6 \pm 0.5) \cdot 10^{-7}$	$K_L^0 \to \pi^0 e^+ e^-$	$< 2.3 \cdot 10^{-6}$ [2]
$K^+ \to \pi^+ \mu^+ \mu^-$	$< 2.4 \cdot 10^{-6}$	$K_L^0 \to \pi^0 \mu^+ \mu^-$	$< 1.2 \cdot 10^{-6}$ [2]
$K^+ \to \pi^\pm e^\mp \mu^\mp$	$< 7 \cdot 10^{-9}$	$K_L^0 \to \pi^+\pi^- e^+ e^-$	$< 5 \cdot 10^{-6}$ [3]
$K^+ \to \pi^+ e^- \mu^+$	$< 5 \cdot 10^{-9}$	$K_L^0 \to e^+ e^- \gamma$	$(1.7 \pm 0.9) \cdot 10^{-5}$ [2]
$K_S^0 \to e^+ e^-$	$< 3.4 \cdot 10^{-4}$	$K_L^0 \to \mu^+ \mu^- \gamma$	$(2.8 \pm 2.8) \cdot 10^{-7}$ [2]
$K_S^0 \to \mu^+ \mu^-$	$< 3.2 \cdot 10^{-7}$	$K_L^0 \to e^+ e^- e^+ e^-$	$< 5.8 \cdot 10^{-5}$ [3]
$K_L^0 \to e^+ e^- (e^\pm \mu^\mp)$	$< 2 \cdot 10^{-9}$ [1][b]	$K_L^0 \to e^+ e^- \mu^+ \mu^-$	$< 1.1 \cdot 10^{-5}$ [3]

[a] The data are given on decays which would be induced by strangeness-nonconserving neutral currents. Some of the decays can be caused by non-leptonic charged currents together with the electromagnetic interaction.

[b] The value of $B(K_L^0 \to \mu^+ \mu^-) < 3 \cdot 10^{-9}$ obtained in the same experiment [1] is in contradiction with the results of other experiments.

References

[1] A. R. Clark et al., Phys. Rev. Lett. 26 (1971) 1667
[2] A. S. Carrol et al., Phys. Rev. Lett. 44 (1980) 525
[3] V. M. Berezin et al., Preprint ITEP-47 (1980) (In Russian)

30.6.4. Absence of processes with $\Delta Q = 0$ for $\Delta S \neq 0$ or $\Delta C \neq 0$

$\sigma(\nu_\mu N \to \nu_\mu \Lambda(\Sigma^0)X)/\sigma(\nu_\mu N \to \nu_\mu X) < 5.4 \cdot 10^{-3}$	[1]
$\dfrac{\sigma(\nu_\mu N \to \nu_\mu CX)}{\sigma(\nu_\mu N \to \nu_\mu X)}\bigg|_{E_X > 1\,\text{GeV}} B(C \to e^+ + \cdots)\big|_{E_{e^+} > 0.2\,\text{GeV}} < 1.3 \cdot 10^{-3}$	[2]
$\dfrac{\sigma(\nu_\mu N \to \nu_\mu CX)}{\sigma(\nu_\mu N \to \nu_\mu X)}\bigg|_{E_X > 100\,\text{GeV}} B(C \to \mu^+ + \cdots) < 3.9 \cdot 10^{-3}$	[3]
$\dfrac{\sigma(\bar\nu_\mu N \to \bar\nu_\mu CX)}{\sigma(\bar\nu_\mu N \to \bar\nu_\mu X)} B(C \to e^+ + \cdots) < 1.4 \cdot 10^{-3}$	[4]
$\sigma(\nu_\mu N \to \nu_\mu CX)/\sigma(\nu_\mu N \to \mu^- CX) \leq 0.08$	[5]

Notations: σ and B denote the cross section and the relative decay rates, respectively; C denotes a charmed particle, and X all other hadrons.

References
[1] J. Blietschau et al., Phys. Lett. 71B (1977) 231
[2] H. Deden et al., Phys. Lett. 67B (1977) 474
[3] M. Holder et al., Phys. Lett. 74B (1978) 277
[4] V. Efremenko et al., Phys. Lett. 88B (1979) 181
[5] C. Baltay, Proc. Neutrino '78 Conf., p. 533

SUBJECT INDEX

abelian symmetry 164, 165, 180, 309
Ademollo-Gatto theorem 41, 43, 270
analogy of the weak and electromagnetic
    interaction 1, 23–25, 29, 31, 104, 105,
    108, 110, 145, 146, 149, 151, 152, 159,
    191, 192
annihilation e^+e^- see $e^+e^- \to \ldots$
annihilation-type quark graphs 49, 68, 79
antiquark sea 146, 150, 155, 220
asymptotic freedom 49, 119, 146, 242
astrophysical processes 140, 141, 259, 260, 293
axial charge 32, 34, 46, 47, 70, 211, 270, 339
—— current see current, axial
—— ——, conservation of see
    conservation of axial current

baryon octet 44, 65, 66, 316
baryonic asymmetry of the universe 246, 263, 294, 306
—— number 164, 166, 182, 243–246, 249, 251, 263, 294
—— "photons" 166, 294
Bjorken scaling 155, 278
B-mesons 3, 5, 299
b-quark 3–5, 12, 13, 89, 123, 195, 215, 216, 227, 228, 277
β-decay 12, 22, 25, 32, 127, 267, 282, 297

Cabibbo angle 13, 22, 39, 89, 116–118, 126, 127, 148, 153, 195–197, 215, 235, 270, 331
Callan-Gross relation 155
Callan-Treiman relation 43, 270
charged current see current, charged
charmed particles 3, 6, 117, 152, 275, 298, 299, 324, 326, 336, 338, 354

chiral invariance 25, 34, 75, 115, 232, 269
C invariance 1, 12, 48, 66, 85, 98, 99, 202, 272, 273
classification of leptons see lepton classification
—— —— quarks see quark classification
colliding beam projects 220, 285, 307
color symmetry 4, 239, 249
confinement 51, 55, 147, 198, 231, 235, 253, 254, 260, 306
conservation of axial current 26, 34, 114, 159; see also PCAC
—— —— electric charge 181, 186, 296, 307; see also U(1)
—— —— vector current 24, 26, 29, 31, 116, 159–161, 164, 182, 186, 269
cosmological effects 181, 246, 254, 290, 305, 306; see also baryonic asymmetry
CP invariance 1, 12, 13, 32, 35, 48, 63, 66, 76, 77, 85, 86, 90, 94, 128, 213, 263, 273, 274, 275, 278, 293, 294, 297, 339, 350
CPT invariance 35, 63, 95, 100, 274, 333, 334, 350, 351
c-quark 3–6, 52, 53, 88, 89, 117–121, 123–125, 127–129, 153, 275
crossing symmetry 133, 134, 169, 203
CVC see conservation of vector current
current, axial 10, 26, 32, 34, 105, 110, 113, 114, 190, 208–211, 226; see also conservation of axial current and PCAC
——, charged 1–6, 9, 10, 12, 52, 89, 104, 105, 123, 135, 138, 157, 186, 192, 195–197, 213, 268, 278; see also selection rule
——, isoscalar 24, 31, 39, 110, 208
——, isovector 23, 24, 29, 31, 39, 110, 208

Index

———, left-handed 10, 14, 26, 39, 123, 138, 154, 186, 196, 208, 209, 278
———, lepton 2, 5, 6, 12, 139, 193
———, neutral 4–6, 9, 13, 40, 139, 140, 185, 190, 196, 199, 213, 259, 275, 279, 299, 331, 354, 355
———, quark 3–6, 12, 23–26, 195–197, 208
———, right-handed 14, 26, 138, 186, 196, 208, 209
———, strange 5, 6, 38–40, 48, 275; *see also* selection rule
———, vector 10, 26, 29, 31, 32, 110, 113, 208–211, 226; *see also* conservation of vector current

decays *see* respective particle names, *see also* form factors
deep inelastic processes 145, 151, 154, 278, 279
density matrix 16, 19, 107, 328
dileptonic events 118, 129, 153, 275
dipole moment of the neutron 100, 274, 338
discrete symmetry 175; *see also* C, CP, CPT, G, P and T invariance
DUMAND 221, 285

$E_{6,7,8}$ groups 249, 250, 288, 289, 303
$e^+e^- \to$ hadrons 104, 110, 111, 113, 276, 277
$e^+e^- \to \mu^+\mu^-$ 202, 205
$e^+e^- \to \nu\bar{\nu}$ 140, 157
$e^+e^- \to W^+W^-$, Z^0 217, 218, 285, 287, 300
effective pseudoscalar 32, 35, 269
——— scalar 32
electric charge 165, 167, 240; *see also* conservation of electric charge
"electrism" *see* weak "electrism"
electromagnetic interaction *see* analogy of the weak and electromagnetic interaction *and see* quantum electrodynamics
electroweak interaction 7, 14, 138, 171, 185, 199, 214, 223, 241, 284, 296, 300
exceptional groups 249, 288, 289, 303

fermion generation 195, 236, 242, 247, 248, 252, 257–259; *see also* lepton and quark classification
——— mass *see* masses

Fierz transformation 15, 53, 57, 70, 71, 80, 139, 201, 204, 312, 315, 321, 323
flavour *see* quark flavours
form factors, in hyperon decays 36, 37, 44–47, 57, 69, 70, 72–74, 269, 270
——— ———, ——— K-meson decays 40–44, 81, 82, 87, 270, 271
——— ———, ——— neutrino reactions 144, 145, 208, 209
——— ———, ——— neutron decay 30–35, 45–47, 144, 145
——— ———, ——— pion decays 27–29, 33, 34, 105, 269–271
four-fermion interaction *see* weak interaction

gauge fields 166, 168, 170, 171, 181, 185–188, 191, 192, 283, 289
——— invariance 164, 174, 177, 180, 183, 189, 214, 246, 265, 282
Gedanken experiments 85, 101
Glashow-Iliopoulos-Maiani mechanism 88, 128, 157
global symmetry 164, 165, 170, 177, 179
gluonic effects 48–50, 52, 56–57, 82, 118–120, 146, 151, 155, 220, 228, 230–232, 240, 271, 272, 287, 297, 298
——— monopole 271, 272
Goldberger-Treiman relation 34, 269
Goldstone bosons 34, 177, 179, 180, 269
G-parity 23, 32, 110
grand unification 171, 236, 252, 288, 295, 302, 303, 306; *see also* Higgs bosons, intermediate bosons, superunification, *and* unification of interactions
gravitational interaction 167, 170, 171, 236, 251, 252, 255, 307, 330
gravitino 238, 251
gravitons 168–170, 189, 232, 251, 252

hadronic weak current *see* current, quark
heavy quarks *see* b-, c-, and t-quark
helicity 11, 21, 27, 56, 137–139, 149, 207
Higgs bosons 101, 194–198, 223, 274, 286, 297, 300
——— ———, in grand unification 237, 246, 250
——— mechanism 174, 181, 182, 184, 186–191, 235, 283
hypercharge 186–189, 191, 195, 240, 241, 317

hyperon decays 4, 6, 27, 36, 38, 44, 48, 49, 58, 67, 102, 270–272, 297, 337, 338, 340, 346, 353, 354
hypothetical particles and forces 306

interaction, electromagnetic *see* analogy of the weak and electromagnetic interaction and *see* quantum electrodynamics
———, electroweak *see* electroweak interaction
———, gravitational 167, 170, 171, 236, 251, 252, 255, 307, 330
———, milliweak 94, 98, 129
———, strong *see* strong interaction, quantum chromodynamics
———, superweak 98, 101, 274, 350
———, weak *see* weak interaction
intermediate bosons, in grand unification 237, 239, 242–246, 250
——— ———, weak interaction 1–5, 50, 51, 53, 88, 105, 106, 158, 185, 186, 189, 199, 200, 202, 203, 205, 225–227
——— ———, masses of $W^{\pm}$, Z^0 171, 174, 183–186, 189, 193, 194, 214, 286, 300
——— ———, production and decay of $W^{\pm}$, Z^0 214, 216–218, 220–222, 285, 287, 300
——— ———, properties of $W^{\pm}$, Z^0 158, 160, 162; *see also* weak charge
isoscalar current *see* current, isoscalar
isotopic invariance 24, 25, 29, 39, 40, 46, 57, 120, 170, 187, 189, 198; *see also* SU(2)
isovector current *see* current, isovector

J/ψ family 3, 103, 217, 276

K^0-$\overline{K}^0$ system 76, 85, 94, 95, 127, 272, 273, 323, 297–299
K-meson decays 6, 13, 38, 40, 41, 43, 48, 49, 76, 86, 90, 94, 98, 270, 271, 274, 275, 322, 323, 327, 331, 332, 338, 342, (28.7.2)
K^0-meson regeneration 91, 272, 273
Kobayashi-Maskawa matrix *see* quark mixing

Lee-Sugawara relation 66, 271, 349
left-handed weak current *see* current, left-handed

lepton classification 2, 13, 186, 236, 248, 249
——— mixing 13
leptonic number 149, 164, 166, 167, 243, 244, 251, 295, 297, 352
——— photons 166, 294
——— processes *see* weak processes
——— weak current *see* current, leptonic
local symmetry 164, 165, 167, 170, 174, 180, 181, 183–186, 189, 251

magnetic monopole 284, 295, 306
masses of Higgs bosons 223, 232, 235
——— ——— intermediate bosons 171, 174, 180, 181, 183, 186, 189, 193, 194, 214, 286, 300
——— ——— leptons 187, 188, 194, 195, 223, 235, 247, 248
——— ——— neutrinos 103, 188, 195, 247, 249, 258, 291, 301, 302, 305, 342
——— ——— nucleons 230, 232
——— ——— quarks 25, 52, 72, 75, 89, 119, 128, 187, 196, 197, 229, 235, 248
mass, non-diagonal 65, 97, 99, 181
——— of photon 161, 166, 171, 182, 183, 190, 293–294, 332
meson decays *see* K-meson *and see* pion decays
——— octet 65, 66, 316
Michel parameter 18, 104, 268
milliweak interaction 94, 98, 129
muon capture 6, 27, 32, 34
——— decay 2, 6, 7, 15, 126, 268, 282, 297, 300, 332, 339, 353
———-electron universality 43, 142

neutral current *see* current, neutral
neutrino masses 103, 188, 195, 247, 249, 258, 291, 301, 302, 305, 342
——— oscillations 295, 301, 302, 352
——— reactions 2, 6, 14, 30, 118, 130, 143, 157, 158, 199, 201, 206, 278, 282, 285, 299–301, 355
———, right-handed 11, 12, 186, 188, 189, 195, 237, 238, 249, 292
———, relic 254, 258
neutron, decay of 3, 4, 6, 22, 24, 27, 30–36
———, dipole moment of 100, 274, 338
neutron-antineutron oscillations 249, 302
non-abelian photons 169–172, 183

_____ symmetry 165, 168, 170, 183, 282
non-diagonal mass 95, 97, 99, 181
_____ neutral current 4, 13, 39, 196, 299, 355
non-leptonic processes *see* weak processes

octet enhancement 121, 271
_____ of baryons 44, 65, 66, 316
_____ _____ currents 39, 40, 101
_____ _____ mesons 66, 67, 316
operator expansion 50, 271
orthogonal groups 239, 248, 249, 288, 302, 303
oscillations of neutrinos 295, 301, 302, 352
_____ _____ strangeness 89, 273
_____, neutron ↔ antineutron 249, 302

parity non-conservation in atoms 211, 280, 281, 299
partons 130, 145–153, 155, 218, 220, 278
PCAC 26, 28, 32, 34, 44, 186, 269; *see also* conservation of axial current
perturbation theory 39, 49, 52, 88, 97, 101, 127, 156–159, 169, 225, 226, 252
photon, baryonic and leptonic 166, 294
_____ mass 161, 166, 171, 182, 183, 190, 293, 294, 332
_____, non-abelian 169–172, 183
_____, relic 255, 290
P invariance 1, 12, 21, 48, 59, 66, 77, 85, 98–101, 205, 208, 209, 211, 213, 268, 272, 273, 280, 281, 296, 298
pion decays 6, 22, 25, 27, 29, 40, 41, 268–270, 271, 297, 333
_____ pole 33, 34
Planck mass 252, 256
P-odd nuclear forces 208, 213, 272, 296
Pomeranchuk theorem 138
proton non-stability 240, 243, 248, 249, 288, 294, 295, 302

quantum chromodynamics 155, 171, 220, 298
_____ electrodynamics 54, 157, 165, 171, 265
quark classification 3, 13, 195, 237, 238, 248, 249, 269, 275, 297
_____ color 4, 50, 69, 70, 104, 113, 118–120, 215, 218, 219, 228, 237–239, 249

_____ current *see* current, quark
_____ flavours 3, 52, 57, 104, 196–198, 228, 231, 242, 249, 252
_____ masses 25, 52, 72, 75, 89, 119, 128, 187, 196, 197, 229, 235, 248
_____ mixing 12, 13, 89, 101, 123–129, 195–198, 277
_____, relic 254, 258, 260, 292
_____, right-handed 25, 56, 57, 69, 71–73, 196, 199, 206, 208, 209, 236, 237
quintons 238, 239

regenerations of K^0 mesons 91, 272, 273
relic monopoles 293
_____ neutrinos 254, 258
_____ photons 255, 290
_____ quarks 254, 258, 260, 292
renormalizability 156, 171, 172, 180, 182, 185, 189, 227, 252, 284
right-handed current *see* current, right-handed
_____ _____ neutrino *see* neutrino, right-handed
_____ _____ quark *see* quark, right-handed
running coupling constants 51, 52, 55, 104, 228, 240, 243

selection rule, $\Delta S = 0$ 22, 36, 45, 46, 101, 269
_____ _____, $\Delta S = 1$ 38, 45, 46, 48, 49, 56, 94, 98, 99, 269, 354, 355
_____ _____, $\Delta S = 2$ 39, 85–89, 97, 101, 127, 273, 354
_____ _____, $\Delta S = \Delta C$ 117, 118, 354
_____ _____, $\Delta S = \Delta Q$ 38, 89, 96, 270, 339, 353
_____ _____, $\Delta T = \frac{1}{2}$ 40, 49, 50, 52, 56, 57, 61–63, 65, 67, 69, 74, 78, 82, 122, 270, 271, 297, 343
_____ _____, $\Delta T = \frac{3}{2}$ 49, 50, 52, 57, 61–63, 67, 69, 74, 78, 84, 271, 345, 348
semileptonic processes *see* weak processes
sextet enhancement 119, 120, 121, 276, 299
Shifman-Vainshtein-Zakharov lagrangian 56, 271
S-matrix 54, 63–65, 325
SO(8) 252
SO(10) 239, 248, 249, 288, 302, 303
spectral density 110–115

spin *see* U- and V-spin
spontaneous symmetry breaking 26, 34, 174, 246, 248, 269, 283, 293
spurion 61, 62, 65, 78
strange current *see* current, strange
strangeness oscillations 89, 273
strong interaction 26, 33, 51, 240, 266; *see also* quantum chromodynamics *and* SU(3)
sub-quarks 252, 300
supergravity 238, 251, 252, 290, 304, 305
superunification 251, 288, 290, 304
supersymmetry 251, 289, 290, 304
superweak interaction 98, 101, 274, 350
_____ mixing 95, 97, 129
SU(2) 165, 167, 174, 179, 183, 186, 192, 239, 240, 242, 246, 310; *see also* isotopic invariance
SU(3) 39–44, 46, 47, 57, 65, 66, 70, 116, 120, 171, 238–240, 242, 246, 249, 251, 270, 314, 315
SU(4) 275, 304
SU(5) 236, 237, 239, 240, 242, 243, 246–248, 288, 302, 303, 305, 306
SU(6) 47, 249
SU(8) 252, 303–305
SU(9) 303
SU(n) 198, 242, 310
symmetry, abelian 164, 165, 180, 309
_____, chiral 25, 34, 75, 115, 232, 269
_____, color 4, 239, 249
_____, discrete 174; *see also* C, CP, CPT, G, P, and T invariance
_____, gauge 164, 174, 177, 180, 183, 189, 214, 246, 265, 282
_____, global 164, 165, 170, 177, 179
_____, local 164, 165, 167, 170, 174, 180, 181, 183, 186, 189, 251
_____, non-abelian 165, 168, 170, 183, 282

tachyon 175, 177
technicolor 235, 248, 287, 300, 303
T invariance 35, 36, 61, 63, 64, 100, 350
t-quark 3, 5, 12, 101, 123, 195, 215, 249, 250, 277, 299
transversality 24, 29, 32, 34, 54, 107, 110, 161, 163, 166, 172
triangle anomaly 231
_____ relation 62, 347

τ-lepton decays 6, 103, 276, 332, 333, 342, 353

U(1) 164, 166, 174, 177, 182, 186, 192, 239, 242, 247, 252
unification of interactions 171, 251, 289; *see also* grand unification *and see* superunification
unified electroweak interaction *see* electroweak interaction
unitarity limit 157, 227, 281
universality, μ-e 43, 142
_____ of charged current 12, 105, 157, 268
U- and V-spin 41, 121, 317

vacuum 174–181, 183, 189, 192, 224, 225, 246, 286, 293
$V - A$ interaction *see* weak interaction
vector current *see* current, vector

W-bosons 214, 218; *see also* intermediate bosons
weak charge 24, 29, 31, 35, 41, 160, 187, 191, 192, 199, 206, 211, 216; *see also* axial charge
_____ "electrism" 32
weak interaction constants 7, 9, 18, 88, 89, 97, 98, 101, 126, 127, 142, 156, 157, 185, 191, 192, 236, 240, 269, 270, 331
_____ _____, four-fermion 7, 9, 53, 136, 138, 157, 190, 200, 206, 209, 281
_____ _____ in atoms 211, 279, 280, 281, 299
_____ _____ _____ nuclear forces 208, 213, 272, 296
_____ _____, lagrangian of 9, 48, 50, 51, 53, 55, 56, 67, 69, 81, 89, 187, 200, 271
_____ _____, SU(2)-symmetry *see* isotopic symmetry *and* SU(2)
_____ _____, $V - A$ 10, 12, 43, 53, 103, 104, 201, 268
_____ magnetism 31, 32, 269
_____ processes, leptonic 6, 15, 104, 116, 130, 133–135, 138, 140, 141, 199, 201, 202, 268, 279, 280, 300, 332, 353
_____ _____, non-leptonic 6, 48, 58–63, 69, 73, 74, 76, 77, 79, 82, 90, 94, 96, 98, 101, 102, 118, 121, 122, 270, 274, 276, 277, 297, 299, 332–340, 343–349, 354

——— ———, semileptonic 6, 27, 29, 30, 36, 38–41, 43, 44, 90, 104, 105, 110, 114, 116, 117, 119, 129, 143, 144, 152, 153, 206, 209, 211, 213, 269–271, 277–279, 297, 332–340, 343, 350, 353–355

Weinberg angle 14, 138, 189–192, 199, 214, 235, 236, 241, 243, 249, 250, 331

——— sum rule 115, 116

———-Salam model *see* electroweak interaction

Wilson operator expansion 50, 271

X, Y-bosons 239, 243–246

Yang-Mills theory 164, 167, 168, 170, 282

ϒ-mesons 3, 126, 227, 228, 277

Z-boson 216, 217; *see also* intermediate bosons